Vanilla

Medicinal and Aromatic Plants — Industrial Profiles

Individual volumes in this series provide both industry and academia with in-depth coverage of one major genus of industrial importance.

Series Edited by Dr. Roland Hardman

Volume 1
Valerian, edited by Peter J. Houghton
Volume 2
Perilla, edited by He-ci Yu, Kenichi Kosuna and Megumi Haga
Volume 3
Poppy, edited by Jenö Bernáth
Volume 4
Cannabis, edited by David T. Brown
Volume 5
Neem, edited by H.S. Puri
Volume 6
Ergot, edited by Vladimír Křen and Ladislav Cvak
Volume 7
Caraway, edited by Éva Németh
Volume 8
Saffron, edited by Moshe Negbi
Volume 9
Tea Tree, edited by Ian Southwell and Robert Lowe
Volume 10
Basil, edited by Raimo Hiltunen and Yvonne Holm
Volume 11
Fenugreek, edited by Georgios Petropoulos
Volume 12
Ginkgo biloba, edited by Teris A. Van Beek
Volume 13
Black Pepper, edited by P.N. Ravindran
Volume 14
Sage, edited by Spiridon E. Kintzios
Volume 15
Ginseng, edited by W.E. Court
Volume 16
Mistletoe, edited by Arndt Büssing
Volume 17
Tea, edited by Yong-su Zhen
Volume 18
Artemisia, edited by Colin W. Wright
Volume 19
Stevia, edited by A. Douglas Kinghorn
Volume 20
Vetiveria, edited by Massimo Maffei
Volume 21
Narcissus and Daffodil, edited by Gordon R. Hanks
Volume 22
Eucalyptus, edited by John J.W. Coppen
Volume 23
Pueraria, edited by Wing Ming Keung
Volume 24
Thyme, edited by E. Stahl-Biskup and F. Sáez
Volume 25
Oregano, edited by Spiridon E. Kintzios
Volume 26
Citrus, edited by Giovanni Dugo and Angelo Di Giacomo
Volume 27
Geranium and Pelargonium, edited by Maria Lis-Balchin
Volume 28
Magnolia, edited by Satyajit D. Sarker and Yuji Maruyama
Volume 29
Lavender, edited by Maria Lis-Balchin
Volume 30
Cardamom, edited by P.N. Ravindran and K.J. Madhusoodanan

Volume 31
Hypericum, edited by Edzard Ernst
Volume 32
Taxus, edited by H. Itokawa and K.H. Lee
Volume 33
Capsicum, edited by Amit Krish De
Volume 34
Flax, edited by Alister Muir and Niel Westcott
Volume 35
Urtica, edited by Gulsel Kavalali
Volume 36
Cinnamon and Cassia, edited by P.N. Ravindran, K. Nirmal Babu and M. Shylaja
Volume 37
Kava, edited by Yadhu N. Singh
Volume 38
Aloes, edited by Tom Reynolds
Volume 39
Echinacea, edited by Sandra Carol Miller Assistant Editor: He-ci Yu
Volume 40
Illicium, Pimpinella and Foeniculum, edited by Manuel Miró Jodral
Volume 41
Ginger, edited by P.N. Ravindran and K. Nirmal Babu
Volume 42
Chamomile: Industrial Profiles, edited by Rolf Franke and Heinz Schilcher
Volume 43
Pomegranates: Ancient Roots to Modern Medicine, edited by Navindra P. Seeram, Risa N. Schulman and David Heber
Volume 44
Mint, edited by Brian M. Lawrence
Volume 45
Turmeric, edited by P. N. Ravindran, K. Nirmal Babu, and K. Sivaraman
Volume 46
Essential Oil-Bearing Grasses, edited by Anand Akhila
Volume 47
Vanilla, edited by Eric Odoux and Michel Grisoni

Vanilla

Edited by Eric Odoux and Michel Grisoni

Medicinal and Aromatic Plants — Industrial Profiles

CRC Press
Taylor & Francis Group
Boca Raton London New York

CRC Press is an imprint of the
Taylor & Francis Group, an **informa** business

Cover photos: Left: Curing of mature pods of *Vanilla planifolia* © Michel Grisoni, CIRAD. **Right:** *Vanilla planifolia* in full bloom © René Carayol, Région Réunion.

CRC Press
Taylor & Francis Group
6000 Broken Sound Parkway NW, Suite 300
Boca Raton, FL 33487-2742

International Standard Book Number: 978-1-4200-8337-8 (Hardback)

Library of Congress Cataloging-in-Publication Data

Vanilla / editors: Eric Odoux, Michel Grisoni.
p. cm. -- (Medicinal and aromatic plants--industrial profiles ; 47)
ISBN 978-1-4200-8337-8
1. Vanilla. 2. Vanilla industry. I. Odoux, Eric. II. Grisoni, Michel. III. Title. IV. Series: Medicinal and aromatic plants--industrial profiles ; v. 47.

SB307.V2V35 2010
633.8'2--dc22 2010001438

This book is dedicated to the memory of
Dr. Miguel Angel Soto Arenas (1963–2009)

Contents

Series Preface

There is increasing interest in industry, academia, and the health sciences in medicinal and aromatic plants. In passing from plant production to the eventual product used by the public, many sciences are involved. This series brings together information that is currently scattered through an ever-increasing number of journals. Each volume gives an in-depth look at one plant genus about which an area specialist has assembled information ranging from the production of the plant to market trends and quality control.

Many industries are involved, such as forestry, agriculture, chemical, food, flavor, beverage, pharmaceutical, cosmetic, and fragrance. The plant raw materials are roots, rhizomes, bulbs, leaves, stems, barks, wood, flowers, fruits, and seeds. These yield gums, resins, essential (volatile) oils, fixed oils, waxes, juices, extracts, and spices for medicinal and aromatic purposes. All these commodities are traded worldwide. A dealer's market report for an item may say "drought in the country of origin has forced up prices."

Natural products do not mean safe products, and account of this has to be taken by the above industries, which are subject to regulation. For example, a number of plants that are approved for use in medicine must not be used in cosmetic products.

The assessment of "safe to use" starts with the harvested plant material, which has to comply with an official monograph. This may require absence of, or prescribed limits of, radioactive material, heavy metals, aflatoxin, pesticide residue, as well as the required level of active principle. This analytical control is costly and tends to exclude small batches of plant material. Large-scale, contracted, mechanized cultivation with designated seed or plantlets is now preferable.

Today, plant selection is not only for the yield of active principle, but for the plant's ability to overcome disease, climatic stress, and the hazards caused by mankind. Methods such as *in vitro* fertilization, meristem cultures, and somatic embryogenesis are used. The transfer of sections of DNA is leading to controversy in the case of some end uses of the plant material.

Some suppliers of plant raw material are now able to certify that they are supplying organically farmed medicinal plants, herbs, and spices. The Economic Union directive CVO/EU No. 2092/91 details the specifications for the obligatory quality controls to be carried out at all stages of production and processing of organic products.

Fascinating plant folklore and ethnopharmacology lead to medicinal potential. Examples are the muscle relaxants based on the arrow poison curare from species of *Chondrodendron*, and the antimalarials derived from species of *Cinchona* and *Artemisia*. The methods of detection of pharmacological activity have become increasingly reliable and specific, frequently involving enzymes in bioassays and avoiding the use of laboratory animals. By using bioassay-linked fractionation of crude plant juices or extracts, compounds can be specifically targeted which, for example, inhibit blood platelet aggregation, or have antitumor, antiviral, or any other required activity. With the assistance of robotic devices, all the members of a genus may be readily screened. However, the plant material must be fully authenticated by a specialist.

The medicinal traditions of ancient civilizations such as those of China and India have a large armamentarium of plants in their pharmacopoeias that are used throughout Southeast Asia. A similar situation exists in Africa and South America. Thus, a very high percentage of the world's population relies on medicinal and aromatic plants for their medicine. Western medicine is also responding. Already in Germany all medical practitioners have to pass an examination in phytotherapy before being allowed to practice. It is noticeable that medical, pharmacy, and health-related schools throughout Europe and the United States are increasingly offering training in phytotherapy.

Multinational pharmaceutical companies have become less enamored of the single compound, magic-bullet cure. The high costs of such ventures and the endless competition from "me-too" compounds from rival companies often discourage the attempt. Independent phytomedicine companies have been very strong in Germany. However, by the end of 1995, 11 (almost all) had been acquired by the multinational pharmaceutical firms, acknowledging the lay public's growing demand for phytomedicines in the Western world.

The business of dietary supplements in the Western world has expanded from the health store to the pharmacy. Alternative medicine includes plant-based products. Appropriate measures to ensure their quality, safety, and efficacy either already exist or are being answered by greater legislative control by such bodies as the U.S. Food and Drug Administration and the recently created European Agency for the Evaluation of Medicinal Products based in London.

In the United States, the Dietary Supplement and Health Education Act of 1994 recognized the class of phytotherapeutic agents derived from medicinal and aromatic plants. Furthermore, under public pressure, the U.S. Congress set up an Office of Alternative Medicine, which in 1994 assisted the filing of several Investigational New Drug (IND) applications required for clinical trials of some Chinese herbal preparations. The significance of these applications was that each Chinese preparation involved several plants and yet was handled as a *single* IND. A demonstration of the contribution to efficacy of each ingredient of each plant was not required. This was a major step toward more sensible regulations with regard to phytomedicines.

My new book series "Traditional Herbal Medicines for Modern Times" (CRC Press) has included some important examples of Chinese and Japanese formulae, commonly of three to six dried herbs and now available as tablets or water soluble granules for the treatment of cardiovascular disease (Vol. 1) or liver disease (Vols. 3 and 7) or to relieve the adverse effects of Western anticancer drugs (Vol. 5). Other books have covered Ayurvedic herbs and *Rasayana* (Vol. 2); antimalarial plants (Vol. 4); antidiabetic plants (Vol. 6); cosmetic plants (Vol. 8) and figs (Vol. 9). More are in preparation.

To return to the present series and particularly the topic of vanilla, James A Duke, in his *Handbook of Medicinal Plants of Latin America* (CRC Press, 2009) has given the medicinal uses of the tinctures and decoctions of the pods, stems and roots of this plant (pages 733–735).

This volume, *Vanilla*, edited by Eric Odoux and Michel Grisoni, is outstanding in that it is the first comprehensive volume on the subject in English. I am very grateful to them for all their hard work and to the contributors of the 24 chapters for their authoritative information. My thanks are also due to Barbara Norwitz and her staff, including production coordinator Jessica Vakili, for their unfailing help.

Roland Hardman, BPharm, BSc (Chem), PhD (London), FR Pharm S
Head of Pharmacognosy (Retired), School of Pharmacy & Pharmacology
University of Bath, United Kingdom

Preface

Vanilla is a legacy of Mexico, and like chocolate, another major global delicacy, it is the basis of many sweets, ice cream, and cola drinks. Vanilla flavor is appreciated in any concentration by most people all over the world. It represents a large market of almost a half billion Euros per year, with only a few countries producing the pods of this tropical orchid. An orchid with special demands for soil and climate, sensitive to pests and diseases, and because of its vegetative propagation it has little genetic variation in the producing areas. In addition, several of the major growing regions, such as Madagascar, are regularly hit by tropical storms. This makes vanilla a vulnerable crop, resulting in large yearly changes in price. Moreover, the green beans need an elaborate curing procedure, which results in the final product: the dark colored pods which contain a high amount of vanillin. This process is still not well understood, though of crucial importance for the vanilla flavor.

The supply issue obviously resulted in efforts to start production in other regions and even in greenhouses, or to alternatively look for other sources of vanillin. With the food and beverage industry as the major users, the preferred source is a natural one, which means production by other plants or microorganisms, including the microbial bioconversion of vanillin precursors. Vanillin is thus available as a pure chemical entity both of natural and synthetic origin, but the pure compound does not give the same flavor as obtained with vanilla pods, or extracts thereof. Because of the large differences in price between the different commodities, adulteration is not uncommon.

This very brief sketch of vanilla explains the diverse research in this field. This includes biotechnology aimed at finding novel production methods of vanillin, horticultural studies for improving yields and increasing the resistance of the plants, entomology for finding possible pollinators required in areas outside of the original habitat, studies on the chemistry and biochemistry of the curing process, and unfortunately also advanced analytical chemistry to be able to identify adulterations such as vanillin-spiked pods, and synthetic vanillin instead of natural vanillin.

This book gives an excellent overview of this field. All chapters are written by experts, each with many years of experience in their respective fields. This book shows the past, present, and future of vanilla, and with no doubt will serve for many years to come as the major comprehensive source of information on vanilla, the standard reference source for all who have interest in vanilla, such as producers, flavorists, researchers, and consumers.

Rob Verpoorte
Department of Pharmacognosy/Metabolomics
Leiden University
Leiden, the Netherlands

Editors

Eric Odoux graduated in biochemistry and has had a career with CIRAD (French Agricultural Research Centre for International Development) since 1988. He successively worked on coffee, cocoa, aromatic plants, and tropical fruits processing in Cameroon (and other countries in West Africa) and in France before he developed a research project on vanilla curing in Reunion Island and Madagascar in 1996. His research has mainly focused on vanilla aroma development related to curing practices. He received his PhD in food sciences from University of Montpellier II (France) in 2004.

His work led to research in partnership with industry, to consultancy reports, and to scientific articles.

cirad

Michel Grisoni graduated in agronomy and holds a PhD in plant pathology from Montpellier SupAgro, France. He has pursued a career as agro-virologist for CIRAD since 1984 in Colombia, French Polynesia, and Reunion Island. His research on vanilla has focused primarily on virus diseases and then moved towards the characterization, preservation, and development of genetic resources, particularly to improve the resistance of vanilla plants to diseases.

He is presently in charge of the Vanilla Genetics and Certification Research Program of CIRAD and curator of the vanilla collection at the Center for Biological Resources (Vatel) on Reunion Island. He is the author or coauthor of many scientific articles, consultancy reports, and conference communications related to vanilla.

cirad

Contributors

K. Nirmal Babu
Division of Crop Improvement and Biotechnology
Indian Institute of Spices Research
Calicut, Kerala, India

Pascale Besse
Unité Mixte de Recherche—Peuplement Végétaux et Bioagresseurs en Milieu Tropical
Université de la Réunion
Saint Denis, La Réunion, France

Séverine Bory
Unité Mixte de Recherche—Peuplement Végétaux et Bioagresseurs en Milieu Tropical
Université de la Réunion
Saint Pierre, La Réunion, France

Spencer Brown
Institut des Sciences du Végétal
Centre National de la Recherche Scientifique
Gif-sur-Yvette, France

Christel Brunschwig
Département Recherche et Développement
Etablissement Vanille de Tahiti
Uturoa, Raiatea, French Polynesia

and

Laboratoire de Biodiversité Terrestre et Marine
Université de la Polynésie Française
Tahiti, French Polynesia

Kenneth M. Cameron
Department of Botany
University of Wisconsin
Madison, Wisconsin

Dexin Chen
Hainan Bright Fragrance Co. Ltd.
Haikou, China

François-Xavier Collard
Département Recherche et Développement
Etablissement Vanille de Tahiti
Uturoa, Raiatea, French Polynesia

Bertrand Côme
Provanille / La Vanilleraie
La Réunion, France

Geneviève Conéjéro
Unité Mixte de Recherche—Biochimie et Physiologie Moléculaire des Plantes
Institut National de la Recherche Agronomique
Montpellier, France

Minoo Divakaran
Department of Botany
Providence Women's College
Calicut, Kerala, India

Michel Dron
Institut de Biotechnologie des Plantes
Université Paris Sud
Orsay, France

Marie-France Duval
Unité Propre de Recherche—Multiplication Végétative
Centre de Recherche Agronomique Pour le Développement
Montpellier, France

Karin Farreyrol
School of Biological Sciences
The University of Auckland
Auckland, New Zealand

Clemens Fehr
Gourmet Gardens Ltd.
Kampala, Uganda

Michel Grisoni
Unité Mixte de Recherche—Peuplement Végétaux et Bioagresseurs en Milieu Tropical
Centre de Recherche Agronomique Pour le Développement
Saint Pierre, La Réunion, France

Franz-Josef Hammerschmidt
Symrise GmbH & Co
Holzminden, Germany

Juan Hernández Hernández
Instituto Nacional de Investigaciones Forestales
Agrícolas y Pecuarias
Martínez de la Torre, Veracruz, México

Jens-Michael Hilmer
Symrise GmbH & Co
Holzminden, Germany

Fabienne Lapeyre-Montes
Unité Mixte de Recherche—Développement et Amélioration
Centre de Recherche Agronomique Pour le Développement
Montpellier, France

Sandra Lepers-Andrzejewski
Département Recherche et Développement
Etablissement Vanille de Tahiti
Uturoa, Raiatea, French Polynesia

Kenny Le Roux
Unité Mixte de Recherche--Peuplement Végétaux et Bioagresseurs en Milieu Tropical
Centre de Recherche Agronomique Pour le Développement
Saint Pierre, La Réunion, France

Edward C.Y. Liew
Royal Botanic Gardens Sydney
Botanic Gardens Trust
Sydney, New South Wales, Australia

Gerd Lösing
Symrise GmbH & Co
Holzminden, Germany

Pesach Lubinsky
Department of Botany and Plant Sciences
University of California
Riverside, California

Raoul Lucas
Faculté des Lettres et Sciences Humaines
Université de la Réunion
Saint Denis, La Réunion, France

Eric Odoux
Unité Mixte de Recherche—Qualisud
Centre de Recherche Agronomique Pour le Développement
Montpellier, France

Michael Pearson
School of Biological Sciences
The University of Auckland
Auckland, New Zealand

Marc Pignal
Unité Mixte de Recherche—Origine, Structure et Evolution de la Biodiversité
Muséum National d'Histoire Naturelle
Paris, France

Serge Quilici
Unité Mixte de Recherche—Peuplement Végétaux et Bioagresseurs en Milieu Tropical
Centre de Recherche Agronomique Pour le Développement
Saint Pierre, La Réunion, France

Patricia Rain
The Vanilla Company
Santa Cruz, California

Agathe Richard
Unité Mixte de Recherche—Peuplement Végétaux et Bioagresseurs en Milieu Tropical
Centre de Recherche Agronomique Pour le Développement
Saint Pierre, La Réunion, France

Michel Roux-Cuvelier
Unité Mixte de Recherche—Peuplement Végétaux et Bioagresseurs en Milieu Tropical
Centre de Recherche Agronomique Pour le Développement
Saint Pierre, La Réunion, France

Y.R. Sarma
Indian Institute of Spices Research
Calicut, Kerala, India

Samira Sarter
Unité Mixte de Recherche—Qualisud
Centre de Recherche Agronomique Pour le Développement
Antananarivo, Madagascar

B. Sasikumar
Indian Institute of Spices Research
Calicut, Kerala, India

Joseph Thomas
Indian Cardamom Research Institute
Spice Board
Idukki, Kerala, India

Mesak Tombe
Entomology and Phytopathology Divison
Indonesian Medicinal and Aromatic Crops Research Institute
Bogor, Indonesia

S. Varadarasan
Indian Cardamom Research Institute
Spice Board
Idukki, Kerala, India

Jean-Luc Verdeil
Unité Mixte de Recherche—Développement et Amélioration
Centre de Recherche Agronomique Pour le Développement
Montpellier, France

Hongyu Wang
The National Center for Agricultural Biodiversity Research and Development
Yunnan Agricultural University
Kunming, China

Yunyue Wang
The National Center for Agricultural Biodiversity Research and Development
Yunnan Agricultural University
Kunming, China

Xurui
Flavors and Fragrances Research Institute
Yunnan Agricultural University
Kunming, China

Robber Zaubin
Eco-Physiology Division
Indonesian Medicinal and Aromatic Crops Research Institute
Bogor, Indonesia

Hengcang Zhou
Flavors and Fragrances Research Institute
Yunnan Agricultural University
Kunming, China

1 Vanilloid Orchids
Systematics and Evolution

Kenneth M. Cameron

CONTENTS

INTRODUCTION

Vanilla and its relatives are surviving members of what is likely an ancient lineage of flowering plants. Many are restricted to remote localities, and some are threatened with extinction. We certainly know a great deal about *Vanilla planifolia*—methods of cultivation, diseases that affect the domesticated vines, and techniques of fruit processing—but the fundamental natural history of the entire genus *Vanilla* and its closest relatives is still poorly known. The systematic study of these plants has been and continues to be surrounded by controversies. For these reasons it is encouraging to witness the increased level of knowledge in recent years regarding their classification and evolution, which has come about primarily thanks to the increased use of DNA-based data in systematic studies (e.g., see Cameron, 2003, 2004, 2006).

Until the end of the twentieth century, the vanilloid orchids had proven difficult to classify within any particular subtribe, tribe, or subfamily of the family Orchidaceae. On the one hand, they share the presence of a fully bent, single, fertile anther with various advanced orchid lineages. On the other hand, they exhibit a variety of characters considered primitive among orchids. Botanists now consider the single fertile anther at the apex of the vanilla flower's column to have risen by way of a different evolutionary process than that of nearly all other orchids (i.e., those classified within the Epidendroideae and Orchidoideae subfamilies). For this reason and others mentioned below, vanilla and related orchids are now classified within their own unique subfamily, Vanilloideae, as shown in Figure 1.1.

As we move further into the twenty-first century and the genomics era, there is little doubt that plant breeders will endeavor to improve vanilla as a crop plant using genetic

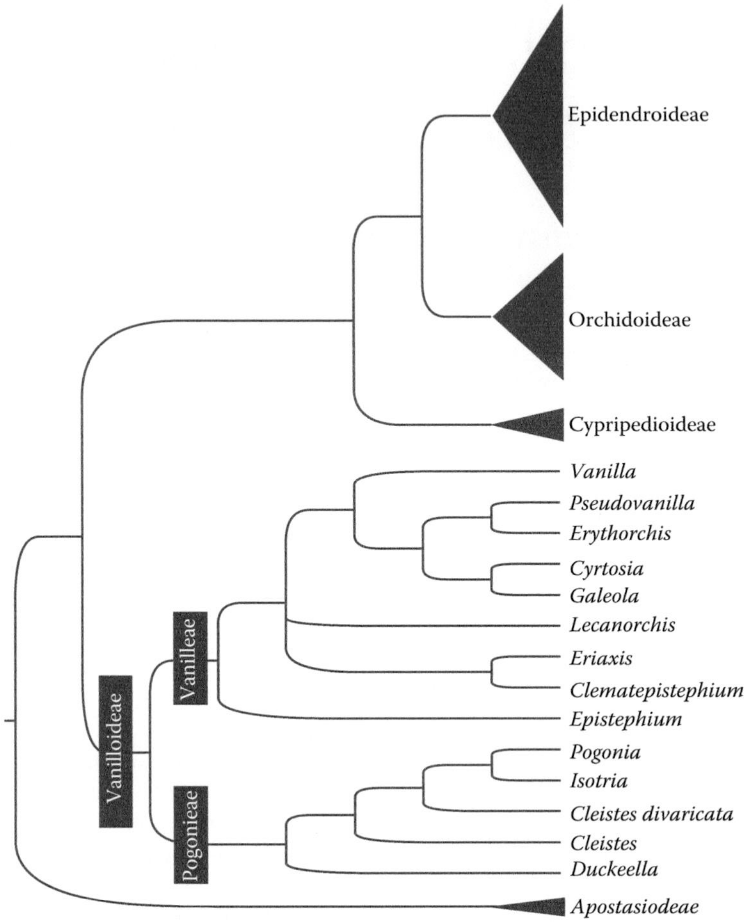

FIGURE 1.1 Cladogram depicting the phylogenetic relationships among subfamilies of Orchidaceae and among genera within Vanilloideae based on a combination of nuclear, mitochondrial, and plastid DNA sequence data. The subfamily is divided into two tribes: Pogonieae and Vanilleae. Note that *Vanilla* shares a common ancestor with a clade of four genera including *Galeola* and *Pseudovanilla*.

modification. Any future genetic studies into the structure and development of the vanilla flower and/or fruit should consider looking closely at other genera of Vanilloideae with shared ancestry, rather than making direct comparisons only to more distantly related orchids or other flowering plants. Such comparisons could be misleading in their assumptions of homology. This point is best appreciated by considering that over the course of more than 65 million years, vanilloid orchids have become adapted to a variety of specialized habitats, pollinators, and seed-dispersal strategies. They all share a fundamental genome in common, based on a now extinct ancestor, and yet differences in gene expression and regulation ultimately determine whether a given vanilloid orchid grows in the tropics or survives temperatures well below freezing, whether it grows as an erect herb or as a vine, and whether it will produce a dry flavorless capsule or an aromatic fleshy fruit. As genomic and proteomic technology is eventually applied to crop plants of lesser economic value (compared to cereals and legumes, for example) studies targeting the improvement of vanilla may also wish to consider other genera of tribe Vanilleae or subfamily Vanilloideae. For example, it might be possible to develop more cold- and shade-tolerant vanilla vines by first studying the physiology and genetic makeup of *Cyrtosia*, a close relative that survives in the deciduous forests of Japan and China. On the basis of these arguments, a review of vanilloid orchid systematics (the scientific study of the diversity and classification of organisms) is presented here in order to set the stage for a more comprehensive understanding of the biology of *V. planifolia* and these exceptional orchids.

EVOLUTION OF VANILLOID ORCHIDS

An unsubstantiated hypothesis has persisted among biologists that the orchid family is only recently evolved relative to other flowering plants. To support this opinion, botanists cite the relatively low levels of genetic diversity among orchid genera and species, many of which can be hybridized easily with one another. They provide evidence in the fact that the geologically young Andes of South America and Highlands of New Guinea are centers of greatest orchid diversity. The close relationship between orchids and social bees, which are thought to have evolved much later than other insects, is also given as proof, and the fact that most orchid genera are found in either the Paleotropics or the Neotropics, but rarely are pantropical, indicates to some that Orchidaceae evolved only recently and certainly long after the separation of today's continents.

Molecular phylogenetic studies of Vanilloideae challenge the notion that the entire orchid family is recently evolved, however, and new perspectives on the systematics of Orchidaceae downplay or even contradict some of the facts mentioned above. For example, *Vanilla* is one of a few orchid genera with a transoceanic distribution that may not be due entirely to long-distance dispersal. Extant species are native to North America, South America, Africa, and Asia (see Figure 1.2). The fact that vanilloid orchids survive in the Guyana Shield region of South America, tropical Australia and Africa, Madagascar, and on the island of New Caledonia (a nonvolcanic Pacific island

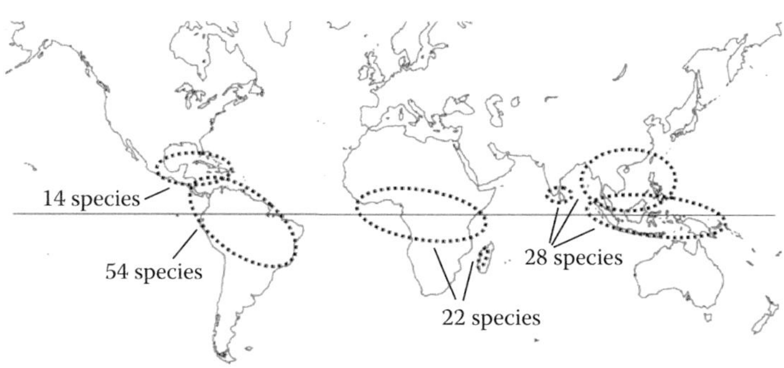

FIGURE 1.2 Paleotropical distribution of *Vanilla*, and estimates of species diversity within each geographic region.

with a peculiar ancient flora that separated from Gondwana around 65 million years ago) may also provide evidence of their considerable age and possible status as ancient relicts (Cameron, 1999, 2000).

Furthermore, subfamily Vanilloideae is positioned near the base of the orchid family tree, and Orchidaceae is the basal family within the large monocot order Asparagales (including onions, agaves, hyacinths, and the iris family, among others). Molecular clock estimates of the evolutionary age of these plants have calculated that Orchidacaeae may trace their origins back at least 76–119 million years (Janssen and Bremer, 2004; Ramirez et al., 2007). Vanilloid orchids, in turn, are at least 62 million years old. Molecular clocks can only provide minimum ages, so these plants are probably even older. Critical to this approach is the use of a calibration point for the "clock," which, in the case of Orchidaceae, has been provided by a 15–20-million-year-old fossil specimen of orchid pollen attached to an extinct bee preserved in amber (Ramirez et al., 2007).

SUBFAMILY VANILLOIDEAE AMONG ORCHIDS

As mentioned already, the vanilloid orchids, Vanilloideae, have been recognized as a subfamily of Orchidaceae only in the past decade, as DNA data have been used to reevaluate relationships among all orchids. Cameron (2007) has provided a detailed review of this DNA-driven revolution in orchid taxonomy from 1997 to 2007. The current systems of orchid classification (e.g., Chase et al., 2003) divide Orchidaceae into five subfamilies. The largest, with approximately 650 genera and 18,000 species, is Epidendroideae, which is dominated by tropical epiphytes and those orchids most highly prized as ornamentals. Orchidoideae, the second largest subfamily, is made up almost exclusively of terrestrial species classified within approximately 200 genera. Both subfamilies are characterized by monandrous flowers (meaning they have only one anther). All species within the subfamily Vanilloideae also possess flowers with just a single fertile anther, but this condition is considered to have evolved independently from Orchidoideae and Epidendroideae, and is the result of a unique mode of floral development (Freudenstein et al., 2002). In other words, the reduction in stamen/anther number from several (probably from six down to three and eventually down to one) occurred at least two times within Orchidaceae. Through the process of evolution, orchid flowers are thought to have undergone significant structural modifications resulting in flowers with pronounced bilateral symmetry, loss of stamens, and fusion of the remaining stamen(s) with the pistil to form a central column. A clue to explain the beginnings of this hypothetical evolutionary continuum can be found today by examining living members of the fourth orchid subfamily, Apostasioideae, which contains two genera: *Neuwiedia* and *Apostasia*. Species of *Neuwiedia* are triandrous, possessing flowers with three fertile anthers. These are only partially fused with the base of the pistil, and the perianth of the flower is only slightly bilateral in symmetry. Apostasioid orchids in many ways may be viewed as the most "primitive" of all orchids in that they show the least number of modifications from the basic blueprint of a hypothetical pre-orchid monocot ancestor. Diandrous flowers (i.e., with two fertile anthers) define the fifth orchid subfamily, Cypripedioideae. This group of about 120 species is commonly called "lady's slipper orchids." In terms of relative size, Cypripedioideae is more diverse than Apostasioideae (15 species), but less diverse than Vanilloideae (200 species), which will be considered further below.

Before they were classified as their own subfamily of Orchidaceae, most of the vanilloid orchids were considered to be primitive members of the monandrous subfamily Epidendroideae, but somewhat unconvincingly so. In fact, Dressler's (1993) pre-molecular system of orchid classification listed many of the vanilloid orchids under the category *insertae sedis* (meaning "of uncertain status"). At one time, it was even suggested that they might be best treated as a separate family all their own, Vanillaceae, closely related to, but separate from, Orchidaceae (Lindley, 1835). Why the uncertainty? A mix of what are assumed to be both primitive and advanced floral features among vanilloid orchids can be claimed to be the source of greatest confusion. Their precise position among orchids was eventually laid to rest using comparisons of DNA sequence information, and among the most unexpected results of the first molecular phylogenetic studies of orchids was the relocation of vanilla and its relatives from a position among the other orchids with a single fertile anther to a placement near the base of the orchid family tree (Cameron et al., 1999). Recognition of Vanilloideae as a monophyletic subfamily helped in solving one of the more perplexing enigmas of orchid systematics.

SPECIES DIVERSITY WITHIN *VANILLA*

Within Vanilloideae are no fewer than 15 genera, but *Vanilla* is the most diverse of these. There is yet to be published a formal monograph of the genus, but there does exist a taxonomic treatment of *Vanilla* that considered all the species known at the time. Unfortunately, this treatment was written more than 50 years ago (Portères, 1954).

Very recently, a taxonomic synopsis for *Vanilla* was published posthumously based upon the work of the late Mexican botanist Miguel A. Soto Arenas (Soto Arenas and Cribb, 2010). Within this important preliminary work are presented keys to the species, information about geographic distribution, and lists of select specimens. It serves as a significant step toward updating the systematic treatment of the genus. The 15 Mexican and Central American species were treated more completely in a posthumously published work by Soto Arenas and Dressler (2010). Within this paper one will find detailed descriptions, illustrations, and information on the molecular characterization of the Mesoamerican species.

The current worldwide checklist of all orchid species today recognizes 110 species of *Vanilla* (Govaerts et al., 2008). Most of these (61 species) are Neotropical natives of South America, Central America, Caribbean islands, and southern Florida. Africa claims 23 native species, with at least five of these restricted to Madagascar. The remaining species of *Vanilla* are found on the Indian subcontinent and throughout tropical Southeast Asia. No species of *Vanilla* are native to Australia. Likewise, Polynesia and other oceanic islands of the Pacific lack native species of *Vanilla*. This is perplexing to some since "Tahitian Vanilla" is cultivated throughout the Pacific, and its scientific name, *Vanilla tahitensis*, implies that it is indigenous to the French Polynesian island of Tahiti. What was described more than 75 years ago

(Moore, 1933) as a new "species" of *Vanilla*, however, has been proven recently by Lubinsky et al. (2008) to be nothing more than a primary hybrid between Neotropical *V. planifolia* (the maternal parent) and *V. odorata* (the paternal parent).

In terms of classification of species within the genus *Vanilla*, these were formally placed into one of two possible sections by Rolfe (1896). The first, *Vanilla* section *Aphyllae*, was erected to accommodate all of the leafless species in the genus (e.g., *V. aphylla*, *V. barbellata*, *V. roscheri*, and others). Species within this section grow on the African mainland, Madagascar, Southeast Asia, and also on islands in the Caribbean. Although some of these species produce fleshy fruits, there is no evidence that any of them are aromatic. Rolfe's classification of these species together implies that they share a recent common ancestor, but molecular studies have demonstrated that this is not the case (Cameron, 2005). Instead, there appears to be at least three independent cases of probably leaf loss in *Vanilla*—once in Africa, once in the Caribbean, and at least once in Asia. The section, therefore, is not monophyletic, but an artificial grouping of species with shared vegetative morphology derived by convergent evolution. According to modern rules of natural classification, it should not be recognized formally.

For the remaining species not classified in *Vanilla* section *Aphyllae*, Rolfe created section *Foliosae*. As the name indicates, all of these are leafy. This is a large group of species, and so Portères (1954) further divided the section into subsections. *Vanilla* section *Foliosae* subsection *Membranaceae* is a small cluster of species characterized by thin stems, thin leaves, short aerial roots, and flowers in which the labellum is not fused with the column. The labellum also lacks the complex bristles, hairs, and scales characteristic of other *Vanilla* species, and the fruits tend to dry on the vines and split lengthwise. *Vanilla mexicana* exemplifies this section, and molecular systematic studies have demonstrated that the group is the most primitive of all *Vanilla* species. These plants are very difficult to cultivate, probably because they have close relationships with mycorrhizal fungi, and there is no evidence that the fruits produce aromatic vanillin.

The other remaining species of the genus, including *V. planifolia* and *V. pompona*, were classified into either *Vanilla* section *Foliosae* subsection *Lamellosae* or subsection *Papillosae*. The former group is so named because species within this section are characterized by flowers with flattened scale-like appendages (lamellae), hairs, bristles, and complex ornamentation on their labella, which is always fused to the column along its margins to form a floral tube. The latter subsection was proposed for those species characterized by fleshy leaves and flowers usually with thick trichomes positioned in the center of the labellum, but without lamellate scales. Species within this leafy section are pantropical in distribution, but recent molecular systematic studies have demonstrated that this group is also artificial. Instead, species of *Vanilla* cluster primarily by geographic origin, as can be seen in Figure 1.3. Specifically, all Old World species (from the African and Asian Paleotropics) share a common ancestor together with the leafless New World species. These were probably dispersed from Africa to the Caribbean at some point in the past. All remaining Neotropical species, including *V. planifolia*, share a different common ancestor. It is within this group that aromatic fruits producing significant levels of vanillin are found. As such, the group has informally been named the "Neotropical, fragrant,

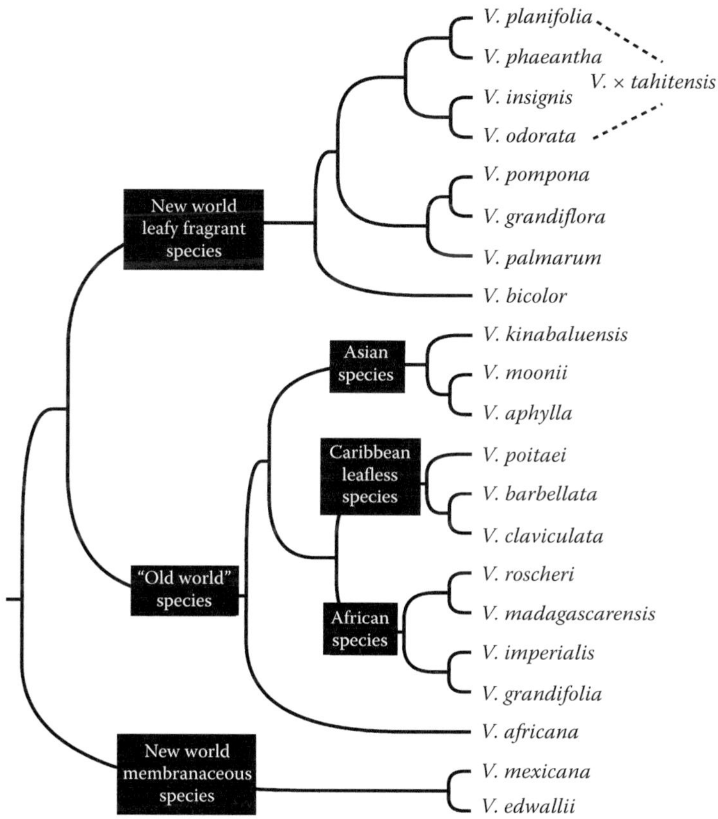

FIGURE 1.3 Phylogenetic relationships among select species of *Vanilla*. The cladogram is based on molecular sequence data from different genes including nuclear ribosomal ITS, plastid *rbc*L, *mat*K, *rpo*C1, and others. The hybrid origin of *V. tahitensis* from a cross between *V. odorata* and *V. planifolia* is highlighted by the dashed lines. Informal clades and subclades are labeled on the branch representing the common ancestor of each major species group.

leafy species." Note that molecular studies position *V. tahitensis* inside this group of Neotropical relatives, thereby confirming the hybrid origin of Old World Tahitian Vanilla, many individuals of which are tetraploid, from New World parents.

In their recent synopsis of the genus, Soto Arenas and Cribb (2010) classify 106 species and offered a new infrageneric classification of *Vanilla* based primarily on molecular phylogenetic reconstructions. The species with membranaceous leaves are classified as *Vanilla* subgenus *Vanilla*, which contains two informal "groups." A second subgenus, *Vanilla* subgen. *Xanatha* was created for the remainder of the species. The name is based on the Mexican Totonac Indian name for *Vanilla*, "xanath." This subgenus is further divided into a pair of sections: *Xanatha* and *Tethya*. The former corresponds to mostly leafy neotropical species and is divided into six informal groups (e.g., the *V. palmarum group* and *V. pompona group*). The latter is almost entirely paleotropical in distribution, except that it also includes the Caribbean leafless species. Those taxa are clustered into an informal unit (the

V. barbellata group), along with 11 other groups that are included within the section (e.g., the *V. phalaenopsis* group and *V. africana* group).

GENUS DIVERSITY WITHIN VANILLOIDEAE: TRIBE VANILLEAE

Having examined higher-level relationships among subfamilies of Orchidaceae, and lower-level relationships among species within *Vanilla*, let us now consider relationships among the genera of Vanilloideae. Examples of these genera are shown in Figure 1.4. The subfamily is divided into two tribes, the first of which is Vanilleae. In addition to *Vanilla* itself, this tribe contains eight other tropical genera. Two of these, *Eriaxis* and *Clematepistephium*, are endemic to the isolated Pacific island of New Caledonia. Both genera are monotypic, meaning that they contain only a single species each. An unusual aspect of one of these two species is that *Clematepistphium smilacifolium* grows in the dense shade of the New Caledonian rainforests as a climbing vine. Unlike species of *Vanilla*, however, *Clematepistphium* vines produce no aerial roots. Instead they climb by twisting around the trunks of small trees. Its large, leathery leaves exhibit prominent venation patterns that are reticulate (net-like) rather than exclusively parallel as we see in most orchids and other monocotyledons (Cameron and Dickison, 1998).

The two New Caledonian endemics described above were once classified as species of the genus *Epistephium*, but that genus of 20 species is now considered to be exclusively South American in distribution. Most of these species are erect herbs native to open savanna habitats, and they are most commonly found in nutrient-poor areas of Brazil and Venezuela. Some have been described as scrambling loosely through surrounding vegetation, but none are true climbers. The leaves of *Epistephium* exhibit reticulate venation like their New Caledonian relatives, and the stunning flowers are mostly dark pink or violet. Like most vanilloid orchids, however, they are almost impossible to cultivate. The fruits of these orchids are capsules that dehisce to release distinctive seeds with circular wings, a feature in Orchidaceae found only among Vanilloideae (Cameron and Chase, 1998).

Winged seeds are also found in three other genera of vanilloid orchids: *Pseudovanilla*, *Erythrorchis*, and *Galeola*. These are all closely related, and are native to Southeast Asia, Northeast Australia, and a few Pacific islands. All three of these genera are leafless climbing vines, two of which (*Erythrorchis* and *Galeola*) completely lack chlorophyll. These nonphotosynthetic genera are exclusively parasitic on fungi, a lifestyle technically known as mycoheterotrophy. The leafless genus *Pseudovanilla* is similar to the other two in most aspects, but does eventually develop green pigment within its stems even if it may persist in a presumably nonphotosynthetic state during the juvenile stages of its life cycle. Recent studies have shown that these orchids are the closest living relatives of vanilla (Cameron and Molina, 2006). They climb by means of aerial roots produced at each node of the stem, just like vanilla, and their flowers are remarkably similar to those of *Vanilla* species. Their fruits, however, are designed to accommodate the winged seeds within and so are dry, dehiscent, and nonaromatic at maturity.

There are two other genera of Vanilloideae that grow as nonphotosynthetic mycoheterotrophs: *Cyrtosia* and *Lecanorchis*. Both grow as erect herbs within

FIGURE 1.4 (See color insert following page 136.) Representative genera of subfamily Vanilloideae, the "vanilloid orchids." (a) *Pogonia ophioglossoides* from the United States; (b) *Pseudovanilla foliata* from Queensland, Australia; (c) *Epistephium elatum* from Ecuador; (d) *Erythrorchis cassythoides* from New South Wales, Australia; (e) *Clematepistephium smilacifolium* vine and leaf with reticulate venation from New Caledonia; and (f) *Eriaxis rigida* from New Caledonia.

forested areas of southeast Asia, and both share a number of floral features with *Vanilla*, which has made them difficult to be classified within the subfamily. For example, the fruits of *Cyrtosia* are like those of *Vanilla* in being fleshy and contain small, black, spherical, crustose seeds, but are typically bright red to attract bird or mammal dispersers (Nakamura and Hamada, 1978). The small flowers of *Lecanorchis* are similar in structure to many species of *Vanilla* in that the labellum is fused with the column along its margins to produce a floral tube. Also, like many species of *Vanilla*, the labellum of *Lecanorchis* is ornamented with characteristic bristles and hairs, but *Lecanorchis* fruits are dry capsules lacking odor and containing numerous dust-like seeds with long slender appendages. Further study of the natural history of all these genera is warranted.

GENUS DIVERSITY WITHIN VANILLOIDEAE: TRIBE POGONIEAE

The second tribe within subfamily Vanilloideae is Pogonieae, which contains tropical members but also half a dozen temperate species as well. The tribe is divided into four or possibly five genera. *Pogonia* is one of the temperate genera, and is unusual in that its species are in disjunction between eastern North America (one species, *P. ophioglossoides*) and eastern Asia (3–5 species). These plants are found most commonly in acidic bogs, around the edges of lakes, and within wet savannas. Also native to North America, specifically the eastern United States, is the genus *Isotria*. There are two species in the genus, both of which are characteristic among orchids for having leaves arranged into a whorl of five or six. These plants are spring ephemerals that emerge and reproduce quickly within their deciduous forest habitat before the tree canopy closes fully during the summer months. One other genus, *Cleistes*, has members in temperate North America, and this is the genus *Cleistes*. Most species of this genus (>30 species) are native to tropical South America where they are most commonly found in open savannas that experience seasonal periods of drought. They are equipped with underground tubers that presumably allow them to survive by entering an annual state of dormancy. However, one species of this genus, *Cleistes divaricata*, is native to the southeastern United States. Detailed systematic studies of Pogonieae and vanilloid orchids indicate that this species might be better treated as a separate genus (Cameron and Chase, 1999). The final genus of Pogonieae is *Duckeella*, which contains one or possibly two species indigenous to Venezuela and northern Brazil. The genus produces long linear leaves and bright yellow flowers that rise above wet grassland and savanna habitats. It may occasionally be found rooted in mats of floating vegetation.

FINAL THOUGHTS

The vanilloid orchids are a tremendously diverse group of flowering plants. Whereas the greatest amount of research has been focused on *V. planifolia*, it is important to realize and to appreciate that this is only one species of a lineage that has become adapted to a variety of habitats, lives in greater or lesser partnerships with fungi, exhibits a variety of growth habits, relies on different pollinators, and develops flowers of diverse form (see Figure 1.5). In other words, *V. planifolia* may be the only

FIGURE 1.5 **(See color insert following page 136.)** Representative species of *Vanilla*. (a) *Vanilla phaeantha*; (b) *Vanilla kinabaluensis*; (c) *Vanilla aphylla*; (d) *Vanilla mexicana*; (e) *Vanilla mexicana* in fruit with seeds that are visible; and (f) *Vanilla odorata*.

orchid species of significant agricultural value (out of more than 25,000 naturally occurring species), but it is not entirely unique in the family. Rather, it is just one of approximately 110 species in the genus *Vanilla*, all of which are similar to and yet different from one another. Furthermore, *Vanilla* is only one genus out of 15 genera that are classified within the orchid subfamily Vanilloideae (the "vanilloid orchids"), and some of these are remarkable like vanilla in terms of their growth patterns, floral structure, and fruit dispersal mechanisms. Unfortunately, these orchids are generally overlooked by biologists and those in the vanilla industry, who know only of *V. planifolia*. Many of the genera and species discussed in this chapter are rare and in great danger of extinction primarily due to habitat destruction. By further appreciating and studying their diversity, there is offered a hope of their survival and evolution for another 70 million years.

REFERENCES

Cameron, K. 1999. Biogeography of Vanilloideae (Orchidaceae). *XVI International Botanical Congress, Abstracts*, p. 749. St. Louis, Missouri.

Cameron, K. 2000. Gondwanan biogeography of Vanilloideae (Orchidaceae). *Southern Connections Congress, Programme and Abstracts*, pp. 25–26. Lincoln, New Zealand.

Cameron, K. 2003. Vanilloideae. In: A. Pridgeon, P. Cribb, M. Chase, and F. Rasmussen, eds. *Genera Orchidacearum Vol. 3*. Oxford University Press, Oxford, UK, 281–334.

Cameron, K. 2004. Utility of plastid *psaB* gene sequences for investigating intrafamilial relationships within Orchidaceae. *Molecular Phylogenetics and Evolution* 31:1157–1180.

Cameron, K. 2005. Recent advances in the systematics biology of *Vanilla* and related orchids (Vanilloideae, Orchidaceae). In: *Vanilla: First International Congress*. Allured Publishing, Carol Stream, Illinois.

Cameron, K. 2006. A comparison of plastid *atpB* and *rbcL* gene sequences for inferring phylogenetic relationships within Orchidaceae. In: J. T. Columbus, E. A. Friar, J. M. Porter, L. M. Prince, and M. G. Simpson, eds. *Monocots: Comparative Biology and Evolution*. Rancho Santa Ana Botanic Garden, Claremont, California, 447–464.

Cameron, K. 2007. Molecular phylogenetics of Orchidaceae: The first decade of DNA sequencing. In: K. Cameron, J. Arditti, and T. Kull, eds. *Orchid Biology Reviews and Perspectives Vol. IX*. The New York Botanical Garden Press, Bronx, New York, 163–200.

Cameron, K. and M. Chase. 1998. Seed morphology of the vanilloid orchids. *Lindleyana* 13:148–169.

Cameron, K. and W. Dickison. 1998. Foliar architecture of vanilloid orchids: Insights into the evolution of reticulate leaf venation in monocotyledons. *Botanical Journal of the Linnean Society* 128:45–70.

Cameron, K. and M. W. Chase. 1999. Phylogenetic relationships of Pogoniinae (Vanilloideae, Orchidaceae): An herbaceous example of the eastern North America-eastern Asia phytogeographic disjunction. *Journal of Plant Research* 112:317–329.

Cameron, K. and M. C. Molina. 2006. Photosystem II gene sequences of *psbB* and *psbC* clarify the phylogenetic position of *Vanilla* (Vanilloideae, Orchidaceae). *Cladistics* 22:239–248.

Cameron, K., M. Chase, M. Whitten, P. Kores, D. Jarrell, V. Albert, T. Yukawa, H. Hills, and D. Goldman. 1999. A phylogenetic analysis of the Orchidaceae, evidence from *rbcL* nucleotide sequences. *American Journal of Botany* 86:208–224.

Chase, M. W., K. M. Cameron, R. Barrett, and J. F. Freudenstein. 2003. DNA data and Orchidaceae systematics: A new phylogenetic classification. In: K. W. Dixon, S. P. Kell,

R. L. Barrett, and P. J. Cribb, eds. *Orchid Conservation*. Natural History Publications, Kota Kinabalu, Sabah, 69–89.

Dressler, R. L. 1993. *Phylogeny and Classification of the Orchid Family*. Dioscorides Press, Portland, Oregon.

Freudenstein, J., E. Harris, and F. Rasmussen. 2002. The evolution of anther morphology in orchids: Incumbent anthers, superposed pollinia, and the vandoid complex. *American Journal of Botany* 89:1747–1755.

Govaerts, R., M. A. Campacci, D. Holland Baptista, P. Cribb, A. George, K. Kreuz, and J. Wood. 2008. *World Checklist of Orchidaceae*. The Board of Trustees of the Royal Botanic Gardens, Kew. http://www.kew.org/wcsp/ (accessed 26 February 2008).

Janssen, T. and K. Bremer. 2004. The age of major monocot groups inferred from 800+ *rbcL* sequences. *Botanical Journal of the Linnaean Society* 146:385–398.

Lindley, J. 1835. *Key to Structural, Physiological and Systematic Botany*. Longman, London.

Lubinsky, P., K. Cameron, S. Lepers-Andrzejewski, M. Wong, M. del Carmen Molina, A. Gómez-Pompa, and S.-C. Kim. 2008. Neotropical roots of a Polynesian spice: The hybrid origin of Tahitian vanilla, *Vanilla tahitensis* (Orchidaceae). *American Journal of Botany* 95:1040–1047.

Moore, J. W. 1933. New and critical plants from Raiatea. *Bernice P. Bishop Museum Bulletin* 102:1–53.

Nakamura, S. J. and M. Hamada. 1978. On the seed dispersal of an achlorophyllous orchid, *Galeola septetrionalis. Journal of Japanese Botany* 53:260–263.

Portères, R. 1954. Le genere *Vanilla* et ses especes. In: G. Bouriquet, ed. *Le Vanillier et la Vanille dans le Monde*. Paul Lechevalier, Paris, 94–290.

Ramirez, S. R., B. Gravendeel, R. B. Singer, C. R. Marshall, and N. E. Pierce. 2007. Dating the origin of the Orchidaceae from a fossil orchid with its pollinator. *Nature* 448:1042–1045.

Rolfe, R. A. 1896. A revision of the genus *Vanilla. Journal of the Linnaean Society* 32:439–478.

Soto Arenas, M. A. and P. Cribb. 2010. A new infrageneric classification and synopsis of the genus *Vanilla* Plum. ex Mil. (Orchidaceae: Vanillinae). *Lankesteriana* 9:355–398.

Soto Arenas, M. A. and R. L. Dressler. 2010. A revision of the Mexican and Central American species of *Vanilla* plumier ex Miller with a characterization of their ITS region of the nuclear ribosomal DNA. *Lankesteriana* 9:285–354.

2 Evolutionary Processes and Diversification in the Genus *Vanilla*

Séverine Bory, Spencer Brown, Marie-France Duval, and Pascale Besse

CONTENTS

INTRODUCTION

The diversity of the genus *Vanilla* Plumier ex Miller appears complex at many levels. First, its taxonomy is confused and species delimitation is unclear. Second, at the intraspecific level, genetic diversity is often not correlated with phenotypic diversity. At the moment, a considerable amount of data is available, providing new insights on the possible evolutionary processes responsible for the evolution and diversification of the genus. These processes are detailed and their implication for vanilla conservation and improvement are discussed.

A CONFUSED TAXONOMY

Vanilla is an ancient genus within the Orchidaceae family, the Vanilloidae subfamily, Vanilleae tribe, and Vanillinae subtribe, as demonstrated by recent molecular phylogenetic studies (Bory et al., 2008c; Cameron, 2004, 2005; see Chapter 1). *Vanilla* species are naturally distributed throughout America, Africa, and Asia-Oceania between the 27th north and south parallels (Portères, 1954). Portères (1954) described 110 species in the genus *Vanilla*, a number that reduced to 90 (Cameron and Chase, 1999) and to 107 (Soto Arenas, 2003). New species have also been added, such as the seven additional American species proposed (Soto Arenas, 1999, 2006, 2010) or

V. shenzenica recently described in China (Liu et al., 2007). Altogether, there are more than 200 *Vanilla* species described to date but numerous synonymies remain (Bory et al., 2008c). Taxonomic classification is based on morphological variations (Portères, 1954) and such vegetative and floral characters are strongly influenced by the environment. In particular, vegetative traits (leaves, stems) display considerable variations at the intraspecific level making taxonomic identification difficult (Figure 2.1). This is exemplified by the lack of reliable herbarium vouchers and often the nonavailability of flowers (see Chapter 4). Taxonomy of *Vanilla* will therefore greatly benefit from the development of molecular phylogenetics, which already showed that the sections and subsections used in the taxonomic description of species by Portères do not have a phylogenetic value (Bouetard, 2007; Soto Arenas, 2003). As such, based on cladistic analysis of morphological and molecular data, a new infrageneric classification of *Vanilla* was recently proposed (Soto Arenas and Cribb, 2010) for 106 species examined, dividing genus *Vanilla* in two sub-genera: *Vanilla* and *Xanata* (further divided into sect. *Xanata* and *Tethya*). New keys for 15 Mexican and Central American species (Soto Arenas and Dressler, 2010) and more largely for the infrageneric taxonomic identification within *Vanilla* are also proposed (Soto Arenas and Cribb, 2010). This recent work represents a crucial and major step towards a complete taxonomic revision of the genus.

Intraspecific Diversity

The aromatic species *Vanilla planifolia* G. Jackson, syn. *V. fragrans* (Salisb.) Ames, the main source of commercial vanilla, was disseminated from its area of origin (Mexico) following the discovery of the Americas by Christopher Columbus. Plantations were easily established by cuttings but pod production was unsuccessful in the absence of natural pollinators in the areas of introduction. In 1841, a simple method to hand-pollinate vanilla was discovered by Edmond Albius, a slave, in Reunion Island, and vanilla cuttings rapidly spread from Reunion Island to the Indian Ocean area and worldwide (Bory et al., 2008c; Kahane et al., 2008; see Chapter 17). As a consequence of this dissemination history, extremely low levels of genetic diversity are observed in vanilla plantations worldwide as shown by recent molecular genetic studies (Besse et al., 2004; Bory et al., 2008b, 2008d; Lubinsky et al., 2008a; Minoo et al., 2007; Sreedhar et al., 2007) suggesting a single clonal origin for the vanilla crop. This clone could correspond to the lectotype that was introduced, early in the nineteenth century, by the Marquis of Blandford into the collection of Charles Greville at Paddington (Portères, 1954). Cuttings were sent to the botanical gardens of Paris (France) and Antwerp (Belgium) from where these specimens were disseminated worldwide (Bory et al., 2008c; Kahane et al., 2008). It is thus surprising to observe an important morphological diversity in *V. planifolia* in the areas of introduction such as Reunion Island (Bory et al., 2008b, 2008c, 2008d) for a crop with a clonal origin and vegetatively propagated by cuttings.

All these observations raise important questions regarding the processes that might be involved in the evolution and diversification of vanilla. Some of the key processes that have been identified so far and the explanations that these can provide for the genetic and taxonomic complexity observed in *Vanilla* are discussed.

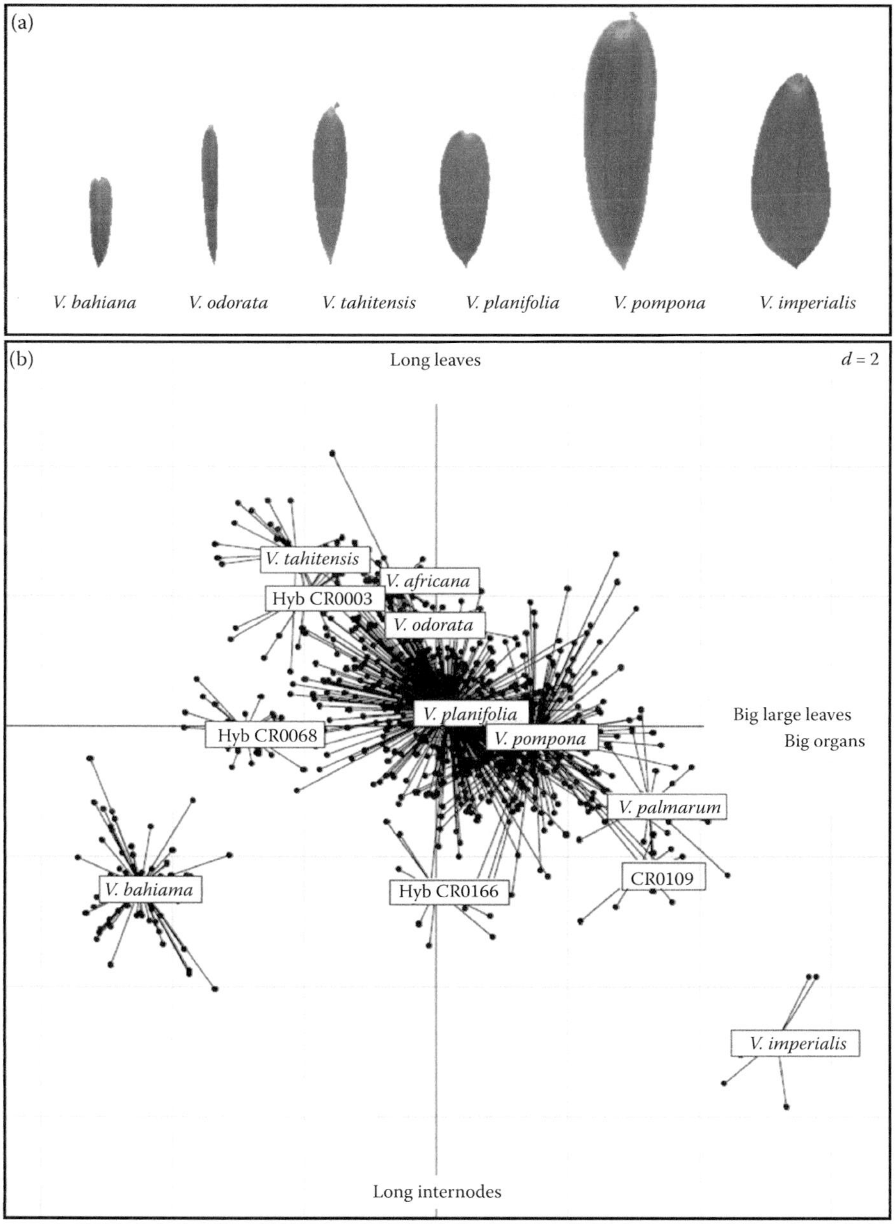

FIGURE 2.1 Morphological vegetative traits in *Vanilla* species from the CIRAD collection in Reunion Island (see Chapter 3): (a) typical leaf specimen for some species; (b) principal component analysis of vegetative traits (leaf and stem) measured in different species showing the importance of intraspecific variations leading to overlapping of species.

VEGETATIVE VERSUS SEXUAL REPRODUCTION

For most *Vanilla* species, vegetative growth occurring naturally from stem cuttings (Portères, 1954) is the predominant reproductive mode, and appears as an efficient strategy adopted by the plant to develop settlements (Figure 2.2). Stems running on the ground are frequently observed, giving new roots and creating new individuals when the stem is cut, as reported for species such as *V. bahiana* Hoehne (Pignal, 1994) and *V. chamissonis* Klotzsch (Macedo Reis, 2000) in Brazil, *V. barbellata* Reichenbach f., *V. claviculata* (W. Wright) Swartz and *V. dilloniana* Correll (Nielsen and Siegismund, 1999) in Puerto Rico or *V. madagascariensis* Rolfe in Madagascar (P. Besse, pers. obs.). In Mexico, with reference to *V. planifolia*, in natural conditions, the same individual can cover up to 0.2 ha (Soto Arenas, 1999).

In *Vanilla* species, a rostellum membrane separates the female and male parts of the flower, and pollination therefore depends on the intervention of external pollinators. A notable exception is the species *V. palmarum* (Salzm. ex Lindl.) Lindl., which spontaneously self-pollinates (Bory et al., 2008c; Soto Arenas, 2006). Consequent, due to the need for pollinators, sexual reproduction is rarely observed in natural conditions. For *V. planifolia*, rates of 1% to 1‰ are reported (Bory et al., 2008c; Soto Arenas, 1999) with possible natural pollinators in America being orchid bees from the *Euglossa* and perhaps from the *Eulaema* genera (Lubinsky et al., 2006; Soto Arenas, 2006). Sexual reproduction rates reported for the species *V. chamissonis* (6% autogamy and 15% allogamy) are also relatively low (Macedo Reis, 2000).

FIGURE 2.2 Typical vegetative growth observed in *Vanilla* species. Left: *V. madagascariensis* in Madagascar. Right: *V. pompona* in Guadeloupe. (Courtesy of P. Besse.)

However, even rare sexual reproduction events can generate an important genetic diversification because a single sexual reproduction event is able to generate numerous genotypes that can vegetatively propagate rapidly. Heterozygosity observed in *V. planifolia* was reported to be 0–0.078 using isozymes (Soto Arenas, 1999), 0.154 using SSR markers (Bory et al., 2008b) and 0.293 using AFLPs (Bory et al., 2008d). Given these heterozygosity levels, even selfing can generate genetic diversity, as demonstrated through manual self-pollination experiments (Bory et al., 2008d) leading to increased diversity estimates (D_{max} from 0.106 to 0.140) through novel allelic combinations (Figure 2.3). This is well illustrated in the case of *V. planifolia* in areas of introduction, where natural pollinators are absent. In these areas, such as in Reunion Island, traditional cultivation practices involve vine propagation by cuttings, and manual self-pollination to produce pods. This resulted in the appearance of novel vanilla varieties such as the "Aiguille" type observed in Reunion Island, which most likely resulted from accidental seed germination in the field from a forgotten pod, and subsequent vegetative propagation of the individual (Bory et al., 2008d) (Figure 2.3). Such a novel type can rapidly spread in plantations given the vegetative propagation used to multiply vines. This must also happen in the wild. A combination of sexual and vegetative reproduction, where one creates diversity and the other helps settlement, has already been suggested for the species *V. pompona* Schiede and *V. bahiana* in tropical America based on AFLP patterns (Bory et al., 2008d). Sexual reproduction is therefore a key evolutionary process for most species of the genus despite its low rates and because of their major vegetative reproduction. A few species of *Vanilla* appear to rely solely on sexual reproduction for propagation. This is the case for *V. palmarum*, which is entirely epiphytic on a palm tree with a

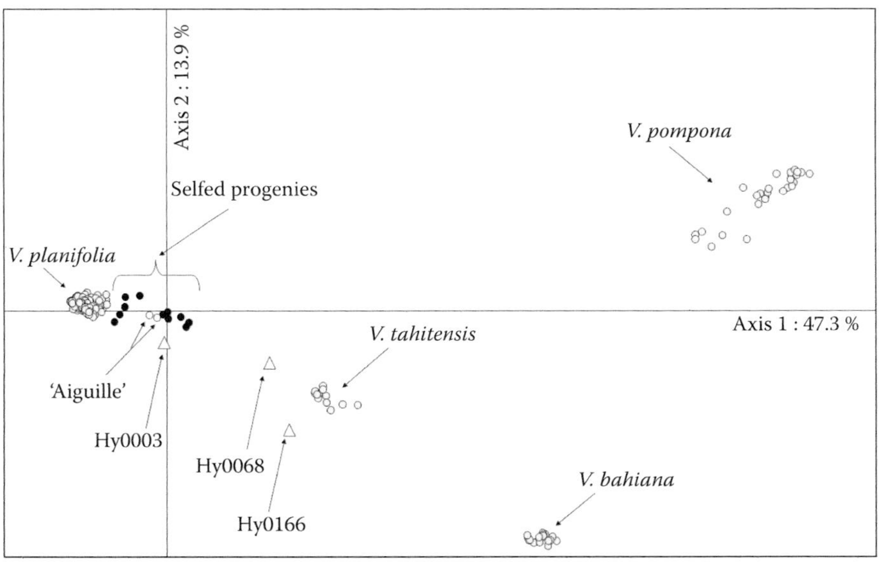

FIGURE 2.3 Factorial analysis from AFLP markers on different American *Vanilla* species illustrating the increased diversity for *V. planifolia* selfed progenies.

short lifecycle (Pignal, 1994) and for *V. mexicana* Mill. (syn *V. inodora* Shiede) in which even artificial vegetative propagation is unsuccessful (P. Feldmann, pers. com.) (Figure 2.4).

INTERSPECIFIC HYBRIDIZATION

The main factors preventing interspecific hybridization in the Orchidaceae family are pre-pollination mechanisms such as pollinator specificity, flowering phenologies, or mechanical barriers in flowers (Dressler, 1981; Gill, 1989; Grant, 1994; Paulus and Gack, 1990; Van Der Pijl and Dodson, 1966). On the contrary, genetic incompatibility between closely related species is rarely observed (Dressler, 1993; Johansen, 1990; Sanford, 1964, 1967). This is also the case for *Vanilla*. Indeed, most interspecific artificial crosses attempted to date in *Vanilla* have been successful showing the lack of genetic incompatibility between the species involved. Interspecific hybrids were successfully obtained between closely related American species (*V. planifolia* × *V. tahitensis* J.W. Moore—accession Hy0003 in Figures 2.1 and 2.3, *V. planifolia* × *V. pompona*) in breeding programs in Madagascar (Bory et al., 2008c), and even between distantly related species such as the Indian *V. aphylla* Blume and the American *V. planifolia* in breeding programs in India (Minoo et al., 2006).

FIGURE 2.4 Exclusive sexually reproducing species. Left: *V. palmarum* in the CIRAD collection. Fruits were spontaneously obtained in an insect proof quarantine glasshouse. (Courtesy of M. Grisoni.) Right: *V. mexicana* in Guadeloupe. (Courtesy of P. Besse and P. Feldmann.)

There is a growing evidence for the occurrence of natural interspecific hybridization in *Vanilla*. A study on three native species of *Vanilla*, *V. claviculata*, *V. barbellata*, and *V. dilloniana* in the western part of the island of Puerto Rico, showed the possibility of interspecific hybridization between *V. claviculata* and *V. barbellata* in sympatric areas (Nielsen, 2000; Nielsen and Siegismund, 1999). This was demonstrated by using isozyme markers, and floral morphological observations confirmed the hybrid status of sympatric populations. On the other hand, *V. dilloniana*, showing a different phenology, did not hybridize with the other two species. Recent work using AFLP and SSR markers also suggested the possibility of interspecific hybrid formation in tropical America, involving species such as *V. bahiana*, *V. planifolia*, or *V. pompona* (Bory, 2007; Bory et al., 2008d) (Figure 2.5). The species *V. tahitensis* was also recently shown using nuclear ITS and cp DNA sequences to result from intentional or inadvertent hybridization between the species *V. planifolia* and

FIGURE 2.5 Flowers, fruits, and leaves from accession CR0068 (a) from Costa Rica, putatively identified as a hybrid species derived from a maternal *V. planifolia* donor species (b) based on AFLP, SSR, and cp DNA markers data. (Courtesy of M. Grisoni.)

V. odorata C. Presl that could have happened during the Late Postclassic (1350–1500) in Mesoamerica (Lubinsky et al., 2008b).

POLYPLOIDIZATION

We recently demonstrated the occurrence of recent polyploidization events (in less than 200 years as *V. planifolia* was introduced in 1822) in Reunion Island (Bory et al., 2008a). Congruent evidences (AFLP markers, genome size, chromosome counts, and stomatal length) showed the formation of auto-triploid self-sterile types ("Stérile"), as well as auto-tetraploid types ("Grosse Vanille"), with genome sizes of 2C = 7.5 and 10 pg, respectively, as opposed to 5 pg for conventional "Classique" varieties (Bory et al., 2008a) (Table 2.1). The most likely formation of these types was suggested to be through manual self-pollination accompanied by the formation of unreduced *2n* gametes (Bretagnolle and Thompson, 1995), seed germination from a forgotten pod, followed by vegetative multiplication of the individuals (Bory et al., 2008a) (Figure 2.6). Polyploidy was also reported for the cultivated species of *V. tahitensis* in Polynesia with the existence of diploid and tetraploid (i.e., "Haapape") types (Duval et al., 2006; see Chapter 13) and this species might have resulted from both polyploidy and sexual regeneration following its *V. planifolia* × *V. odorata* origin (Lubinsky et al., 2008b).

Polyploidization could therefore be a major phenomenon in the evolution of *Vanilla*. In order to put this hypothesis to test, we conducted a preliminary survey on genome size variation in different *Vanilla* species. Thirty-eight accessions were analyzed by flow cytometry according to the protocol detailed by Bory et al. (2008a) using wheat as an internal standard: *Triticum aestivum* L. cv. Chinese Spring, 2C = 30.9 pg, 43.7% GC (Marie and Brown, 1993). These accessions belong to 17 different *Vanilla* species and also included 3 artificial hybrids (*V. planifolia* × *V. planifolia*, *V. planifolia* × *V. tahitensis*, *V. planifolia* × *V. phaeantha* Rchb. f.). The entire leaf samples were collected from vines maintained in the *Vanilla* genetic resources collection of CIRAD in Reunion Island (see Chapter 3 and Grisoni et al., 2007). Details for each accession (species, putative continent of origin, place of sampling, and genome size) are presented in Table 2.1. For each species, fluorescence histograms revealed five endoreplicated peaks and the marginal replication ratio was still irregular (from 1.5 to 1.8 instead of 2), as encountered in *V. planifolia* (Bory et al., 2008a).

2C nuclear DNA content varied from 4.72 (±0.05) pg (*V. tahitensis*) to 12.00 (±0.11) pg (*V. phalaenopsis* Reichb. f. ex Van Houtte) for 34 wild accessions. One accession CR0067 (*Vanilla* sp.) had an extreme value at 22.31 pg (one measure with wheat standard) (Table 2.1), which was confirmed by using another standard (pea, data not shown). Intra-accession variation coefficients did not exceed 5%.

These results indicate that genome size variations exist in *Vanilla*, which could suggest the occurrence of polyploid species, based on what was detected for *V. planifolia* (Bory et al., 2008a). In particular, African accessions displayed bigger genome sizes than American accessions, with 2C DNA content ranging from 6.93 to 22.31 pg and 4.72 to 9.23 pg, respectively. African species may, therefore, have been subjected to more dramatic genomic rearrangements and polyploidization events than their American or Asian counterparts. Finally, as it was observed in *V. planifolia* in Reunion Island (Bory et al., 2008a), intraspecific genome size

TABLE 2.1
Accession Code from the *Vanilla* Genetic Resources Collection of CIRAD Reunion Island

Accession Code	Species	Origin[a]	Place of Sampling	Mean 2C pg (±s.d.)
Group A[b]	*V. planifolia "Classique"*	America	Reunion Island	5.03 (±0.16)
Group B[b]	*V. planifolia "Stérile"*	America	Reunion Island	7.67 (±0.14)
Group C[b]	*V. planifolia "Grosse Vanille"*	America	Reunion Island	10.00 (±0.28)
CR0056	*V. planifolia* × *V. phaeantha*	N/A	Artificial hybrid	4.48 (±0.10)
CR0747	*V. planifolia* × *V. planifolia*	N/A	Artificial hybrid	5.24 (±0.12)
CR0003	*V. planifolia* × *V. tahitensis*	N/A	Artificial hybrid	10.12 (±0.10)
CR0062	*V. bahiana*	America	Unknown	6.70 (±0.32)
CR0072	*V. bahiana*	America	Brazil (Bahia)	6.60
CR0076	*V. bahiana*	America	Brazil (Bahia)	6.52
CR0085	*V. bahiana*	America	Brazil (Bahia)	7.28
CR0097	*V. bahiana*	America	Brazil (Bahia)	7.09
CR0099	*V. bahiana*	America	Brazil (Bahia)	6.91
CR0666	*V. chamissonis*	America	Brazil (Sao Paulo)	8.22 (±0.06)
CR0667	*V. chamissonis*	America	Brazil (Sao Paulo)	8.14 (±0.35)
CR0702	*V. chamissonis*	America	Unknown	7.47 (±0.03)
CR0693	*V. (cf.) grandiflora* Lindl.	America	Guyana	9.23 (±0.25)
CR0109	*V. leprieuri* R. Porteres	America	French Guyana	7.74 (±0.02)
CR0686	*V. odorata*	America	Unknown	4.95 (±0.12)
CR0083	*V. palmarum*	America	Brazil (Bahia)	7.00 (±0.29)
CR0017	*V. tahitensis*	America	French Polynesia	4.72 (±0.05)
CR0063	*V. tahitensis*	America	Unknown	6.88 (±0.13)
CR0069	*V. trigonocarpa* Hoehne	America	Brazil (Alagoinhas)	8.21 (±0.07)
CR0103	*V. africana* Lindl.	Africa	Africa	10.25 (±0.06)
CR0107	*V. africana*	Africa	Africa	10.22 (±0.18)
CR0696	*V. crenulata* Rolfe	Africa	Unknown	9.88 (±0.34)
CR0091	*V. crenulata*	Africa	Africa	10.24 (±0.16)
CR0102	*V. crenulata*	Africa	Africa	9.79 (±0.37)
CR0106	*V. crenulata*	Africa	Africa	10.47 (±0.38)
CR0108	*V. humblotii* Rchb. f.	Africa	Comoros	11.81 (±0.09)
CR0104	*V. imperialis* Kraenzl.	Africa	Africa	6.93
CR0105	*V. imperialis*	Africa	Africa	10.14 (±0.34)
CR0796	*V. imperialis*	Africa	Unknown	7.18 (±0.00)
CR0797	*V. imperialis*	Africa	Unknown	7.06 (±0.24)
CR0141	*V. madagascariensis*	Africa	Madagascar	8.06
CR0142	*V. madagascariensis*	Africa	Madagascar	8.02

continued

TABLE 2.1 (continued)
Accession Code from the *Vanilla* Genetic Resources Collection of CIRAD Reunion Island

Accession Code	Species	Origin[a]	Place of Sampling	Mean 2C pg (±s.d.)
CR0146	*V. phalaenopsis*	Africa	Unknown	12.00 (±0.11)
CR0067	*Vanilla* sp.	Africa	Central Africa	22.31
CR0058	*V. albida*	Asia	Unknown	5.90 (±0.16)
CR0793	*V. albida*	Asia	Thailand	5.15
CR0059	*V. albida*	Asia	Unknown	8.65 (±0.08)
CR0145	*V. aphylla*	Asia	Thailand	9.81 (±0.03)

In the first three lines, are given results for *V. planifolia* according to the study of Bory et al. (2008a).

[a] According to Portères (1954).

[b] From the study of Bory et al. (2008a).

variations were revealed in some species (*V. imperialis* Kraenzl., *V. albida* Blume), which may reflect the occurrence of different ploidy levels even within species. These results need to be explored further by chromosome counts for each species, but they already strongly suggest that polyploidy might be a major phenomenon in the evolution of *Vanilla*.

CONCLUSIONS

There is, therefore, growing evidence that demonstrate the complexity of the processes involved in the evolution and diversification of *Vanilla*. As for many species for which vegetative reproduction is predominant, we observed phenotypic diversity at the intraspecific level in *V. planifolia*, which was not congruent with the observed low genetic diversity of this clonal crop. We demonstrated that this discrepancy was in part due to the occurrence of rare sexual reproduction events, as well as to the occurrence of polyploidization in this species. Given that these variations have happened in Reunion Island in less than 200 years; there is little doubt that such intraspecific variations exist in other species of the genus found in the wild, and might be responsible for the difficulty to correctly identify species solely based on morphological observations. This is exacerbated by the occurrence of interspecific hybridizations in the genus, which makes clear taxonomic designation is even more delicate.

Vanilla can therefore be considered as a TCG, a "Taxonomic Complex Group" *sensu* Ennos et al. (Ennos et al., 2005). Indeed, it exhibits (1) a uniparental reproduction mode (vegetative reproduction), (2) interspecific hybridization in sympatric areas, and (3) polyploidy. These mechanisms have profound effects on the organization of the biological diversity and have been described as being responsible for the difficulty to define discrete, stable and coherent taxa in such TCGs (Ennos et al., 2005). TCGs are widespread in plants and uniparental reproduction can produce a

FIGURE 2.6 Schematic representation of the possible formation of autotriploid and autotetraploid *V. planifolia* types in Reunion Island.

complex mixture of sexual outcrossing and uniparental lineages that can be at different ploidy levels and the whole complex can be involved in reticulate evolution generating novel uniparental lineages by hybridization (Ennos et al., 2005). This has great implications on the way conservation programs should be conducted. In such TCGs, as it is often not possible to classify biodiversity into discrete and unambiguous species, traditional species-based conservation programs are not appropriate. As recommended, *in situ* conservation should focus on the evolutionary processes that generate taxonomic diversity rather than on the poorly defined taxa resulting from this evolution (Ennos et al., 2005). This includes concentrating on species that might be widespread (and thus not concerned by classical conservation efforts) but responsible for the generation of taxonomic diversity (through hybridization, introgression, or polyploidization).

Therefore, not only the mechanisms described in this chapter provide a better understanding of the *Vanilla* genus evolution, but they also are of major importance with regard to future genetic resources management and conservation (Crandall et al., 2000; Moritz, 2002).

These mechanisms are also of major interest with regard to the future improvement of *V. planifolia*. Interspecific hybridizations between *V. planifolia* and other aromatic species have already proved successful. In Madagascar, the production of *V. planifolia* × *V. tahitensis* hybrid variety "Manitra ampotony" led to an increased vanillin content (6.7% vanillin versus 2.5% in *V. planifolia*) and the (*V. planifolia* × *V. pompona)* × *V. planifolia* "Tsy taitry" shows increased resistance to different diseases (Dequaire, 1976; FOFIFA, 1990; Nany, 1996). In India, *V. planifolia* × *V. aphylla* hybrids were produced to increase *Fusarium* resistance (Minoo et al., 2006). At the intraspecific level, self-pollination could also be used to increase diversity in this genetically uniform crop (Bory et al., 2008d; Minoo et al., 2006). Furthermore, the agronomic characterization (vigor, resistance, vanillin production) of autotetraploid types is currently being performed in Reunion Island as a first step for testing for the potentialities of polyploidy breeding strategies in *V. planifolia*.

Unraveling the evolution and acquisition of traits of agronomic interest in the genus will also be of major importance. These traits include fragrance of fruits, which despite its considerable breeding interest has received limited attention particularly with regards to its evolution. Fragrant fruit species are almost exclusively restricted to America (Soto Arenas, 2003), and this character could have been selected as favoring fruit dispersion by bats (Soto Arenas, 1999) or sticky-seed dispersion by bees through fruit fragrance collection as observed in *V. grandiflora* (Lubinsky et al., 2006). This matter was recently addressed by surveying intron length variations in the COMT gene family (Besse et al., 2009), encoding key enzymes in the phenylpropanoid pathway putatively involved in the biosynthesis of vanillin. Further work is also needed to understand the evolution, mechanisms and genetic determinism of spontaneous self-pollination in the genus (*V. palmarum*)—a highly desirable character that would considerably reduce bean production costs. Finally, elucidating how the aphyllous species of the genus have evolved and differentiated might be of great interest in our understanding of adaptation to dry conditions, given the predicted future of great climatic changes.

REFERENCES

Besse, P., D. Da Silva, S. Bory, et al. 2004. RAPD genetic diversity in cultivated vanilla: *Vanilla planifolia*, and relationships with *V. tahitensis* and *V. pompona*. *Plant Science* 167:379–385.

Besse, P., D. Da Silva, S. Bory, M. Noirot, and M. Grisoni, 2009. Variation in intron length in caffeic acid *O*-methyltransferase (COMT) in *Vanilla* species (Orchidaceae). *Plant Science* 176:452–460.

Bory, S. 2007. Diversité de *Vanilla planifolia* dans l'Ocean Indien et de ses espèces apparentées: Aspects génétiques, cytogénétiques et épigénétiques. PhD thesis, Reunion Island University, France.

Bory, S., O. Catrice, S.C. Brown, et al. 2008a. Natural polyploidy in *Vanilla planifolia* (Orchidaceae) *Genome* 51:816–826.

Bory, S., D. Da Silva, A.M. Risterucci, et al. 2008b. Development of microsatellite markers in cultivated *Vanilla*: Polymorphism and transferability to other *Vanilla* species. *Scientia Horticulturae* 115:420–425.

Bory, S., M. Grisoni, M.-F. Duval, and P. Besse, 2008c. Biodiversity and preservation of vanilla: Present state of knowledge. *Genetic Resources and Crop Evolution* 55: 551–571.

Bory, S., P. Lubinsky, A.M. Risterucci, et al., 2008d. Patterns of introduction and diversification of *Vanilla planifolia* (Orchidaceae) in Reunion Island (Indian Ocean). *American Journal of Botany* 95 (7):805–815.

Bouetard, A. 2007. Utilisation de marqueurs chloroplastiques pour l'identification spécifique et l'élucidation des relations phylogénétiques au sein du genre *Vanilla*. Master M2 thesis, Université de Pau et des Pays de l'Adour, France, Pau.

Bretagnolle, F. and J.D. Thompson, 1995. Gametes with the somatic chromosome number: Mechanisms of their formation and role in the evolution of autopolyploid plants. *New Phytologist* 129 (1):1–22.

Cameron, K.M. 2004. Utility of plastid *psa*B gene sequences for investigating intrafamilial relationships within Orchidaceae. *Molecular Phylogenetics and Evolution* 31 (3): 1157–1180.

Cameron, K.M. 2005. Recent advances in the systematic biology of *Vanilla* and related orchids (Orchidaceae: subfamily Vanilloideae). *First International Congress*, November 11–12, 2003, Princeton, NJ, USA.

Cameron, K.M. and M.W. Chase, 1999. Phylogenetic relationships of Pogoniiae (Vanilloideae, Orchidaceae): An herbaceous example of the eastern north America-eastern Asia phylogeographic disjunction. *J. Plant Res.* 112:317–329.

Crandall, K.A. O. R. P. Bininda-Emonds, G. M. Mace, and R. K. Wayne. 2000. Considering evolutionary processes in conservation biology. *Trends Ecol. Evol.* 15:290–295.

Dequaire, J. 1976. L'amélioration du vanillier à Madagascar. *Journal d'Agriculture Tropicale et de Botanique Appliquée* 23 (7–12):140–158.

Dressler, R.L. 1981. *The Orchids: Natural History and Classification*. Cambridge, MA: Harvard University Press.

Dressler, R.L. 1993. *Phylogeny and Classification of the Orchid Family*. Portland, OR: Dioscorides Press.

Duval, M.-F., S. Bory, S. Andrzejewski, et al., 2006. Diversité génétique des vanilliers dans leurs zones de dispersion secondaire. *Les Actes du BRG* 6:181–196.

Ennos, R.A., G.C. French, and P.M. Hollingsworth, 2005. Conserving taxonomic complexity. *Trends in Ecology and Evolution* 20 (4):164–168.

FOFIFA. 1990. Le vanillier. Bilan de la Recherche Agricole à Madagascar. 112–119.

Gill, D.E. 1989. Fruiting failure, pollinator inefficiency and speciation in orchids. In: D. Otte and J. A. Endler (ed.) *Speciation and its Consequences*. Sunderland, MA: Sinauer Associates.

Grant, V. 1994. Modes and origins of mechanical and ethological isolation in angiosperms. *Proceedings of the National Academy of Science USA* 91:3–10.

Grisoni, M., M. Moles, P. Besse, et al., 2007. Towards an international plant collection to maintain and characterize the endangered genetic resources of vanilla. *Acta Horticulturae (ISHS)* 760:83–91.

Johansen, B. 1990. Incompatibility in *Dendrobium* (Orchidaceae). *Botanical Journal of the Linnean Society* 103:165–196.

Kahane, R., P. Besse, M. Grisoni, F. Le Bellec, and E. Odoux, 2008. Bourbon Vanilla: Natural flavour with a future. *Chronica Horticulturae* 48 (2):23–28.

Liu, Z.J., S.C. Chen, and Z.Z. Ru, 2007. *Vanilla shenzhenica* Z.J. Liu & S.C. Chen, the first new species of Orchidaceae in Shenzhen, South China. *Acta Phytotaxonomica Sinica* 45:301–303.

Lubinsky, P., S. Bory, J.H. Hernandez, S.-C. Kim, and A. Gomez-Pompa, 2008a. Origins and dispersal of cultivated vanilla (*Vanilla planifolia* Jacks. [Orchidaceae]). *Economic Botany* 62:127–138.

Lubinsky, P., K.M. Cameron, M.C. Molina, et al. 2008b. Neotropical roots of a polynesian spice: The hybrid origin of tahitian vanilla, *Vanilla tahitensis* (Orchidaceae). *American Journal of Botany* 95 (8):1–8.
Lubinsky, P., M. Van Dam, and A. Van Dam, 2006. Pollination of *Vanilla* and evolution in Orchidaceae. *Orchids* 75 (12):926–929.
Macedo Reis, C.A. 2000. Biologia reprodutiva e propagacao vegetativa de *Vanilla chamissonis* Klotzsch: subsidios para manejo sustentado, Ciencas Florestais, Escola Superior de Agric Luiz de Queiroz, Piracicaba, Sao Paulo, Brasil, Piracicaba, SP - Brasil.
Marie, D. and S.C. Brown, 1993. A cytometric exercise in plant DNA histograms, with 2C values for seventy species. *Biology of the Cell* 78:41–51.
Minoo, D., V.N. Jayakumar, S.S. Veena, et al. 2007. Genetic variations and interrelationships in *Vanilla planifolia* and few related species as expressed by RAPD polymorphism. *Genetic Resources and Crop Evolution* DOI: 10.1007/s10722-007-9252-3.
Minoo, D., K. Nirmal Babu, P.N. Ravindran, and K.V. Peter, 2006. Interspecific hybridization in vanilla and molecular characterization of hybrids and selfed progenies using RAPD and AFLP markers. *Scientia Horticulturae* 108 (4):414–422.
Moritz, C. 2002. Strategies to protect biological diversity and the evolutionary processes that sustain it. *Systematic Biology* 51:238–254.
Nany, F. 1996. Résultats de recherche vanille: Manitra ampotony et Tsy taitra, deux variétés prometteuses. *Les cahiers du CITE "Spécial plantes aromatiques et médicinales"* 4:47–49.
Nielsen, R.L. 2000. Natural hybridization between *Vanilla claviculata* (W.Wright) Sw. and *V. barbellata* Rchb.f. (Orchidaceae): Genetic, morphological, and pollination experimental data. *Botanical Journal of the Linnean Society* 133 (3):285–302.
Nielsen, R.L. and H.R. Siegismund, 1999. Interspecific differentiation and hybridization in *Vanilla* species (Orchidaceae). *Heredity* 83 (5):560–567.
Paulus, H.F. and C. Gack, 1990. Pollination of *Ophrys* (Orchidaceae) in Cyprus. *Plant Systematics and Evolution* 169:177–207.
Pignal, M. 1994. Deux vanilles du Brésil: *Vanilla palmarum* Lindley et *Vanilla bahiana* Hoehne. *L'Orchidophile* 110:23–25.
Portères, R. 1954. Le genre *Vanilla* et ses espèces. In: G. Bouriquet, ed. *Le vanillier et la vanille dans le monde*. Paul Lechevalier, Paris, pp. 94–290.
Sanford, W.W. 1964. Sexual compatibility relationship in *Oncidium* and related genera. *American Orchids Society Bulletin* 33:1035–1048.
Sanford, W.W. 1967. Sexual compatibility relationship in *Oncidium* and related genera: Part II. *American Orchids Society Bulletin* 36:114–122.
Soto Arenas, M.A. 1999. *Filogeografia y recursos genéticos de las vainillas de México*. Herbario de la Asociación Mexicana de Orquideología, March 31, 1999 [cited 2004]. Available at http://www.conabio.gob.mx/institucion/proyectos/resultados/InfJ101.pdf.
Soto Arenas, M.A. 2003. *Vanilla*. In: A. M. Pridgeon, P. J. Cribb, M. W. Chase and F. N. Rasmussen, eds. *Genera orchidacearum: Orchidoideae.* Oxford University Press, USA.
Soto Arenas, M.A. 2006. Vainilla: los retos de un cultivo basado en una especie amenazada con una historia de vida compleja. *Congreso Internacional de Productores de Vainilla*, May 26–28, 2006, Papantla, Veracruz, Mexico.
Soto Arenas, M.A. 2010. A new species of *Vanilla* from south America. *Lankesteriana* 9(3):281–284.
Soto Arenas, M.A. and P. Cribb, 2010. A new infrageneric classification and synopsis of the genus *Vanilla* Plum. ex Mil. (Orchidaceae: Vanilinae). *Lankesteriana* 9(3):355–398.
Soto Arenas, M.A. and R.L. Dressler, 2010. A revision of the Mexican and Central American species of *Vanilla* plumier ex Miller with a characterization of their ITS region of the nuclear ribosomal DNA. *Lankesteriana* 9(3):285–354.

Sreedhar, R.V., L. Venkatachalam, K. Roohie, and N. Bhagyalakshmi, 2007. Molecular analyses of *Vanilla planifolia* cultivated in India using RAPD and ISSR markers. *Orchid Science and Biotechnology* 1 (1):29–33.

Van Der Pijl, L. and C.H. Dodson, 1966. *Orchid Flowers, Their Pollination and Evolution*. University of Miami Press, Coral Cables, FL.

3 Conservation and Movement of Vanilla Germplasm

Michel Roux-Cuvelier and Michel Grisoni

CONTENTS

IMPORTANCE OF PLANT GENETIC RESOURCES

Plant genetic resources are a major strategic challenge for all activities linked to agriculture and the agro-food industry, especially in the present context of climate change. Globally, since the 1950s, population increase and the development of intensive agriculture have contributed to a reduction in the diversity of plant species. Today, the protection of these resources is of vital importance in achieving sustainable food security for the populations.

From the 1970s, a number of initiatives were developed to safeguard the diversity of cultivated species. During 1971, the initiation of the CGIAR (Consultative Group on International Agricultural Research) provided an initial response to the problem of the loss of genetic diversity for the major agricultural species. At present, 11 of the 15 CGIAR centers are responsible for maintaining international gene banks for the preservation and dissemination of the plant genetic resources that provide the basis of world food security (http://www.cgiar.org/index.html). More recently, initiatives aimed at securing collections of genetic resources have been launched. These include the creation of the OECD's BRCs (Biological Resource Centres) or of the Global Crop Diversity Trust (http://www.croptrust.org/main/), that is behind operational projects such as the Svalbard Global Seed Vault that currently conserves almost 660 different genera and 3300 species from all continents (http://www.croptrust.org/main/arctic.php?itemid=211).

While the major agricultural species are covered by recognized conservation systems, few large-scale initiatives have been taken to preserve underutilized and orphan species. However, to meet the challenges of the future, the study and protection of agricultural biodiversity must not be based solely on a limited number of species, but must be envisaged from a broad, open perspective, presenting each species as being interdependent on the others and a representative of its own specific evolutionary process.

Today, a conservation program cannot be implemented without first defining the objectives for the use of resources and for acquiring knowledge of intra- or interspecies genetic diversity. Once the objectives have been established, questions should be asked about the representativeness of genetic diversity (what should be preserved and in what quantities?) and the cost of setting up and maintaining these collections.

In situ collections consist of preserving species in their natural or human-modified ecosystems, in other words, the place in which they developed their distinctive characteristics. These collections mainly concern wild species and are often represented by the national or regional parks that protect the ecosystems as a whole. While this type of conservation represents the ideal model in that it maintains the selection pressure of its environment, its effectiveness in terms of conservation often depends on the degree to which local populations are involved in the management of the region and its resources. Moreover, it provides no protection against climate or biological risks (pathogens, plant pests, or invasive species).

Ex situ collections consist of preserving genetic resources outside of their natural habitat. This conservation method freezes the resources at the genetic level. The degree of protection of resources and their independence in relation to the natural environment is highly variable depending on the type of conservation. Whole plants may be kept in the form of field collections, arboreta, in greenhouses, or in tissue culture. Fragments of plants (meristems, embryos, tissues), which can be used to regenerate a whole plant, may be preserved at very low temperature in liquid nitrogen. Seeds are commonly held in cold storage, under controlled humidity conditions. Conserving pollen at low temperature means that a stock of haploid material that can be maintained. By conserving as much diversity as possible at lower cost, DNA banks can be used for intra- and interspecies genetic studies. Within *ex situ* collections, a distinction can be made between safeguarding and active or working collections, which give rise to specific activities that involve the resources stored, such as genetic diversity studies and plant breeding research.

These *ex situ* conservation methods are, nevertheless, reliant on human activity and on the smooth functioning of the conservation facilities and structures.

Finally, herbaria (see Chapter 4)—the first method for conserving plant species (sixteenth century)—and spirit collections are essential, especially for conducting taxonomic studies.

Although the economic cost of maintaining collections is linked to the number of accessions conserved, the issue of the number of individuals that represent diversity and ensure the security of the accession is crucial. The principle of the "core collection" may provide a response. Noirot et al. (1996) developed a method for creating a core collection based on quantitative data. This method (Principal Component Scoring) aims to include the maximum diversity from the base collection in a sample of minimum size, while avoiding redundancies.

For plant genetic resources, each conservation system has its advantages and disadvantages and often only a combination of these different systems will make it possible to optimize and to secure the conservation of resources.

In addition to the quantitative aspect (the number of species and accessions conserved), the quality of data and research programs linked to resources is now a key element in justifying the material and human investments. The value and accuracy of these data also contribute to the development of resources.

THE CASE OF VANILLA

The genus *Vanilla*, which belongs to the Orchid family, includes 90–100 species depending on the author (Bory et al., 2008c). Most of these species are wild; only two of them, *V. planifolia* and *V. tahitensis*, are grown to produce commercial vanilla, with *V. planifolia* providing 95% of the world production. The taxonomy of vanilla is very old (Portères, 1954; Rolfe, 1896), incomplete, and still imprecise (see Chapter 1). It must therefore be thoroughly revised, especially in light of the findings of recent molecular genetics and cytogenetics research (Bory et al., 2008a, 2008b; Cameron, 2003, 2009; Schluter et al., 2007; Soto Arenas, 2003; Verma et al., 2009). For example, several recent research studies on the origin of *V. tahitensis* suggest that this species is the result of cross-breeding between *V. planifolia* and *V. odorata* (Besse et al., 2004; Bory et al., 2008d; Lubinsky et al., 2008b).

The species of the genus *Vanilla* are mainly found in natural habitats in tropical and subtropical regions of the American, African, and Asian continents. Most are threatened by the destruction of their original habitats, which is accentuated by climate change. The species *V. planifolia* is particularly endangered, as the primary gene pool in its region of origin (southern Mexico) is subject to considerable pressure linked to deforestation and the overexploitation of natural resources (Soto Arenas, 1999). In secondary diversification zones such as the islands of the Indian Ocean, and especially Réunion—the point of entry for the species into this region in the nineteenth century—there is considerable intraspecies homogeneity, indicating that cultivation may rely on a very restricted genetic base, which probably developed from one single individual through vegetative reproduction. Molecular studies have confirmed the very low level of genetic diversity in the vanilla plants cultivated throughout the world (*V. planifolia*) (Bory et al., 2008d; Lubinsky et al., 2008a; Minoo et al., 2008a). The vegetative reproduction process, which is predominant for the species of vanilla grown, does not make it possible to maintain and extend the gene pool. Nevertheless, an interesting phenotypic diversity is observed, which may be explained by the accumulation of somatic mutations, by the possibility of natural seed germination in the case of sexual reproduction, but also by the variable ploidy level that can be found in cultivated vanilla species (see Chapter 2). However, due to varietal homogeneity, vanilla cultivation is particularly vulnerable to environmental hazards, such as climate change and the emergence of plant pests.

The secondary gene pool of vanilla includes some 100 species that have diversified in America, Africa, and Asia. These species have individual properties that may be of particular interest for the genetic improvement of cultivated vanilla, such as autofertility, resistance to disease (fusariosis, viruses), the capacity to bear a large amount of fruit, a lower dependence on the photoperiod for the induction of flowering, a higher vanillin content, the presence of other aromatic or medicinal metabolites, and resistance to drought. The establishment of gene banks, in the form of collections, is thus extremely urgent for vanilla in order to safeguard the endangered endemic and patrimonial genetic resources (Grisoni et al., 2007; Lubinsky et al., 2008a; Pandey et al., 2008; Soto Arenas, 2006). Their interesting characteristics could thus be used in genetic crop improvement programs, as has begun in India (Minoo et al., 2006, 2008b; Muthuramalingam et al., 2004).

OVERVIEW OF THE CONSERVATION OF VANILLA GENETIC RESOURCES

The most effective means of protecting the diversity of vanilla species ought to be *in situ* protection in their areas of origin. However, the natural areas where the primary gene pools are found are very often subject to strong demographic pressure that endangers the different species. This type of conservation can therefore only be envisaged if it is associated with a global conservation strategy at the level of a territory (Soto Arenas, 1999). In Mexico, the success and experience of the CONABIO (*Comisión Nacional para el Conocimiento y Uso de la Biodiversidad*) may enable the implementation of this conservation method. In South Africa, the iSimangaliso reserve focuses particularly on conserving *V. rosheri*, which is second in the list of rare and endangered endemic species (Combrink and Kyle, 2006).

However, the diversity of vanilla, including more than 100 species distributed over three continents and living in varied biotopes, makes it difficult to set up *in situ* conservation systems. The establishment of *ex situ* collections therefore appears to be a necessary strategy for protecting vanilla genetic resources.

Initiatives of varying degrees of importance have been taken in the past in certain vanilla-producing countries where the genetic resources are found (Puerto Rico, Madagascar, Costa Rica, Mexico). Thus, from the 1940s onward on the Mayaguez station in Puerto Rico, research was conducted to characterize and improve vanilla plants (Childers et al., 1959). At about the same time in Madagascar, the Ambohitsara vanilla station near Antalaha began to collect and hybridize a wide range of vanilla plants (Dequaire, 1976). However, this collection was decimated by repeated cyclones, a lack of maintenance, and the propagation of viruses (Grisoni, 2009). In the late 1970s, the CATIE (*Centro Agronómico Tropical de Investigación y Enseñanza*) in Costa Rica collected and maintained about 32 vanilla accessions from Central America. A part of this collection was safeguarded by passage *in vitro* (Jarret and Fernandez, 1984). In Mexico, alongside the CONABIO program, a collection of clones representing the diversity of the country was established, but unfortunately could not be maintained (M.A. Soto Arenas, pers. comm.).

Today, the most important collections are found in France—Réunion (CIRAD), French Polynesia (EVT), and Cherbourg (council/MNHN), in India (ICRI), and in the United States (Table 3.1). Several botanical gardens and research institutions also have varying quantities of vanilla specimens in their greenhouses (Les Serres d'Auteuil, Jardin du Luxembourg and Jardin Botanique de Nancy in France, the Royal Botanic Gardens, Kew and the Copenhagen Botanical Garden in Europe, or

TABLE 3.1
Information on the Major Collections of Vanilla

Name of the Collection	CRB VATEL	American Vanilla and Wild Relatives Collection	Myladumpara Collection Vanilla (MCV)	EVT
Location and address of the collection	Pole de protection des Plantes Saint Pierre 97410 La Reunion—France	University of California, USA	Indian Cardamom Research Institute Myladumpara, Kerala 682 553, India	Hamoa (Taputapuatea) 98735 Raiatea French Polynesia
Organization or company responsible for the collection	Centre de Coopération Internationale en Recherche Agronomique pour le Développement (CIRAD)	Dr. Pesach Lubinsky, University of California, USA	Indian Cardamom Research Institute Spices Board, Govt. of India	Etablissement Vanille de Tahiti (EVT)
Curator of the collection and email	Michel Grisoni michel.grisoni@cirad.fr	Pesach Lubinsky plubi@hotmail.com sdoneill@ucdavis.edu	Dr. K.J. Madhusoodanan dirres_spices@yahoo.com; icrimyla@eth.com; dirres_spices@rediffmail.com	Sandra Lepers-Andrzejewski sandra.andrzejewski@labo-vanilledetahiti.pf
Purpose of the collection	Research, preservation	Conservation, breeding, research	Research, preservation	Vanilla improvement
Date of creation of the collection	1984	Not official; based on collections since 2004	1981	1999
Number of Vanilla species in the collection	25 (800)	30	6	3

continued

TABLE 3.1 (continued)
Information on the Major Collections of Vanilla

Name of the Collection	CRB VATEL	American Vanilla and Wild Relatives Collection	Myladumpara Collection Vanilla (MCV)	EVT
Number of accessions	*800*	*100*	*21*	*231*
Living material	*400*	*All live*	*21*	*181*
Dehydrated or spirit samples, nucleic acid extracts	*400*			*50*
Morphological and molecular descriptors	Yes RAPD, AFLP, SSR, plastid DNA sequences	Some ITS/psBA markers and AFLP	Yes No molecular descriptors	Yes AFLP, chromosome counts, 2C content
Availability of material for export	Yes for some accessions, submitted to conditions (MTA, CITES)		No	Beans or DNA extract only
List of species in the collection	*V. africana, V. albida, V. aphylla, V. bahiana, V. chamissonis, V. crenulata, V. ensifolia, V. grandiflora, V. humblotii, V. imperialis, V. insignis, V. leprieuri, V. lindmaniana, V. madagascariensis, V. odorata, V. palmarum, V. phalaenopsis, V. planifolia, V. polylepis, V. pompona, V. tahitensis, V. trigonocarpa, Vanilla* spp.	Focus on New World species	*V. planifola, V. aphylla, V. pompona, V. tahitensis, V. andamanica, V. wightiana*	Majority of *V. tahitensis* + *V. planifolia, V. pompona* and hybrids

Data kindly provided by the curators of the collections.

Rutgers University, the New York Botanical Garden, and the Rio de Janeiro Botanical Garden in Brazil, the Secretariat of the Pacific Community in Fiji, to name but a few). Private collections belonging to orchid lovers are another source of sometimes rare specimens. In total, around 50% of global diversity (in terms of the number of species) is thus conserved in these collections, essentially in the form of whole plants, *in vivo* or sometimes *in vitro*.

However, in most cases, the material is poorly identified in terms of taxonomy and the conservation methods used do not guarantee complete security for the resources. *In vivo* collections are not shielded from plant health risks (introduction of parasitic fungi or viruses and vectors of viral disease in greenhouses) or climate risks (storms, cyclones). Viral indexing has shown that 40% of the vanilla plants conserved in botanical gardens are infected by the *Cymbidium mosaic virus* (Grisoni et al., 2007). *In vitro* culture techniques for vanilla plants have been mastered (see Chapter 5) and ensure protection of plant material from contamination. However, *in vitro* conservation of collections requires regular maintenance operations that imply a large number of qualified workers. Furthermore, the succession of subcultures means that somaclonal variants may appear and the original genetic resources may be lost. For these reasons, methods to secure the collections must be developed as a matter of urgency. Cryopreservation could be an option for the future. In India, cryopreservation of pollen has been successfully conducted as a part of interspecies hybridization research. This technique solves the problem of the synchronization of flowering in different species. Likewise, a protocol for the cryopreservation of the apex of vanilla plants has been standardized for the storage of genetic resources (see Chapter 5). Protocols for the cryopreservation of meristems are being studied, particularly in Mexico (Gonzalez-Arnao et al., 2009) and France.

In Reunion Island the collection of vanilla genetic resources is the central element of the VATEL Biological Resource Centre, which was accredited in 2009. This recognition implies resource management that follows a quality process similar to the ISO 9001 international standard, and requires compliance with procedures on conservation technologies, the introduction of biological material, and the dissemination of genetic resources.

RULES ON THE TRANSFER OF VANILLA GENETIC RESOURCES

Similar to the rest of the Orchid family, the genus *Vanilla* is protected by the Convention on International Trade in Endangered Species of Wild Fauna and Flora (CITES), also known as the Washington Convention (http://www.cites.org/index.html). The aim of this international convention, signed by 175 countries, is to ensure that international trade in specimens of wild animals and plants does not endanger the survival of the species to which they belong. Indeed, although we can consider that cultivated vanilla is not endangered, almost all the species grow in natural forests located in high-risk zones. As the Orchid family is listed in Appendix 2 of the CITES convention, which includes over 28,000 plant species, vanilla is subject to the application of strict rules on the transfer of and trade in plant material between States (Figure 3.1). This protection concerns all parts of plants and all by-products,

Article IV

Regulation of Trade in Specimens of Species Included in Appendix II (extract)

For full text of the convention see http://www.cites.org/eng/disc/text.shtml#IV and http://www.cites.org/eng/app/appendices.shtml#hash1 (appendices I, II and III of the convention)

Main rules

2. The export of any specimen of a species included in Appendix II shall require the prior grant and presentation of an export permit. An export permit shall only be granted when the following conditions have been met:
   (a) a Scientific Authority of the State of export has advised that such export will not be detrimental to the survival of that species;
   (b) a Management Authority of the State of export is satisfied that the specimen was not obtained in contravention of the laws of that State for the protection of fauna and flora; and
   (c) a Management Authority of the State of export is satisfied that any living specimen will be so prepared and shipped as to minimize the risk of injury, damage to health or cruel treatment.

4. The import of any specimen of a species included in Appendix II shall require the prior presentation of either an export permit or a re-export certificate.

   For the Orchidaceae, these rules concern all parts and derivatives, except:
   (a) seeds, spores and pollen (including pollinia);
   (b) seedling or tissue cultures obtained in vitro, in solid or liquid media, transported in sterile containers;
   (c) cut flowers of artificially propagated plants; and
   (d) fruits and parts and derivatives thereof of artificially propagated plants of the genus *Vanilla*.

Article VII

Exemptions and Other Special Provisions Relating to Trade (extract)

6. The provisions of Articles III, IV and V shall not apply to the non-commercial loan, donation or exchange between scientists or scientific institutions registered by a Management Authority of their State, of herbarium specimens, other preserved, dried or embedded museum specimens, and live plant material which carry a label issued or approved by a Management Authority.

FIGURE 3.1 Extracts from articles IV and VII of the Convention on International Trade in Endangered Species of Wild Fauna and Flora (CITES).

with a few exceptions, especially concerning fruits and their parts and products of cultivated vanilla.

However, point 6 of Article VII of the convention ("Exemptions and Other Special Provisions Relating to Trade") facilitates exchanges of noncommercial plant material for scientific purposes.

Each member country of CITES must implement a legislation to guarantee compliance with the convention at the national level. Countries have the right to adopt more binding legislation.

The second level that regulates access to plant genetic resources in general and to vanilla in particular is the Convention on Biological Diversity (CBD), or Rio Convention (http://www.cbd.int/). At the level of the plant kingdom, the CBD covers all species that are not included in the International Treaty on Plant Genetic Resources for Food and Agriculture (http://www.planttreaty.org/index_en.htm). The CBD came into force on December 29, 1993. It has 191 parties (member countries) and its three main objectives are: to conserve biological diversity, to use biological diversity in a sustainable fashion and to share the benefits of biological diversity fairly and equitably.

Compliance with the rules for exchanging plant genetic resources between countries, set out in the Convention, translates in practical terms into the drafting and signing of an MTA (Material Transfer Agreement) by parties. This document defines, of a common accord, the rights and obligations of the parties concerned by the transfer, especially with regard to the issue of the use of resources and the sharing of benefits and advantages arising from this use. It now seems vital to use an MTA in any exchange of plant material in order to prevent disputes between suppliers and users of genetic resources.

Contrary to the CITES convention, there are no exemptions from the rules of the CBD for the use of plant genetic resources for scientific purposes.

The genetic resources held by a country prior to December 29, 1993 are not subject to the rules of the Convention.

Over and above the CBD and the rules on transfers, any exchange of plant material between countries must comply with the obligations of the International Plant Protection Convention (http://www.ippc.int/IPP/En/default.jsp), especially concerning the phytosanitary certificate that is required for any transfer of plant material. An initial guide was drawn up by the IBPGR, now Biodiversity International (Pearson et al., 1991), but the knowledge acquired over the last 15 years regarding vanilla viruses (see Chapter 7) calls for this guide to be updated.

CONCLUSION AND PROSPECTS

Similar to numerous other plant species, vanilla genetic resources are threatened with extinction or genetic erosion in many areas of origin and diversification. Any *in situ* initiatives for conserving species must therefore be encouraged, but the creation of *ex situ* collections is essential for protecting the diversity. This conservation method means resources are more secure, thanks to *in vitro* or cryopreservation mechanisms, and also makes it easier to promote resources and to acquire scientific data. Indeed, despite many research studies that have been made possible particularly by the widespread use of molecular biology tools, knowledge of vanilla genetics, in terms of the taxonomy of the genus and the properties of different species, is still incomplete. This knowledge gap is even more striking when it comes to usage and customs linked to the vanilla that are grown by local communities in regions where the genetic resources are found.

The creation of a global network of *in situ* and *ex situ* collections of genetic resources, based on branches on the three continents (America, Africa, and Asia) that hold most of these resources, could result in considerable progress in terms of the conservation and the scientific and economic improvement of vanilla. The development of increasingly effective genomics, associated with biotechnology techniques, means plant breeding programs can be set up. Exploiting the specific characteristics of wild species, such as resistance to drought in aphyllous vanilla, could provide a means of diversifying vanilla production areas and anticipating future climate change. The creation and development of vanilla plants that are more disease resistant, more productive, and richer in vanillin and other aromatic compounds could improve the living conditions of small-scale growers in vanilla production areas.

To ensure that these research studies and conservation network initiatives are more effective, plant material exchanges between collections and research programs should be facilitated, especially by relaxing the rules of the CBD in line with the existing exemption in the CITES convention.

REFERENCES

Besse, P., D. Da Silva, S. Bory, M. Grisoni, F. Le Bellec, and M.-F. Duval. 2004. RAPD genetic diversity in cultivated vanilla: *Vanilla planifolia*, and relationships with *V. tahitensis* and *V. pompona*. *Plant Science* 167:379–385.

Bory, S., O. Catrice, S. Brown, I. Leitch, R. Gigant, F. Chiroleu, M. Grisoni, M.-F. Duval, and P. Besse, 2008a. Natural polyploidy in *Vanilla planifolia* (Orchidaceae). *Genome* 51:816–826.

Bory, S., D. Da Silva, A.-M. Risterucci, M. Grisoni, P. Besse, and M.-F. Duval, 2008b. Development of microsatellite markers in cultivated vanilla: Polymorphism and transferability to other vanilla species. *Scientia Horticulturae* 115:420–425.

Bory, S., M. Grisoni, M.-F. Duval, and P. Besse, 2008c. Biodiversity and preservation of vanilla: Present state of knowledge. *Genetic Resources and Crop Evolution* 55:551–571.

Bory, S., P. Lubinsky, A.-M. Risterucci, J.-L. Noyer, M. Grisoni, M.-F. Duval, and P. Besse, 2008d. Patterns of introduction and diversification of *Vanilla planifolia* (Orchidaceae) in Reunion Island (Indian Ocean). *American Journal of Botany* 95:805–815.

Cameron, K.M. (ed.) 2003. *First International Congress on the Future of the Vanilla Business*. Princeton, NJ, USA.

Cameron, K.M. 2009. On the value of nuclear and mitochondrial gene sequences for reconstructing the phylogeny of vanilloid orchids (Vanilloideae, Orchidaceae), *Annals of Botany* 104:377–385.

Childers, N.F., H.R. Cibes, and E.M. Hernández. 1959. Vanilla—The orchid of commerce. In: C. L. Withner, ed. *The Orchids. A Scientific Survey*. Ronald Press Co., New York, 477–508.

Combrink, A.S. and R. Kyle. 2006. *A Handbook on the Rare, Threatened & Endemic Species of the Greater St Lucia Wetland Park*. A product of the Greater St Lucia Wetland Park-Rare, Threatened & Endemic Species Project. Unpublished internal report. 191pp. (accessed on Internet March, 15th 2010, http://www.travellersworldwide.com/downloads/species-survival-report.pdf).

Dequaire, J. 1976. L'amélioration du vanillier à Madagascar. *Journal d'Agriculture Tropicale et de Botanique Appliquée (FRA)*. 1976/07–12. 23 (7–12):139–158: 39 réf. - inter.: V.

Gonzalez-Arnao, M.T., C.E. Lazaro-Vallejo, F. Engelmann, R. Gamez-Pastrana, R. Martinez-Ocampo, Y. Martinez-Ocampo, M.C. Pastelin-Solano, and C. Diaz-Ramos. 2009. Multiplication and cryopreservation of vanilla (*Vanilla planifolia* 'Andrews'). *In Vitro Cellular & Developmental Biology-Plant* 45:474–582.

Grisoni, M. 2009. Mission d'expertise en pathologie du vanillier dans la SAVA. Rapport provisoire. CTHT/FED-Stabex, Toamasina, Madagascar.

Grisoni, M., M. Moles, P. Besse, S. Bory, M.-F. Duval, and R. Kahane. 2007. Towards an international plant collection to maintain and characterize the endangered genetic resources of vanilla. *Acta Hort. (ISHS)* 760:83–91.

Jarret, R.L. and R. Fernandez. 1984. Shoot-tip *Vanilla* culture for storage and exchange. *Plant Genetic Resources Newsletter (IBPGR/FAO),* 57:25–27.

Lubinsky, P., S. Bory, J. Hernandez, S. Kim, and A. Gomez-Pompa. 2008a. Origins and dispersal of cultivated vanilla (*Vanilla planifolia* Jacks. [Orchidaceae]). *Economic Botany* 62:127–138.

Lubinsky, P., K.M. Cameron, M.C. Molina, M. Wong, S. Lepers-Andrzejewski, A. Gomez-Pompa, and S.-C. Kim, 2008b. Neotropical roots of a Polynesian spice: The hybrid origin of Tahitian vanilla, *Vanilla tahitensis* (Orchidaceae) 10.3732/ajb.0800067. *American Journal of Botany* 95:1040–1047.

Minoo, D., V. Jayakumar, S. Veena, J. Vimala, A. Basha, K. Saji, K. Nirmal Babu, and K. Peter. 2008a. Genetic variations and interrelationships in *Vanilla planifolia* and few related species as expressed by RAPD polymorphism. *Genetic Resources and Crop Evolution* 55:459–470.

Minoo, D., K. Nirmal Babu, P.N. Ravindran, and K.V. Peter. 2006. Interspecific hybridization in vanilla and molecular characterization of hybrids and selfed progenies using RAPD and AFLP markers. *Scientia Horticulturae* 108:414–422.

Minoo, D., G. Pillai, S.K. Babu, and K. Peter. 2008b. Isolation and fusion of protoplasts in *Vanilla* species. *Current Science* 94:115–120.

Muthuramalingam, S., K.V. Velmourougane, N. Ramamurthy, and R. Naidu. 2004. *Vanilla*—A golden crop in coffee plantations [Online]. Available at Coffee board, India www.indiacoffee.org/newsletter/2004/may/cover_story.html (verified March 2005).

Noirot, M., S. Hamon, and F. Anthony. 1996. The principal component scoring: A new method of constituting a core collection using quantitative data. *Genetic Resources and Crop Evolution* 43:1–6.

Pandey, A., A. Tomer, D. Bhandari, and S. Pareek. 2008. Towards collection of wild relatives of crop plants in India. *Genetic Resources and Crop Evolution* 55:187–202.

Pearson, M.N., G.V.H. Jackson, F.W. Zettler, and E.A. Frison. 1991. Technical guidelines for the safe movement of vanilla germplasm. Technical Report. FAO/International Board for Plant Genetic Resources.

Portères, R. 1954. Le genre Vanilla et ses espèces. In: G. Bouriquet, ed. *Le vanillier et la vanille dans le monde*. Editions Paul Lechevalier, Paris, 599.

Rolfe, R.A. 1896. A revision of the genus *Vanilla. Journal of the Linnaean Society* 32:439–478.

Schluter, P., M. Arenas, and S. Harris. 2007. Genetic variation in *Vanilla planifolia* (Orchidaceae). *Economic Botany* 61:328–336.

Soto Arenas, M.A. 1999. Filogeografia y recursos genéticos de las vainillas de México [Online] http://www.conabio.gob.mx/institucion/proyectos/resultados/InfJ101.pdf (verified March 31, 1999).

Soto Arenas, M.A. 2003. Vanilla. In: A.M.C. Pridgeon, P.J. Chase, M.W. Ramunsen, eds. *Genera orchidacearum*, Vol. 3, Orchidoideae (Part 2). Oxford University Press, Oxford, UK.

Soto Arenas, M.A. 2006. Vainilla: los retos de un cultivo basado en una especie amenazada con una historia de vida compleja. In *Congreso Internacional de Productores de Vainilla*. Consejo Veracruzano de la Vainilla (ed.) Papantla, Veracruz, Mexico.

Verma, P., D. Chakrabarty, S. Jena, D. Mishra, P. Singh, S. Sawant, and R. Tuli, 2009. The extent of genetic diversity among *Vanilla* species: Comparative results for RAPD and ISSR. *Industrial Crops and Products* 29:581–589.

4 Vanilla in Herbaria

Marc Pignal

CONTENTS

Abundant in tropical regions of the globe, the genus *Vanilla* has challenged morphologists. Although the molecular approach provides a glimpse of the phylogenetic structure of the group, the clear definition of species is fraught with difficulties. This chapter opens the "Vanilla File" and lists some technical aspects that hinder the study of the only orchid cultivated for food.

LACK OF TAXONOMIC REVISION

Although cultivated vanilla has been the subject of many studies, the wild species, and more particularly those that have no agronomic qualities (being odorless or too capricious for cultivation) have not been studied in a synthetic way. Admittedly, studies on many taxa were published, as seen among members of the Orchidaceae family. The Kew Index lists more than 250 species (Anonymous, 2009), but the taxonomic revisions are missing.

The first complete taxonomic treatment of the genus *Vanilla* was carried out by Rolfe in 1896. It recognizes 50 species. The last revision by Portères (1954) in Bourriquet's book, *La vanille et le vanillier dans le monde,* includes 110 species. Portères' work, although rich in biogeographic data, is unfortunately not exhaustive because several taxa published by the Brazilian botanist Hoehne between 1941 and 1944 (Hoehne, 1941, 1944) are not mentioned. These works date back to the years of the Second World War, and European libraries in the 1950s had difficulty in supplementing their collections.

The lack of adequate treatment is directly related to the state of herbarium specimens in collections, which are the basis of any taxonomic and nomenclatural reasoning.

COLLECTIONS STILL POOR

Collections by past botanists are scarce and illustrations have often been used in place of specimens. This is the case for the "historical herbaria" of the *Muséum National d'Histoire Naturelle* in Paris. In Lamarck's herbarium, for example, only one figure illustrates vanilla, taken from the *Encyclopédie Méthodique*. The legend is in French: *Petite vanille ou vanille musquée* (small vanilla or vanilla musk). The drawing is sufficiently precise to identify *Vanilla palmarum*. The Jussieu herbarium is hardly any richer, with only one specimen of *Vanilla aromatica*. Are these illustrations derived from herbarium specimens? This is doubtful. Drawing from live specimen or from another illustration is common. This significantly contributed to confusion about the genus upon its official publication in the botanical nomenclature (Miller, 1768), even though vanilla was already widely cultivated.

Miller (1768) at first distinguishes "two or three varieties which differ in the color of their flowers and the length of their pods." Then he recognized them as two species: *V. mexicana* and *V. axillaris*. It seems today that the specimens described by Miller included at least our existing species *V. planifolia* and *V. pompona*.

Modern collections suffer from the same flaws as their predecessors: rare collections that are often fruitless and flowerless. When flowers are present, they are poorly dried or unprepared, making determinations uncertain. Moreover, species are often represented by a single specimen.

Some major herbaria in the world possess vanilla specimens. In North America, the New York Botanical Garden is very active in New World tropical research. In Brazil, the Hoehne collection is conserved in the Rio de Janeiro Botanical Gardens. But many other structures also conserve vanilla.

This is the reason for many databases and Web sites to allow the remote consultation of specimens. Particular mention should be made of the GBIF, Global Biodiversity Information Facility (http://data.gbif.org), which brings together several hundred natural history collections and offers a gateway for the consultation of specimen data. Since 2003, the Web site of the Muséum National d'Histoire Naturelle in Paris has been providing data and photographs of all the Orchidaceae included in its collections (http://www.mnhn.fr/base:sonnerat.html). The Aluka Foundation proposes a partially paying service for consulting all types from Africa and America (http://www.aluka.org). The Swiss Orchid Foundation at the Jany Renz Herbarium, Basel (http://orchid.unibas.ch/site.sof.php), offers a very comprehensive Web site.

The poor condition of the material makes morphological approaches particularly difficult to conduct. Vanilla herbaria are therefore currently insufficient to resolve the genus taxonomy for technical and biological reasons. The difficulty of preparing suitable samples is a major handicap.

TIPS FOR PREPARING GOOD HERBARIUM SAMPLES

Kill the Plant Quickly

Like many orchids, vanilla is particularly resistant to dehydration. The thick skin prevents water from escaping easily. Samples of Malagasy leafless vanilla have been seen to remain green for more than one year on paper. It is therefore necessary to kill samples rapidly. The most common method in the field is to immerse the plant in boiling water, then dry it and put it in a dryer. Sulfur dioxide fumes can also be used, with the advantage that they avoid oxidation, which deteriorates the pigments.

Prepare the Flowers

The flowers are often thick and fleshy. Failure to prepare them in the herbarium usually leads to the deterioration of structures. It is therefore necessary to dissect fresh flowers and to dry the perianth parts separately (Figure 4.1). Petals and sepals are detached from the flower; one of the two petals and one of the two lateral sepals are presented on the lower face. The lip, which is always fused lengthwise with the column, is cut on the side in order to be able to spread it out and to present the higher face with its scales and its gibbosities. The large inflorescences must imperatively be simplified, retaining only two or three flowers.

Monitor Dehydration

Drying should be monitored and in tropical conditions, the paper should be renewed.

Prioritize Collections in Alcohol

Collections in 70% alcohol are valuable additions to the herbarium. Higher levels that make the tissue brittle should be avoided. Ensuring that the structures remain flexible and unpressed makes them easier to study (Figure 4.2). Alcohol also makes it easier to observe vascularization. The alcohol collections are, however, difficult to maintain since large volumes of alcohol require strict safety conditions and the level

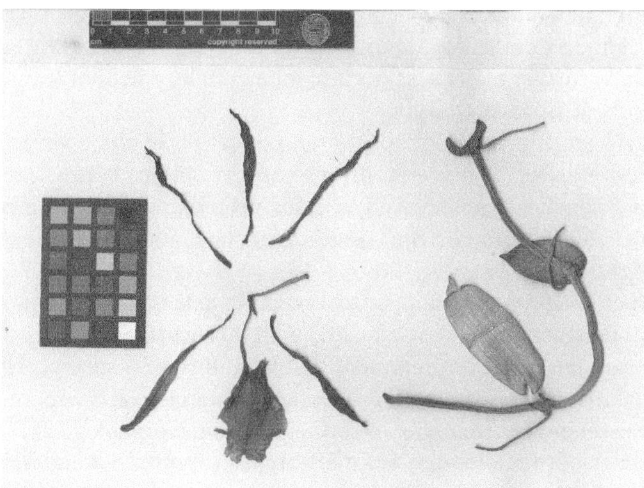

FIGURE 4.1 Dissection of flower and presentation of the leafy stem of *Vanilla trigonocarpa* Hoehne.

FIGURE 4.2 Dissection of a flower of *Vanilla bahiana* Hoehne preserved in 70% ethanol.

of the liquid must be constantly monitored owing to evaporation. Furthermore, alcohol specimens cannot replace dry samples, but they make a very useful addition.

Photographs taken on the field and the fine details of flowers can also be very useful. However no picture, even the best possible one, can replace a collection specimen. Botanists observe many minute details, such as pilosity on the labellum, the structures on the column or organ vascularization, which photography cannot capture.

Phylogenetic analysis, phenetic, and then cladistic methods helped considerably with the comprehension of the plant groups. The use of molecular characters, today very widespread, made it possible in numerous cases to confirm the morphological hypothesis. A combination of molecular genetics and taxonomy, is a very powerful tool. Vanilla herbaria collections must therefore be accompanied by specimens in silica gel to allow further molecular studies, for example, pieces of leaf rapidly dehydrated in absorbent silica crystals.

The herbaria are used as tools in the identification of the species. For all the aforementioned reasons, this is not always easy to obtain. When a regional flora exists, or if a genus has been revised, species determination can be based on the keys of identification defined. But some characters nonetheless require certain knowledge of the group.

The specimen used for the reference to a species name is the holotype (Figure 4.3). In theory, any specimen should be compared with this holotype in order to be suitably identified. It does not, however, represent the variability of a species. Holotypes are used for the stability of names. But it is necessary to examine all the specimens to have an idea of the taxonomy, the number of species, and the morphological relationships.

The duplicates of the holotypes are the isotypes (Figures 4.4 and 4.5). These are different parts of the same individual collected on the same day by the same botanist. They are often deposited in other institutions. Isotypes are important for ease of consultation and for the security of the material. An unhappy incident was observed

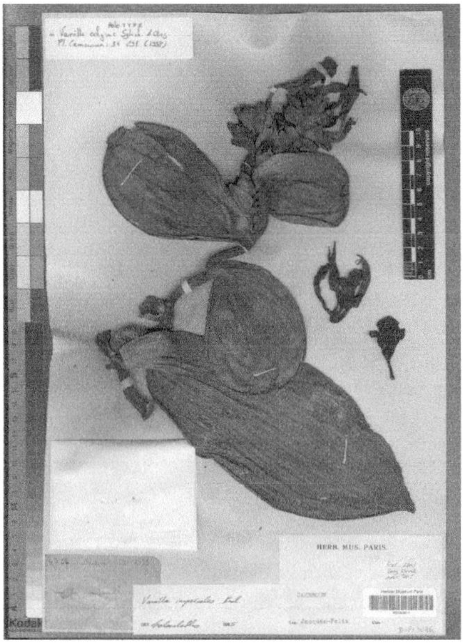

FIGURE 4.3 *Vanilla ochyrae* Szlach. et Olsz. (holotype). [Reproduced from the Herbier National (MNHN) Web site. With permission.]

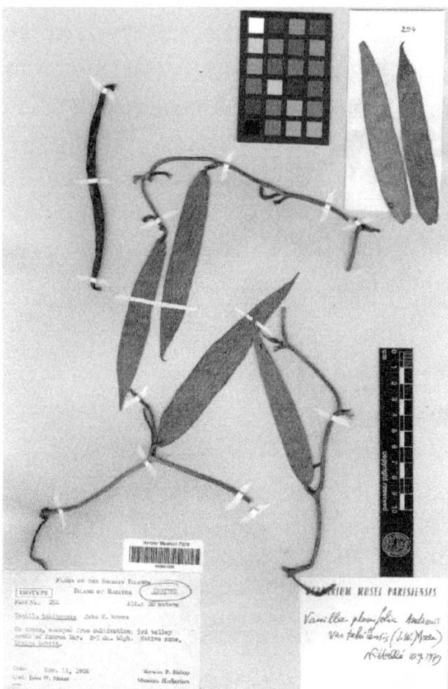

FIGURE 4.4 *Vanilla tahitiensis* Moore (isotype). [Reproduced from the Herbier National (MNHN) Web site. With permission.]

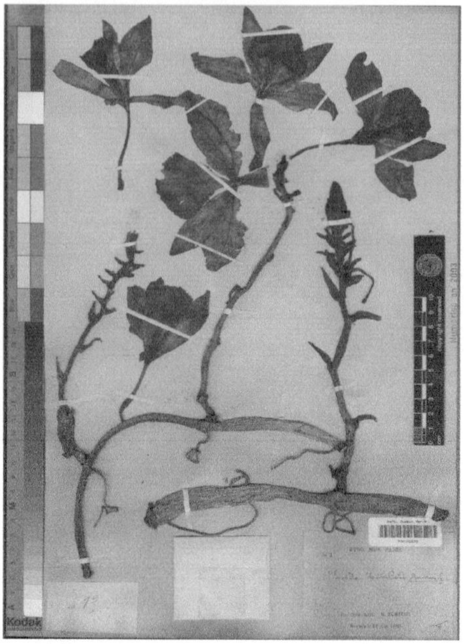

FIGURE 4.5 *Vanilla humblotii* Rchb.f. (isotype). [Reproduced from the Herbier National (MNHN) Web site. With permission.]

after the Second World War, when the Berlin herbarium was largely destroyed, including the type collections. The isotypes conserved in other herbaria are the only reference material available to the scientific community. The International Code of Botanical Nomenclature (McNeill et al., 2006), which sets the rules for giving and using plant names, envisaged the replacement of a holotype that disappeared with an isotype. If there is no isotype, one of the other specimens (paratypes) quoted in the original publication will be used. This replacement specimen is called the lectotype. If all the original material is lacking, the botanist who revised a species can choose a specimen (neotype) that best corresponds to the description published.

SCARCITY OF FLOWERS IN NATURE

Vanillas are seldom observed as flowers in the field. In floristic inventories, it is much more common to observe stems and leaves than inflorescences or fruits.

For these reasons, *ex situ* collections must be created to make it possible to monitor the bloom and properly prepare the samples. If information on initial collections in the field is suitably preserved, the specimens will appear rich and ready to be studied.

Herbarium specimens are difficult to prepare, expensive to maintain, and require specific skills for analysis. They are, however, absolutely indispensable in order to revise the taxonomy of the genus as long as molecular tools, based on DNA analyses in particular, have not been established.

REFERENCES

Anonymous, 2009. *The International Plant Names Index (2009)*. Published on the Internet http://www.ipni.org [accessed October 28, 2009].

Hoehne, F.C. 1941. *Arq. Bot. Estado Sao Paulo* new ser. 1, fasc. 3: 61, tab. 81.

Hoehne, F.C. 1944. *Arq. Bot. Estado Sao Paulo* n. ser. form. maior, 1: 125, tab. 136.

McNeill, J., Barrie, F. R., Burdet, H. M., Demoulin, V., Hawksworth, D. L., Marhold, K., Nicolson, D. H., et al. (eds.). 2006. *International Code of Botanical Nomenclature (Vienna Code) Adopted by the Seventeenth International Botanical Congress Vienna*, Austria, July 2005. Gantner Verlag, Ruggell, Liechtenstein.

Miller, P. 1768. *Gard. Dict*., ed. 8. n. 2.

Portères, R. 1954. Le genre *Vanilla* et ses espèces. In: G. Bouriquet, ed. *Le vanillier et la vanille dans le monde*. Editions Paul Lechevalier, Paris, 94–290.

Rolfe, A. 1896. A revision of the genus *Vanilla*. *Journal of the Linnean Society (Botany)* 32:439.

5 Biotechnological Applications in Vanilla

Minoo Divakaran, K. Nirmal Babu, and Michel Grisoni

CONTENTS

INTRODUCTION

Vanilla planifolia G. Jackson (syn. *V. fragrans* Andrews), is a tropical climbing orchid (Figure 5.1) known for producing the delicate popular flavor, vanilla (Purseglove et al., 1981) and is the second most expensive spice traded in the global market after saffron (Ferrão, 1992). The major vanilla-producing countries are Madagascar, Indonesia, Uganda, India, and the Comoros with Madagascar ranking first. Following the discovery of the New World by Columbus in 1492, the earliest vanilla dissemination record from Mexico is the one by Father Labat who imported three *V. planifolia* vines into Martinique in 1697. The lack of natural pollinators in the areas of introduction prevented sexual reproduction and pod production until the discovery of artificial pollination in the first half of the nineteenth century (Bory et al., 2008d). Continuous vegetative propagation, lack of natural seed set, and insufficient variations in the gene pool all hamper crop improvement programs.

In a short span of time, biotechnology has had a significant impact on the pattern of development and quality of life globally (Bhatia, 1996). The later part of the

FIGURE 5.1 *V. planifolia* vine—in full bearing.

twentieth century saw the rise of new industries based on discoveries made in the field of biological sciences and the progress made over recent years in molecular biology, genetic engineering, and plant tissue culture have provided a new dimension to crop improvement.

In vitro culture is one of the key tools of plant biotechnology, which makes use of the totipotent nature of the plant cells, a concept proposed by Haberlandt (1902) and unequivocally demonstrated for the first time by Steward et al. (1958). It can be employed for the production of disease-free clones, mass cloning of selected genotypes, gene pool conservation, selection of mutants, raising of hybrids between sexually incompatible taxa through somatic hybridization, incorporating the desired traits by genetic engineering, and in the production of secondary metabolites in cultured cells or tissues (Thorpe, 1990). However, the realization of these objectives necessitates prior standardization and optimization of tissue culture procedures.

PROBLEMS TO BE TARGETED

As a cash crop, vanilla plays a major role in the economy of countries such as Madagascar, the Comoros, Indonesia, and Uganda. Continuous clonal propagation of *V. planifolia* leads to monoculture, exposing the crop to severe damage (Gopinath, 1994) as vanilla is affected by a large number of pests and diseases (see Chapters 7, 8, 9, and 20). The introduction of new genetic material is greatly constrained by factors such as its asexual propagation, the fact that the flowers are mostly self-pollinated, and the threat posed to wild populations of vanilla by land pressures (Lubinsky, 2003). The lack of sufficient variability in the gene pool, the threat of destructive diseases that wipe out vanilla plantations, as well as the destruction of its natural habitats, make the search for alternative methods to introduce variability into the gene pool vital. The narrow gene pool can be broadened by using interspecific hybridization to combine the available primary gene pool of the genus *Vanilla*, with the secondary gene pool, that is, the close relatives of *V. planifolia*, which is an important source of desirable traits such as self-pollination, a lower dependence of flower induction on the photoperiod, a higher fruit set, indehiscence of the fruits, and disease resistance (Lubinsky, 2003).

Different species of vanilla are found in various geographical regions and their flowering seasons are not synchronized, bringing about difficulties in the movement of pollen to the receptor species to enable pollination between species. It is in such instances that the development of methods for storing viable pollen for longer periods becomes significant.

Improvements in quality characteristics, such as higher vanillin content, larger bean size, improved aroma and taste, and so on would benefit vanilla processors and consumers. To perfect the germination of immature embryos into a complete plant, embryo rescue techniques can be used for retrieval and regeneration of nonviable hybrid seeds. Cell culture or protoplast culture is useful for creating somatic hybrids for the transfer of characters from alien sources. Protoplasts can be used as target organs for transformation, provided they are made to regenerate into a complete plantlet. Clonal propagation of elite lines, *in vitro* conservation, and international germplasm exchange are possible using micropropagation techniques. Molecular markers such as DNA markers [random amplified polymorphic DNA (RAPD), restriction fragment length polymorphism (RFLP), amplified fragment-length polymorphism (AFLPs)] and biochemical markers (isozyme, protein) can be used for the characterization of germplasm and somaclonal variants.

GENETIC DIVERSITY IN VANILLA

V. planifolia is a crop that differs a little from its wild progenitors. This can be attributed to limited breeding and to recent domestication (Bory et al., 2008c; Lubinsky et al., 2008a). Several types have been recognized within the cultivated vanilla of Mexico differing in vegetative appearance or reproduction mode (Soto Arenas, 2003). The analysis of isoenzyme data of specimens from the vanilla plantations of northern Veracruz, Oaxaca, and elsewhere in Mexico showed little genetic variation in general (Soto Arenas, 1999), although plants originating from two main areas could be differentiated. Nucleotide sequence variation within introns of two specific protein-coding genes—namely, the calmodulin and the glyceraldehyde 3-phosphate dehydrogenase—were detected but were not able to differentiate among the Mexican types of vanilla. Outside the countries of origin, vanilla is likely to be of clonal origin and very little variation can be expected. The vanilla plantations of Réunion, Madagascar, Mauritius, and Seychelles have derived from a single cutting (Lionnet, 1958) and as per the present information, few differences in cultivated types of *V. planifolia* have been observed. However, recent studies revealed that self-progenies as well as polyploidization events have generated phenotypic diversity in cultivated vanilla in Réunion (Bory et al., 2008a, 2008b, 2008c).

At the genus level, molecular markers such as RAPD, AFLP, and sequence single repeats (SSR) have been developed over the last decade for studying genetic diversity.

RAPD was used to estimate the level of genetic diversity and interrelationships among different clones of *V. planifolia* and related species. The data confirmed the very limited variation within accessions of *V. planifolia*, indicative of its narrow genetic base and its close relationship with *V. tahitensis* J.W. Moore (Besse et al., 2004; Minoo et al., 2008; Schlüter et al., 2007). In a study including both leafy and leafless types, such as *V. planifolia*, *V. tahitensis*, *V. andamanica* Rolfe, *V. pilifera* Holtt., and *V. aphylla* Blume (Figure 5.2), there was reasonable variability indicating the possibility of natural seed set in the wild species. In spite of superficial morphological similarity, *V. andamanica* is not closely related to *V. planifolia* or *V. tahitensis* and its accessions are the most divergent from all other species studied, forming a separate and unique cluster (Minoo et al., 2008). There was considerable variability among the eight different accessions of *V. andamanica*, supporting the probability that this species did originate in the Andaman Islands, where sexual reproduction is likely (Minoo et al., 2008). Earlier, Rao et al. (2000) have reported the occurrence of natural seed set in India for *V. wightiana*.

AFLP profiles were developed to analyze *Vanilla* species, interspecific hybrids, and selfed progenies (Bory et al., 2008c; Lubinsky et al., 2008a; Minoo et al., 2006b). All these analyses converged in showing that most of the *V. planifolia* accessions cultivated outside of Mesoamerica exhibit very low levels of genetic diversity, as they derived from a single accession, possibly the Mexican cultivar Mansa from Papantla.

FIGURE 5.2 (See color insert following page 136.) Flowers of Indian species of *Vanilla*: (a) and (b) *V. andamanica* with varying label colors, (c) *V. pilifera* showing indication of insect visits, and (d) *V. aphylla*.

The patterns of diversification of the cultivated species were also studied and compared with other cultivated (*V. tahitensis*) and wild (*V. aphylla, V. bahiana, V. insignis, V. odorata,* and *V. pompona*) species. Clear polymorphism was detected in these related species, interspecific hybrids, and selfed progenies.

The development of SSR markers (microsatellites) have been reported by Bory et al. (2008b). The isolation and characterization of 14 microsatellite loci from *V. planifolia* have been described. These were monomorphic within cultivated accessions, as expected based on the probable single clonal origin of this crop and previous genetic studies. These markers were transferable to *V. tahitensis* and 11 loci were polymorphic between these two closely related species. Furthermore, some of these markers were transferable and polymorphic across 15 other wild American, African, and Asian species and revealed consistent relationships between species, together with a strong pattern of Old World versus New World differentiation in the genus. Furthermore, the use of microsatellites allowed the first molecular-based estimation of heterozygosity levels in this species, which was not possible when dominant markers such as AFLP or RAPD was used.

Sequencing of neutral genes has been used for reconstructing the evolutionary history of Vanilloid orchids, including a few *Vanilla* species (Cameron, 2000, 2004, 2009; Cameron and Molina, 2006; Cameron et al., 1999). Nuclear (internal transcribed spacer, ITS) and plastid (rbcL gene) DNA sequences were also used for unraveling the origin of the Tahitian vanilla (Lubinsky et al., 2008b). Recently, the length polymorphism of the nonneutral caffeic acid *O*-methyl transferase gene was also used to analyze 20 *Vanilla* species, and confirmed the strong differentiation of Old World versus New World species in the genus (Besse et al., 2009). On the basis of sequencing data for nuclear and plasmidic DNA, Cameron (2005) suggested in setting up a bar code system (Lahaye et al., 2008) for vanilla using the ITS region and the psbA-trnH intergenic spacer. This system may allow routine identification of vanilla specimens to the species level, and perhaps even to the accession level. To build a robust phylogeny for the *Vanilla* genus, reference herbarium specimens will need to be included. For this purpose, the development of plastid mononucleotide microsatellites should be considered for vanilla (particularly when using degraded DNA samples extracted from herbarium material), as have already been successfully used for biogeographical studies of orchids (Fay and Krauss, 2003; Micheneau, 2002).

Given the difficulty in using classical phenotypic markers for perennial crops such as vanilla, molecular markers are powerful tools for studying the variability in cultivated vanilla, unraveling species interrelationships, identifying interspecific hybrids, and fingerprinting important genotypes (Minoo et al., 2006a). They are, therefore very helpful for monitoring and evaluating the achievements resulting from biotechnologies.

PROPAGATION AND BREEDING METHODS

Commercial vanilla is always propagated by stem cuttings of healthy vigorous plants and may be cut from any part of the vine. The length of the cutting is usually determined by the amount of planting material available. Short cuttings, 20 cm in length, will take 3–4 years to flower and fruit. Cuttings of 90–100 cm in length are usually preferable as they tend to flower earlier. When available, with their free ends hanging over supports, these will flower and fruit in 1–2 years. Cuttings are usually planted *in situ*, but they may be started in nursery beds when necessary. Because of their succulent nature, cuttings can be stored or transported for a period of up to two weeks, if required.

Traditionally, vanilla germplasm is conserved in clonal repositories belonging to botanical gardens and scientific institutions. The high costs of traditional conservation systems limit the number of accessions that can be preserved. In order to reduce the losses of biodiversity, attempts to conserve *Vanilla* species, *in vitro*, were made (Jarret and Fernandez, 1984; Minoo et al., 2006b) and have been extended to conserve the endangered species.

For breeding purposes, vanilla can be grown from seeds. Hybridization and the production of plants from seeds have been carried out in Puerto Rico and Madagascar. The seeds should be disinfected, washed in sterile distilled water, and cultured in nutrient medium (Knudson, 1950). The germination of vanilla seeds is better if the cultures are maintained in a dark incubator at 32°C. Seeds of interspecific crosses between *V. planifolia* and *V. pompona* required a higher temperature of 34°C for germination.

IN VITRO SEED GERMINATION

Vanilla produces numerous minute seeds that do not germinate under natural conditions. Tissue culture technique can be used to successfully germinate the seeds. Protocols for seed and embryo culture of vanilla have been standardized (Gu et al., 1987; Knudson, 1950; Minoo et al., 1997; Withner, 1955).

Seed culture in different basal media indicated that vanilla seeds had no stringent nutritional requirements for the initiation of germination unlike some terrestrial orchids of temperate climate (Minoo et al., 1997). The germination of seeds began within four weeks of culture and the initial stages of germination were typical of most orchids, such as swelling of the embryo followed by rupturing of the seed testa, and the subsequent emergence of protocorms (Figure 5.3). Seeds germinated directly into plantlets in the medium supplemented with benzyladenine (BA) (0.5 mg L^{-1}) alone, without any intervening callus phase, and could thus be utilized for the production of selfed progenies/seedlings. The addition of tryptone had a growth-promoting effect on the size and development of protocorm, irrespective of the basal medium to which it was added. In treatments with BA, most of the protocorms remained the same with the scale-like leaf primordial and developing into shoots, whereas treatment with auxin supplements showed the gradual disorganization of the protocorms into callus. Murashige and Skoog's (MS) medium gave a better response than Knudson's medium, for *in vitro* cultures of vanilla. The minimum germination (26%) was observed in MS medium at half strength and the maximum (85%) was recorded in full strength MS medium supplemented with 2 g L^{-1} tryptone (Minoo, 2002).

The requirement of cytokinin for germination is considered to be related to the utilization of lipids that constitute the primary storage material in most orchid seeds

FIGURE 5.3 *In vitro* seed germination.

and it has been observed that unless storage lipid is utilized, germination does not continue (De Pauw et al., 1995).

Vanilla, a cross-pollinated crop, is known to have many meiotic and postmeiotic chromosomal abnormalities (Ravindran, 1979). As a result, it is possible to get various cytotypes in the seed progenies. Culturing of seeds can thus give many genetically varied types. Studies on *in vitro* germination of vanilla seeds and the resultant progeny showed morphological and biochemical variations. Isozyme profiles of superoxide dismutase (SOD) and peroxidase (PRX) were studied in selfed progenies of *V. planifolia*. The profiles clearly indicated differences among progenies as expressed by the presence or absence of specific bands. The maximum similarity that these progenies exhibited was 47.37%, indicating high segregation and level of heterozygosity existing in *V. planifolia* (Minoo et al., 1997). This heterozygosity was further confirmed by AFLP analyses (Bory et al., 2008c). Thus, *in vitro* culture can be used for the germination of seeds and the selection of useful genotypes from segregating progenies that might be mass propagated for obtaining disease-free planting material.

MICROPROPAGATION

In vitro propagation of vanilla is essential to generate uniform, disease-free plantlets and for conserving the genetic resources. *In vitro* propagation using apical meristem has been standardized for the large-scale multiplication of disease-free and genetically stable plants (Cervera and Madrigal, 1981; George and Ravishankar, 1997; Kononowicz and Janick, 1984; Minoo, 2002; Minoo et al., 1997; Philip and Nainar, 1986; Rao et al., 1993b). *In vitro* propagation of *V. tahitensis* (Mathew et al., 2000) and endangered species of vanilla, such as *V. wightiana*, *V. andamanica*, *V. aphylla*, and *V. pilifera* (Minoo et al., 2006b) have been standardized to protect these species from extinction.

Clonal propagation methods for the efficient multiplication of *V. planifolia* by induction of multiple shoots from axillary bud explants (Figure 5.4) using semi-solid MS medium supplemented with BA (2 mg L^{-1}) and α-naphthaleneacetic acid (NAA, 1 mg L^{-1}) have been reported (George and Ravishankar, 1997). The multiple shoots were transferred to agitated liquid MS medium with BA at 1 mg L^{-1} and NAA at 0.5 mg L^{-1} for 2–3 weeks, and subsequently cultured on semi-solid medium. Using this method, an average of 42 shoots was obtained from a single axillary bud explant over a period of 134 days. The use of an intervening liquid medium was found to enhance the multiplication of shoots.

In another study (Minoo et al., 1997), the subculture of the explants onto proliferation MS media containing various levels of cytokinin (BA) and auxin (indole butyric acid, IBA) was evaluated (Table 5.1). The initiation of preexisting buds to grow *in vitro* could be induced in MS medium with low cytokinin. However, a combination of cytokinins and auxin promoted multiple shoot formation. The ideal medium for multiplication was MS supplemented with BA (1 mg L^{-1}) and IBA (0.5 mg L^{-1}). In this medium, an average of 15 multiple shoots were induced in 90 days of culture (Figure 5.4). Nodal segments gave a better response, with a mean of 15 shoots per culture compared to the shoot tips, which gave a mean of seven shoots per culture (Minoo, 2002). The culture media and conditions favorable to micropropagation of *V. planifolia* were suitable for other related species, such as *V. andamanica*, *V. aphylla* (Figure 5.5), and *V. pilifera*. The number of shoots induced in different species varied (Table 5.2). About 12–15 shoots/culture could be induced in *V. planifolia*, followed by *V. aphylla* (8–10 shoots). Among the species studied, the lowest multiplication rate was observed in *V. pilifera*. Elongated shoots from proliferation medium were rooted on MS growth regulator free medium containing 30 g L^{-1} sucrose (Figure 5.6). *In vitro* plantlets with well-developed roots were

FIGURE 5.4 Micropropagation of *V. planifolia*.

acclimated with a survival percentage of more than 70%. The root initiation on microcuttings started between four and six days after culture, reaching 100% of the cultures after two weeks, indicating that the optimal endogenous levels of plant growth regulators required for rooting were already present in the tissue/explants.

Janarthanam et al. (2005) and Kalimuthu et al. (2006) have devised simple and rapid protocols for the mass multiplication of *V. planifolia*. A commercially viable protocol for the mass propagation of *V. tahitensis*, another cultivated species of *Vanilla*, was standardized with a multiplication ratio of 1:4.7 over a culture period of 60–70 days (Mathew et al., 2000). Rao et al. (2000) have reported the occurrence and micropropagation of *V. wightiana* Lindl., an endangered species. Giridhar et al. (2001) and Giridhar and Ravishankar (2004) have studied the effects of other additives, namely, silver nitrate, thidiazuron, zeatin, coconut milk, and so on, on *in vitro* shoot multiplication and root formation in *V. planifolia*.

TABLE 5.1
Effect of Growth Regulators on Multiple Shoot and Root Induction from Shoot Explants of *V. planifolia* on MS Medium (Mean of 20 Replicates)

Growth Regulators Concentration				Multiple Shoots	Average No. of Shoots/	Roots Development/ Culture	
Kin	BA	NAA	IBA	Frequency (%)	Culture ± SD	No.	Type
0.5				0.0		—	—
1.0				0.0		—	—
	0.5			71 ± 3.45	4.18 ± 0.30	—	—
	1.0			20 ± 3.63	1.0	—	—
		0.5		0.0	1.0	1	Velamen
		1.0		0.0	1.0	1	Velamen
			0.5	0.0	1.0	1	Long roots
			1.0	0.0	1.0	1	Long roots
0.5	0.5			0.0	1.0	—	—
1.0	0.5			0.0	1.0	—	—
0.5	1.0			0.0	1.0	—	—
0.5		0.5		0.0	1.0	1	Velamen
1.0		0.5		0.0	1.0	1	Branching
0.5		1.0		0.0	1.0	1	Velamen
0.5			0.5	0.0	1.0	1	—
1.0			0.5	0.0	1.0	1	—
0.5			1.0	0.0	1.0	1	—
	0.5		0.5	0.0	1.0	1	
	1.0		0.5	97 ± 6.5	15.15 ± 3.63	—	—
	0.5		1.0	65 ± 11.4	10.35 ± 3.45	—	—
				0.0	1.0	1	Healthy roots

BA = benzyladenine, IBA = indole-3-butyric acid, Kin = kinetin, NAA = α-naphthaleneacetic acid.

FIGURE 5.5 *In vitro* multiple shoot production in *V. aphylla*.

The conversion of root tips into shoots was observed in *V. planifolia* and *V. aphylla* when cultured on MS medium supplemented with BA (1.0 mg L^{-1}) and IBA (0.5 mg L^{-1}). These shoots developed into plantlets and were hardened and established in soil. The conversion of root meristem into shoots in *in vitro* cultures of vanilla was earlier reported (Philip and Nainar, 1988). These meristematic conversions without callus stage are assumed to minimize the chances of induced epigenetic changes. Earlier studies by Sreedhar et al. (2007) indicated no difference in the AFLP-banding patterns of any of the micropropagated samples for a particular primer, suggesting the absence of variation among the micropropagated plants.

PLANT REGENERATION THROUGH CALLUS CULTURES

Continuous vegetative propagation and lack of sufficient variations in the gene pool hamper crop improvement programs. Introduction of somaclonal variation through callus cultures has been attempted to broaden the narrow genetic base. A callus induction and *in vitro* plant regeneration system has been optimized from both vegetative and reproductive tissues. The best results were obtained using vegetative tissues and over 80% callusing was achieved in MS medium supplemented with 1 mg L^{-1} BA and 0.5 mg L^{-1} NAA. Callus differentiated into shoots that could be multiplied successfully in 1:12 ratio in a combination of 1 mg L^{-1} BA and 0.5 mg L^{-1} IBA, when supplemented with MS medium (Table 5.3). *In vitro* rooting was induced with an efficiency of 100% in basal MS media devoid of any growth regulators. This ability of dedifferentiated tissue to regenerate is a crucial prerequisite for genetic transformation experiments. The protocol was successfully extended to the endangered wild species, *V. aphylla*, offering the potential of applying the protocol for mass multiplication as well as induction of variations in *Vanilla* species, in a limited time.

TABLE 5.2
Comparison of *In Vitro* Responses in Different Species of *Vanilla*[a]

Growth Regulators	*In Vitro* Responses			
	V. planifolia	*V. andamanica*	*V. aphylla*	*V. pilifera*
Kin	Single shoot	Single shoot	Single shoot	Single shoot
BA	Multiple shoots (3–4)	Multiple shoots (3–4)	Multiple shoots	Single shoot
NAA	Root induction	Root induction	Root induction	Root induction
BA + Kin	Single shoot	Single shoot	Single shoot	Single shoot
Kin + IBA	Single shoot	Single shoot	Single shoot	Single shoot
BA + IBA (1.0 + 0.5 mg L^{-1})	Multiple shoot induction (12–15 nos., in 10 days of culture)	Multiple shoot induction (5–7 in 90 days)	Multiple shoot induction (8–10 in 90 days)	Multiple shoot induction (2–4 in 120 days)
BA + NAA	Callusing and plant regeneration	Multiple shoots	Callusing and plant regeneration	Single shoot
Kin + NAA	Single shoot	Single shoot	Single shoot	Single shoot
Basal medium	Single shoot elongation and development of roots	Single shoot elongation and development of roots	Single shoot elongation and development of roots	Single shoot elongation and development of roots

BA = benzyladenine, IBA = indole butyric acid, Kin = kinetin, NAA = α-naphthaleneacetic acid.
[a] All growth regulators were supplemented on MS basal medium at 0.5 to 1.0 mg L^{-1}.

FIGURE 5.6 *In vitro* rooting in *V. planifolia* cultures.

TABLE 5.3
Influence of Growth Regulators on Callus Induction and Plant Regeneration in Seed Cultures of *V. planifolia* on MS Medium (Mean of 20 Replicates)

Growth Regulators (mg L^{-1})	Callusing (%)	Shoot Regeneration (%)	No. of Shoots/Culture
0	0	0	0
NAA (0.5)	80	0	0
BA (1.0)	0	—	—
BA (1.0) + NAA (0.5)	80	90	10
BA (0.5) + NAA (1.0)	0	—	—
BA (1.0) + IBA (0.5)	10	60	6

BA = benzyladenine, IBA = indole butyric acid, Kin = kinetin, NAA = α-naphthaleneacetic acid.

Reports on variability among callus-regenerated plants in vanilla are few. They concern successful plant regeneration from leaf- and seed-derived callus (Davidonis and Knorr, 1991; Davidonis et al., 1996; Janarthanam and Seshadri, 2008; Xju et al., 1987) and studies among indigenous collections of vanilla, through polyacrylamide electrophoretic (PAGE) studies (Rao et al., 1993a). A study comprising randomly selected callus-regenerated progenies showed variability in morphology and RAPD profiles (Figure 5.7) among the callus-regenerated plants in comparison with the control plant *V. planifolia* (Minoo, 2002). It showed that a significant amount of variability can be generated with this protocol and be utilized in vanilla improvement programs for developing variants with desirable agronomic characters.

Callus cultures initiated from leaf explants of *V. planifolia* showed better callus initiation than those from nodal explants with callus biomass production maximal when cultured on MS basal medium containing 2,4-dichlorophenoxy acetic acid and BA. Callus transferred to MS basal medium supplemented with 3 mg L^{-1} BA and 2.5 mg L^{-1} μM NAA showed superior growth response. Davidonis et al. (1996) have

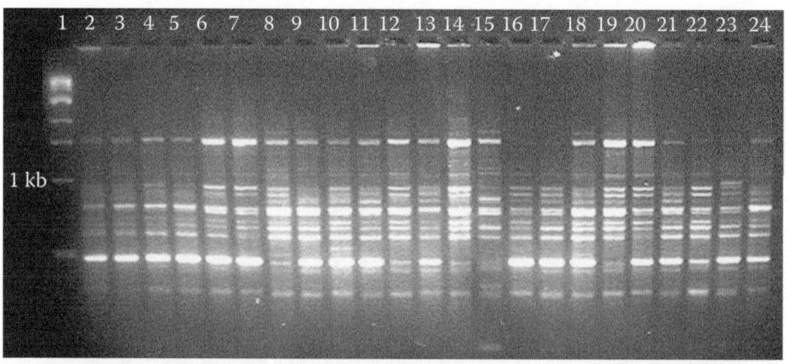

FIGURE 5.7 RAPD profiles of callus-regenerated progenies of vanilla using OPERON primer OPA10. 1: 1 kb ladder, 2–23: callus-regenerated plants, 24: control (*V. planifolia*).

patented the production of callus of *V. planifolia*, extraction of vanillin, and the use of ferulic acid to increase the content of vanillin.

Heritable somaclonal variations with respect to various resistance traits have been reported, namely, resistance to methionine sulfoxime (Carlson, 1973) and *Pseudomonas syringae* (Thanutong et al., 1983), in tobacco, resistance to *Fusarium oxysporum* in tomato (Evans et al., 1984), and resistance to *Helminthosporium sativum* (Chawla and Wenzel, 1987) in wheat. In future attempts to genetically transform vanilla, the ability of transformed tissue to regenerate is a crucial prerequisite. The regeneration protocol optimized (Minoo, 2002) could shorten the length of genetic transformation experiments while inducing a high frequency of regeneration.

EX VITRO ESTABLISHMENT OF SEEDLINGS

Most plant species grown *in vitro* require a gradual acclimatization and hardening for survival and growth in the natural environment. The survival of *in vitro* plants depends upon their ability to withstand water loss and carry out photosynthesis. However in vanilla, the survival rate of transferred plants is currently over 80% during hardening process (Minoo, 2002). Plantlets should be removed from culture vessels (Figure 5.8), washed, treated with fungicide, transferred to polybags containing potting mixture (sand, soil, and vermiculite) and hardened for 30 days under controlled conditions (26–28°C, 80–90% RH). Initiation of new growth occur through development of the axillary branch. These plants are successfully transferred to soil after initial hardening period of three weeks (Figure 5.9) and can be. They were later field planted with proper shade and support.

INTERSPECIFIC HYBRIDIZATION

Interspecific hybridization is an age-old mechanism by which useful genes from wild progenitors and species can be brought into cultivated species. The cultivated

FIGURE 5.8 Hardening of *in vitro* developed plantlets.

FIGURE 5.9 Tissue-cultured plants growing in pots.

types of many crop species were improved through interspecific hybridization and backcrossing. Interspecific hybridization is very common in orchids to produce new and novel varieties of flowering plants.

Natural occurrences of interspecific hybrids have been reported in vanilla by Nielsen and Siegsmund (1999) between *V. claviculata* and *V. barbellata* in localities in Puerto Rico where they coexist. Progenies were discovered having morphological characters intermediate between the two parents.

The cultivated species of *V. planifolia* has been crossed with other American species including *V. pompona* and *V. phaeantha*, which are resistant to *Fusarium* (Purseglove et al., 1981). Interspecific hybridization was also conducted in Java between cultivated and wild vanilla to develop lines resistant to stem rot caused by *Fusarium oxysporum* (Mariska et al., 1997).

In India, *V. aphylla* and *V. pilifera* flower synchronously but *V. aphylla* occurs naturally in South India and *V. pilifera* in Assam, North East India. When cultivated in Kerala, flowers of both species opened sequentially and lasted for one day in *V. pilifera*, whereas it lasted for two days in *V. aphylla*. In the former, signs of fruit set were observed even without manual pollination whereas *V. aphylla* flowers did not set fruit. Since rostellum is present in both species, natural pollination without an aid is ruled out. It can be suspected, that the fragrance of the *V. pilifera* flowers attracts insects (which were found to frequent the flowers often) to visit them and bring about effective pollination (D. Minoo, unpublished data). Self and interspecific hybridizations between the two species were done manually and fruits set was observed.

Successful attempts were made to increase the spectrum of variation of *V. planifolia* by interspecific hybridization with *V. aphylla* which is tolerant to *Fusarium* (Minoo, 2002). Morphological characters and molecular profiles revealed the true hybridity of the interspecific hybrid progenies. Seedling progenies of *V. planifolia*, and interspecific hybrids obtained from crosses between *V. planifolia* (female) and

V. aphylla (male) were evaluated using a number of different loci as markers by using AFLPs and RAPDs loci. The profiles indicated similarity between the parents, selfed progenies, and interspecific hybrids and that all the progenies tested were variable when compared to each other, which can be exploited for crop improvement in vanilla (Minoo et al., 2006a).

Thus, these successful introgressions of male and female characters into the hybrids (Minoo et al., 2006a) by interspecific hybridization, confirmed by molecular profiles are promising to help solve the major bottlenecks in vanilla breeding.

IN VITRO CONSERVATION

Effective procedures for *in vitro* conservation by slow growth in selected species of vanilla have been standardized (Minoo et al., 2006b). The addition of mannitol (10–15 g L^{-1}) and reduction of sucrose to lower levels (15–10 g L^{-1}) induced slow growth and subsequently 80–90% of the cultures could be maintained for a period of 360 days, when the culture vessels were closed with aluminum foil. Supplementing mannitol and sucrose in equal proportions at 10 or 15 g L^{-1}, could help to maintain the cultures for one year and thus were maintained *in vitro* for more than seven years with yearly subculture. The plantlets maintained in this medium showed reduced growth rate and maximum survival. The conserved material was transferred to MS medium fortified with 30 g L^{-1} sucrose and supplemented with 1 mg L^{-1} BA and 0.5 mg L^{-1} IBA, for retrieval of normal shoots and their multiplication. The conserved material was transferred to the multiplication medium (MS + 30 g L^{-1} sucrose and 1 mg L^{-1} NAA) for normal growth. The small-sized plantlets kept in the conservation medium for over one year showed good growth and developed into normal-sized plants with good multiplication rate (1:5). These plantlets were transferred to soil (garden soil:sand:perlite in equal proportions) and established easily with 80% success when kept in a humid chamber for 20–30 days after transfer. They developed into normal plants without any deformities and deficiency symptoms and exhibited apparent morphological similarities to the mother plants. After more than seven years of slow growth storage, involving over five subculture cycles, the genotypic stability of few species was assessed using molecular markers. No changes were observed in DNA fingerprinting vis-à-vis nonconserved controls in the authors' laboratory.

Jarret and Fernandez (1984) have reported storage of *V. planifolia* shoot tips as tissue cultures for 10 months and Philip (1989) has discussed the possibility of using root cultures for conservation of vanilla germplasm for assured genetic stability. *In vitro* conservation of *V. planifolia* (Jarret and Fernandez, 1984) and *V. walkeriae* using slow growth method (Agrawal et al., 1964) has been reported and the effects of polyamines on *in vitro* conservation of *V. planifolia* have been studied by Thyagi et al. (2001).

Conventional and *in vitro* genebanks are complementary as the active and base collections of genetic resources. Although *in vitro* conservation cannot be viewed as a method to replace *in situ* conservation, the advantages of *in vitro* conservation as a component that can be incorporated into an overall vanilla long-term conservation strategy for a safe and economical storage of the germplasm were demonstrated.

CRYOPRESERVATION

Protocols for conservation of gene pools have been developed for slow growth as well as cryopreservation of vanilla accessions as encapsulated shoot tips, pollen, and DNA (Minoo, 2002). Combining the available gene pool in the genus will help in broadening the genetic base and in converging the useful genes into cultivated vanilla from wild species. Interspecific hybridization requires synchronized flowering between the species and availability of viable pollen. Pollen from two asynchronously flowering species of *Vanilla*, namely, cultivated *V. planifolia* and its wild relative *V. aphylla*, were cryopreserved after desiccation, pretreated with cryoprotectant dimethyl sulfoxide (5%) and cryopreserved at –196°C in liquid nitrogen (LN). This cryopreserved pollen was later thawed and tested for their viability both *in vitro* and *in vivo.* A germination percentage of 82.1% and 75.4% in *V. planifolia* and *V. aphylla* pollen, respectively, were observed indicating their viability. These cryopreserved pollens of *V. planifolia* were used successfully to pollinate *V. aphylla* flowers resulting in fruit set. The seeds thus obtained were successfully cultured to develop hybrid plantlets (Minoo, 2002). Viability and fertility assessment of cryopreserved pollen (Figure 5.10) from *Vanilla* species thus showed that it is possible to use cryogenic methods for conservation and management of the haploid gene pool in this species. This is of great importance for the facilitating crosses in breeding programs, for distribution and exchange of germplasm, and for preserving nuclear genes of the germplasm.

A procedure for storage of vanilla germplasm by cryopreservation of shoot tips using encapsulation/dehydration method has been standardized (Figure 5.11). The *in vitro*-grown shoot tips were encapsulated in 4% sodium alginate. The encapsulated beads were subjected to pretreatment by progressive increase of sucrose concentration from 0.1 to 1.0 M, followed by dehydration for 8 h to a moisture content of 22%. This was followed by rapid freezing by plugging into LN. The cryopreserved

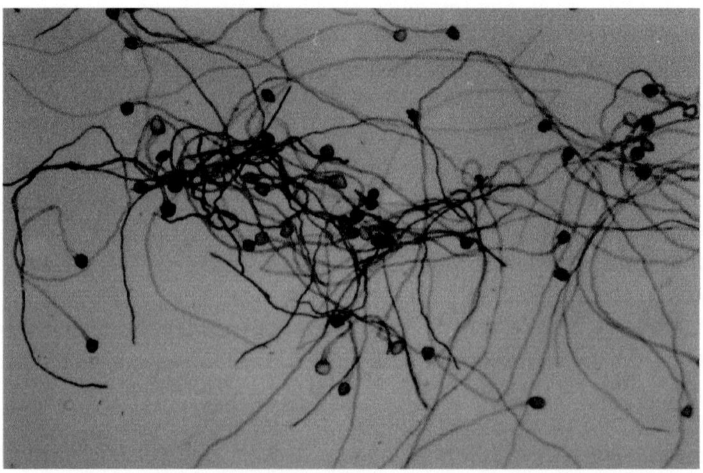

FIGURE 5.10 Germination of cryopreserved pollen.

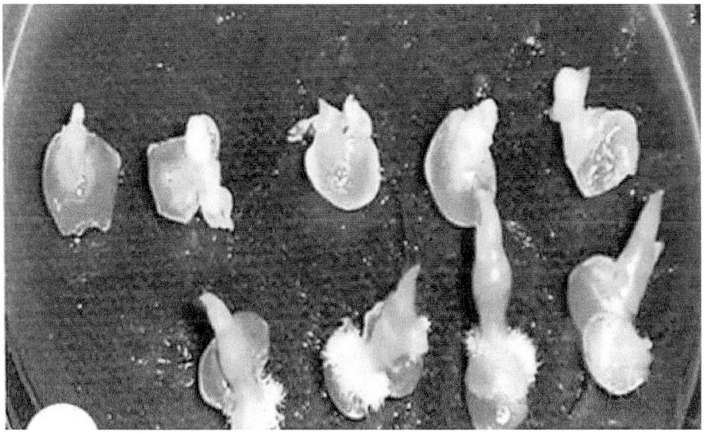

FIGURE 5.11 (See color insert following page 136.) Germination of cryopreserved shoots.

shoot tips were thawed after 12 h in LN by keeping them in water bath at 40°C for 3 min. The thawed propagules were allowed to recover on MS with 3% sucrose, 1 mg L^{-1} BAP, and 0.5 mg L^{-1} IBA in dark for one week and then transferred to light for regrowth and multiplication. Seventy percent of the propagules have for recovered, grown, and multiplied into full-fledged plants (Ravindran et al., 2004).

Cryopreservation, once fully implemented will provide an expeditious and cheaper means to duplicate the base collection for safety reasons, as well as for the distribution of germplasm sets to other countries/continents.

PRODUCTION OF SYNTHETIC SEEDS

Synthetic seed technology was standardized by encapsulating 3–5 mm *in vitro* regenerated shoot buds and protocorms in 4% sodium alginate, to produce good quality rigid beads ideal for withstanding low temperatures and cryopreservation. Higher concentrations of alginate were not suitable as they produced very hard matrix, which hindered the emergence of shoot buds and thereby affecting the rate of germination and recovery, while at lower concentrations of alginate, the beads were difficult to handle during cryopreservation and retrieval. The synthetic seeds were stored at 5°C, 15°C, and 22°C to study the effect of temperature on their storage and viability. Low temperatures (5°C and 15°C) were not suitable for synthetic seed. Shoot buds of 0.4–0.5 cm size were suitable for encapsulation as smaller buds failed to survive the storage and lost their viability within a month. However at 22 ± 2°C, synthetic seeds could be stored for 10 months (Figure 5.12). The plants derived from these encapsulated buds were apparently healthy and developed into normal plants.

Clonal propagation of *V. planifolia* using encapsulated shoot buds have been reported by George et al. (1995). Synthetic seeds are ideal for germplasm conservation and exchange, especially in vanilla, where there is no natural seed set.

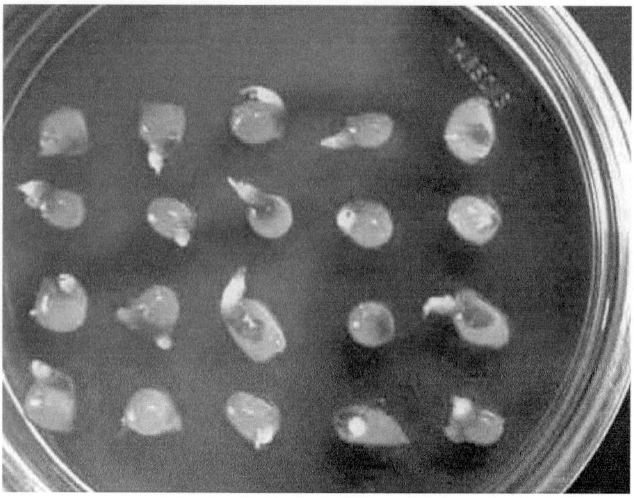

FIGURE 5.12 Synthetic seeds.

PROTOPLAST ISOLATION AND FUSION

The techniques of protoplast isolation and fusion are important because of the far-reaching implications in studies of plant improvement by cell modification and somatic hybridization. The possibility of protoplast systems in spice crops such as cardamom, ginger, and vanilla was studied by Triggs et al. (1995) and Geetha et al. (2000).

Protoplasts were successfully isolated from *V. planifolia* and *V. andamanica* when incubated in an enzyme solution containing macerozyme R10 (0.5%) and cellulase Onozuka R10 (2%) for 8 h at 30°C in dark (Table 5.4). *In vitro* leaves were plasmolyzed in a solution containing cell protoplast washing salts with 9% mannitol before enzymatic digestion. Since it was difficult to peel off the lower epidermis in vanilla, the plasmolyzed leaf tissue was mechanically macerated by scraping the lower surface of the leaf with a sharp blade and incubating in different concentrations and combinations of enzyme solutions. Periodical microscopic observations showed the liberation of cell clusters and individual cells after 2 h of incubation in enzyme solution.

The isolation solution containing 9% mannitol was found necessary for the release and maintenance of viable protoplasts. The isolated protoplasts were round and filled with chloroplasts. Protoplasts of *V. planifolia* were bigger in size (0.031 mm) than those of *V. andamanica* (0.022 mm) and could be distinguished by the arrangement of chloroplasts—peripheral in *V. planifolia* and centrally scattered *V. andamanica*. The visually distinguishable nature of protoplast can be exploited for the purpose of identifying genetic transformation in these species. When subjected to polyethylene glycol (PEG)-mediated fusion, the protoplasts fused forming a heterokaryon. The fusion product was cultured on MS liquid medium with 0.5 mg L^{-1} BA, 0.5 mg L^{-1} IBA supplemented with 3% sucrose and 7% mannitol for 20 days. The cell wall development around the fusion product was observed after 36 h (Minoo et al., 2008). The fusion protoplast technology can be very useful in gene transfer of useful traits to *V. planifolia*, especially the natural seed set and disease tolerance observed in *V. andamanica*.

TABLE 5.4
Effect of Enzyme Concentration and Incubation Conditions on Yield of Protoplasts

Species	Enzyme Solution	Incubation Conditions	Protoplast Yield	Viability (%)
V. planifolia	0.5% macerozyme R10 + 2% onozuka cellulase R10	8 h at 30°C in dark	2.5×10^5/g of leaf	72
V. andamanica	1% macerozyme R10 + 3% hemicellulase + 6% onozuka cellulase R10	8 h at 30°C in dark	1×10^5/g of leaf	55

THE FUTURE OF VANILLA IMPROVEMENT

The landmarks that have been attained in the form of various technologies can be effectively used for the production of a spectrum of genetic variations in vanilla, thus overcoming a major bottleneck in vanilla breeding and crop improvement programs. The protoplast isolation and fusion technology developed can be used in transfer of useful traits through the production of somatic hybrids, thus making way for genetic manipulations in vanilla. The characterization of *Vanilla* species, accessions, seedlings, somaclones, and interspecific hybrids, proved the existence and extent of genetic variations that is available and brought by biotechnological tools. The *in vitro* conservation methods, through synthetic seed, slow growth, and cryopreservation will form an integral and important part of overall conservation strategy in genetic recourses management of vanilla germplasm. Furthermore, the synthetic seed technology forms an ideal means for exchanging disease-free planting material.

REFERENCES

Agrawal, D.C., Morwal, G.C., and Mascarenhas, A.F. 1992. *In vitro* propagation and slow growth storage of shoot cultures of *Vanilla walkeriae* Wight, an endangered orchid. *Lindleyana* 7 (2): 95–99.

Besse, P., Da Silva, D., Bory, S., Grisoni, M., Le Bellec, F., and Duval, M.-F. 2004. RAPD genetic diversity in cultivated vanilla: *Vanilla planifolia*, and relationships with *V. tahitensis* and *V. pompona*. *Plant Science* 167:379–385.

Besse, P., Da Silva, D., Bory, S., Noirot, M., and Grisoni, M. 2009. COMT intron-size variations in *Vanilla* species (Orchidaceae). *Plant Science* 176:452–460.

Bhatia, C.R. 1996. Biotechnology—many application areas. In: Ravi, N. ed. *Hindu Survey of Indian Agriculture*. The Hindu daily, Chennai, India, 173–175.

Bory, S., Catrice, O., Brown, S., Leitch, I., Gigant, R., Chiroleu, F., Grisoni, M., Duval, M.-F., and Besse, P. 2008a. Natural polyploidy in *Vanilla planifolia* (Orchidaceae). *Genome* 51:816–826.

Bory, S., Da Silva, D., Risterucci., A.M., Grisoni, M., Besse, P., and Duval, M.F. 2008b. Development of microsatellite markers in cultivated vanilla: Polymorphism and transferability to other vanilla species. *Scientia Horticulturae* 115:420–425.

Bory, S., Grisoni, M., Duval, M.-F., and Besse, P. 2008d. Biodiversity and preservation of vanilla: Present state of knowledge. *Genetic Resources and Crop Evolution* 55:551–571.

Bory, S., Lubinsky, P., Risterucci, A.-M., Noyer, J.-L., Grisoni, M., Duval, M.-F., and Besse, P. 2008c. Patterns of introduction and diversification of *Vanilla planifolia* (Orchidaceae) in Reunion Island (Indian Ocean). *American Journal of Botany* 95:805–815.

Bouriquet, G. 1954. (ed.) *Le vanillier et la Vanille dans le Monde*. Paris: Editions Paul Lechavalier.

Cameron, K.M. (ed.) 2000. *3rd Southern Connection Congress*, Canterbury, NZ. January 17–22, 2000.

Cameron, K.M. 2004. Utility of plastid psaB gene sequences for investigating intrafamilial relationships within Orchidaceae. *Molecular Phylogenetics and Evolution* 31:1157–1180.

Cameron, K.M. 2005. DNA barcoding as a method for vanilla species identification. In: *Vanilla 2005 Mexico. The International Symposium on the Vanilla Business*, Veracruz, Mexico, November 15–16, 2005.

Cameron, K.M. 2009. On the value of nuclear and mitochondrial gene sequences for reconstructing the phylogeny of vanilloid orchids (Vanilloideae, Orchidaceae) 10.1093/aob/mcp024. Ann Bot:mcp024.

Cameron, K.M., Chase, M.W., Whitten, W.M., Kores, P.J., Jarrell, D.C., Albert, V.A., Yukawa, T., Hills, H.G., and Goldman, D.H. 1999. A phylogenetic analysis of the Orchidaceae: Evidence from rbcL nucleotide sequences. *American Journal of Botany* 86:208–224.

Cameron, K.M. and Molina, C. 2006. Photosystem II gene sequences of psbB and psbC clarify the phylogenetic position of Vanilla (Vanilloideae, Orchidaceae). *Cladistics* 22 (3):239–248.

Carlson, P.S. 1973. Methonine sulfoxamine-resistant mutants of tobacco. *Science* 180: 1366–1368.

Cervera, E. and Madrigal, R. 1981. *In vitro* propagation (*Vanilla planifolia* A.) *Environmental and Experimental Biology* 21:441.

Chawla, H.S. and Wenzel, G. 1987. *In vitro* selection of barley and wheat for resistance against *Helminthosporium sativum. Theoretical and Applied Genetics* 74:841–845.

Davidonis, G. and Knorr, D. 1991. Callus formation and shoot regeneration in *Vanilla planifolia. Food Biotechnology* 5 (1):59–66.

Davidonis, G., Knorr, D., Romagnoli, W., and Lynn, G. 1996. Callus formation *Vanilla planifolia.* US Patent 5573941.

De Pauw, M.A., Remphrey, W.R., and Palmer, C.E. 1995. The cytokinin preference for *in vitro* germination and protocorm growth of *Cyperidium candidum. Annals of Botany* 75:267.

Evans, D.A., Sharp, W.R., and Medina-Filho, A.P. 1984. Somaclonal and gametoclonal variation. *American Journal of Botany* 71:759–774.

Fay, M.F. and Krauss, S.L. 2003. Orchid conservation genetics in the molecular age. In: Dixon, K.W., Kell, S.P., Barrett, R.L., Cribb, P.J. eds. *Orchid Conservation*. Natural History Publications, Sabah, Malaysia, pp. 91–112.

Ferrão, J.E.M. 1992. A Aventura das Plantas e os Descobrimentos Portugueses. Comissão Nacional para a Comemoração dos Descobrimentos Portugueses, Lisboa, Portugal.

Geetha, S.P., Nirmal Babu, K., Rema, J., Ravindran, P.N., and Peter, K.V. 2000. Isolation of protoplasts from cardamom (*Elettaria cardamomum* Maton.) and ginger (*Zingiber officinale* Rosc.). *Journal of Spices and Aromatic Crops* 9 (1):23–30.

George, P.S. and Ravishankar, G.A. 1997. *In vitro* multiplication of *Vanilla planifolia* using axillary bud explants. *Plant Cell Reports* 16:490–494.

George, P.S., Ravishankar, G.A., and Venkataraman, L.V. 1995. Clonal propagation of *Vanilla planifolia* by axillary bud culture and encapsulated shoot buds. In: *Proceedings of the All India Symposium on Recent Advances in Biotechnology and Application of Plant Tissue & Cell Culture.* June 22–26, CFTRI, Mysore.

Giridhar, P., Obul, R., and Ravishankar, G.A. 2001. Silver nitrate influences *in vitro* shoot multiplication and root formation in *Vanilla planifolia* Andr. *Current Science* 81:1166–1170.
Giridhar, P. and Ravishankar, G.A. 2004. Efficient micropropagation of *Vanilla planifolia* Andr. Under influence of thidiazuron, zeatin and coconut milk. *Indian Journal of Biotechnology* 3:113–118.
Gopinath, C. 1994. *Secret of VanillaóFarmer's Notebook on Vanilla.* Indian Spice Associates, Puttur, Karnataka, 31.
Gu, Z.P., Arditti, J., and Nyman, L.P. 1987. *Vanilla planifolia*: Callus induction and plantlet production *in vitro. Lindleyana* 2:48–52.
Haberlandt, G. 1902. Kulturversuche mit isollierten pflanzenzellen. S.B.Weisen Wien *Naturwissenschaften* 111:69–92.
Janarthanam, B., Gopi, C., and Vatsala, T.M.A. 2005. Simple and rapid *in vitro* propagation of *Vanilla planifolia.* Andre. *NSPB, National Symposium on Plant Biotechnology: New Frontiers*, Lucknow, Uttar Pradesh, India. November 18–20.
Janarthanam, B. and Seshadri, S., 2008. Plantlet regeneration from leaf derived callus of *Vanilla planifolia* Andr. *In Vitro Cellular and Developmental Biology—Plant* 44(2):84–89. DOI: 10.1007/s11627-008-9123-4.
Jarret, R.L. and Fernandez, Z. 1984. Shoot tip vanilla culture for storage and exchange. *Plant Genetic Resources Newsletter (IBPGR/FAO)* 57:25–27.
Kalimuthu, K., Senthilkumar, R., and Murugalatha, N. 2006. Regeneration and mass multiplication of *Vanilla planifolia* Andr.—A tropical orchid. *Current Science* 91:1401–1403.
Knudson, L. 1950. Germination of seeds of *Vanilla. American Orchid Society Bulletin* 41:443–445.
Kononowicz, H. and Janick, J. 1984. *In vitro* propagation of *Vanilla planifolia. HortScience* 19:58–59.
Lahaye, R., van der Bank, M., Bogarin, D., Warner, J., Pupulin, F., Gigot, G., Maurin, O., Duthoit, S., Barraclough, T.G., and Savolainen, V. 2008. DNA barcoding the floras of biodiversity hotspots. *Proceedings of the National Academy of Sciences* 105:2923–2928.
Lionnet, J.F.G. 1958. Seychelles vanilla. *World Crops* 10:441–444.
Lubinsky, P. 2003. Conservation of wild vanilla. In: *Proceedings of Vanilla First International Congress*, Princeton, New Jersey, USA, November 11–12.
Lubinsky, P., Bory, S., Hernandez, J., Kim, S., and Gomez-Pompa, A. 2008a. Origins and dispersal of cultivated vanilla (*Vanilla planifolia* Jacks. [Orchidaceae]). *Economic Botany* 62:127–138.
Lubinsky, P., Cameron, K.M., Molina, M.C., Wong, M., Lepers-Andrzejewski, S., Gomez-Pompa, A., and Kim, S.-C. 2008b. Neotropical roots of a Polynesian spice: The hybrid origin of Tahitian vanilla, *Vanilla tahitensis* (Orchidaceae). *American Journal of Botany* 95:1040–1047.
Mariska, I.H., Husni, A., Kosmiatin, M., and Rusyadi, Y. 1997. *In vitro* culture of hybrid seeds cultivated and wild *Vanilla*. In: *Breeding to Increase Competitiveness of Indonesian Agriculture Commodities*, Simposium Nasional dan Kongres III Peripi, Bandung (Indonesia), 24–25 Sep 1997, pp. 471–480.
Mathew, M.K., Rao, Y.S., George, G.L., Lakshmanan, P., and Madhusoodanan, K.J. 2000. *In vitro* propagation of *Vanilla tahitensis* Moore. *Journal of Spices and Aromatic Crops* 9 (2):171–173.
Micheneau, B.C. 2002. *Relations phylogénétiques des Neottieae (Orchidaceae) et éétude de la variabilité génétique de Cephalanthera longifolia à l'aide de marqueurs moléculaires (séquençage et microsatellites chloroplastiques).* Mémoire de DEA, Université Paul Sabatier, Toulouse.
Minoo, D. 2002. Seedling and somaclonal variation and their characterization in Vanilla. PhD thesis. Calicut University, Kerala, India.

Minoo, D., Jayakumar, V.N., Veena, S.S., Vimala, J., Basha, A., Saji, K.V., Nirmal Babu, K., and Peter, K.V. 2008. Genetic variations and interrelationships in *Vanilla planifolia* and few related species as expressed by RAPD polymorphism. *Genetic Resources Crop Evolution* 3:459–470.

Minoo, D., Nirmal Babu, K., and Peter, K.V. 2006a. Conservation of *Vanilla* species, *in vitro*. *Scientia Horticulturae* 110:175–180.

Minoo, D., Nirmal Babu, K., Ravindran, P.N., and Peter, K.V. 2006b. Interspecific hybridization in vanilla and molecular characterization of hybrids and selfed progenies using RAPD and AFLP markers. *Scientia Horticulturae* 108: 414–422.

Minoo, D., Sajina, A., Nirmal Babu, K., and Ravindran, P.N. 1997. Ovule culture of vanilla and its potential in crop improvement. In: Edison, S., Ramana, K.V., Sasikumar, B., Nirmal Babu, K. and Santhosh, J.E. eds. *Biotechnology of Spices, Medicinal and Aromatic Plants*. Indian Society for Spices, Calicut, India, 112–118.

Nielsen, R.L. and Siegismund, H.R. 1999. Interspecific differentiation and hybridization in *Vanilla* species (Orchidaceae). *Heredity* 83:560–567.

Philip, V.J. 1989. Use of tissue culture technology and possibilities of germplasm conservation. *Indian Journal of Plant Genetic Resources* 2 (2):114–121.

Philip, V.J. and Nainar, S.A.Z. 1986. Clonal propagation of *Vanilla planifolia* (Salisb.) Ames using tissue culture. *Journal of Plant Physiol.* 122:211–215.

Philip, V.J. and Nainar, S.A.Z. 1988. *In vitro* transformation of root meristem to shoot and plantlets in *Vanilla planifolia. Annals of Botany* 61:193–199.

Purseglove, J.W., Brown, E.G., Green, C.L., Robbins, S.R.J. (eds) 1981. *Spices*. Vol. 2. Longman Inc., New York, pp. 644–735.

Rao, Y.S., Madhosoodanan, K.J., and Naidu, R. 1993a. Detection of genetic variability in *Vanilla* through PAGE studies. *Journal of Plantation Crops* 21:363–365.

Rao, Y.S., Mary Mathew, K., Madhusoodanan, K.J., and Naidu, R. 1993b. Multiple shoot regeneration in *Vanilla* (*V. planifolia* Andrews). *Journal of Plantation Crops* 21:351–354.

Rao, Y.S., Mathew, K.M., Madhusoodanan, K.J., Sudharshan, M.R., Kumar, V.K., and Potty, S.N. 2000. Natural fruit set in *Vanilla wightiana* Lindl., an endangered species from Andhra Pradesh, India. *Journal of Spices and Aromatic Crops* 9 (1):77.

Ravindran, P.N. 1979. Nuclear behavior in the sterile pollen of *Vanilla planifolia* (Andrews). *Cytologia* 44:391–396.

Ravindran, P.N., Nirmal Babu, K., Saji, K.V., Geetha, S.P., Praveen, K., and Yamuna, G. 2004. *Conservation of Spices Genetic Resources in In Vitro Gene Banks*. ICAR Project report. Indian Institute of Spices Research, Calicut, Kerala, India, 81.

Schlüter, P., Soto Arenas, M., and Harris, S. 2007. Genetic variation in *Vanilla planifolia* (Orchidaceae). *Economic Botany* 61:328–336.

Soto Arenas, M.A. 1999. Filogeografia y recursos genéticos de las vainillas de Mèxico [Online] http://www.conabio.gob.mx/institucion/proyectos/resultados/InfJ101.pdf (verified Marzo 31, 1999).

Soto Arenas M.A. 2003. Vanilla. In: Pridgeon, A.M.C., Chase, P.J., and Ramunsen, M.W.F., ed. *Genera Orchidacearum*, Vol. 3, Orchidoideae (part 2). Oxford University Press, Oxford, UK. 321–334.

Sreedhar, R.V., Venkatachalam, L., and Bhagyalakshmi, N. 2007. Genetic fidelity of long-term micropropagated shoot cultures of vanilla (*Vanilla planifolia* Andrews) as assessed by molecular markers. *Biotechnology Journal* 2 (8):1007–1013.

Steward, F.C., Mapes, M.O., and Mears, K. 1958. Growth and organized development of cultured cells. II. Organization in cultures grown from freely suspended cells. *American Journal of Botany* 45:705–707.

Thanutong, P., Furusawa, I., and Yamamoto, M. 1983. Resistant tobacco plants from tolerant cell lines. *Plant Science* 13:105–111.

Thorpe, T.A. 1990. The current status of plant tissue culture. In: Bhojwani, S.S. ed. *Plant Tissue Culture: Applications and Limitations*. Elsevier, Amsterdam, 1–33.

Thyagi, R.K., Yusuf, A., Jeyaprakash, P., and Dua, P. 2001. Effects of polyamines on *in vitro* conservation of *Vanilla planifolia* (Salisb.) Ames. *Indian Journal of Plant Genetic Resources* 14 (2):300–302.

Triggs, H.M., Triggs, G.S., Lowe, K.C., Davey, M.R., and Power, J.B. 1995. Protoplast systems for spice crops: Cardamom (*Elettaria cardamomum*) and vanilla (*Vanilla planifolia*). *Journal of Experimental Botany* 46 (Suppl.):48–49.

Withner, C.L. 1955. Ovule culture and growth of *Vanilla* seedlings. *American Orchid Society Bulletin* 24:381–392.

Xju, Z., Arditti, J., and Nyman, I.P., 1987. *Vanilla planifolia*—callus induction and plantlet production *in vitro*. *Lindleyana* 2:88–90.

6 Cultivation Systems

Juan Hernández Hernández and Pesach Lubinsky

CONTENTS

INTRODUCTION

Vanilla (*Vanilla planifolia*) is a rare perennial, hemi-epiphytic succulent herb that makes use of forest trees in its natural habitat for support, shade, and natural humus. The forests, where wild *V. planifolia* are found, are classified as *selva alta perennifolia* (tall evergreen tropical forest), the wettest of the tropical forests. Today, *V. planifolia* is cultivated in different production systems that mimic to some degree the agro-ecological parameters that are found in the natural habitat of the species. The cultivation techniques and management practices for vanilla have improved mostly from trial and error by vanilla producers worldwide. On the contrary, empirical agronomic studies of vanilla are largely lacking.

This chapter details the agro-ecological requirements, systems of production, propagation, cultivation/management, and flowering/pollination and harvest of *V. planifolia*. This information is based largely on observation, experiment, and on published articles and cultivation manuals authored by people with personal experience in vanilla cultivation.

AGRO-ECOLOGICAL CONDITIONS

To reach optimal conditions for growth and production, vanilla cultivation requires the following agro-ecological parameters.

Climate

V. planifolia thrives in hot-humid tropical climates.

Temperature

V. planifolia grows best in temperatures ranging from 20°C to 30°C (Childers and Cibes, 1948; Ranadive, 2005), and may tolerate high temperature of 32°C (Purseglove et al., 1981; Anandaraj et al., 2005). Temperatures reaching below 20°C inhibit plant growth and flowering intensity (Ranadive, 2005); temperatures exceeding 32°C cause yellowing of vegetative parts and premature fruit drop (Anandaraj et al., 2005; Hernández Hernández, 2007b).

Precipitation

V. planifolia requires an annual average precipitation from 2000 to 3000 mm (Sasikumar et al., 1992; Soto Arenas, 2003), it is well distributed throughout the year except during flowering/pollination. Since heavy rains may diminish successful pollination and fruit set, it is best to irrigate the plants at their bases during flowering. *V. planifolia* needs 2–3 relatively dry months to stimulate flowering. In areas where average annual precipitation exceeds 3000 mm, plants are more prone to fungal attack (e.g., *Fusarium* sp.). At the other extreme, that is, where precipitation is less than 2000 mm, and where a system of irrigation is not in place, the lack of water greatly compromises the ability of the plant to perform basic physiological functions.

Altitude

The best altitudes for cultivating *V. planifolia* are between the sea level and 600 m (Childers et al., 1959), although cultivation systems do occur as high as 1100 masl in Mexico (Soto, 2003). In India, *V. planifolia* is reported to be cultivated up to 1500 masl (Anandaraj et al., 2005; John, 2005), and in Uganda, cultivation is successfully practiced between 800 and 1200 masl.

Light/Shade

V. planifolia demonstrates most vigorous growth under 50% shade. In dry periods with intense sunlight, it is preferable to use 50–70% shade (Hernández Medina, 1943; Ranadive, 2005) for better conservation of soil and air humidity. In rainy periods, the amount of shade should be reduced to 30–50% to avoid creating favorable conditions for growth of pathogens.

Excess shade causes weak growth and poor flower production, while excess sunlight leads to burning of the leaves and stems, as well as early fruit drop. Plants that suffer from either too much sunlight or shade are the ones most likely to develop diseases.

Site Selection

The first step in designing a successful vanilla cultivation system is to choose a site to plant.

The land destined to be used for the cultivation of vanilla (*vainillal*) should have an excellent drainage, rich humus content, and a pH between 6 and 7 (Childers et al.,

1959; Soto, 2003; Ranadive, 2005); the land should receive morning sunshine and not intense rays during afternoon, which can lead to the sunburn damages on the leaves (Sánchez Morales, 1993; Curti, 1995; Soto, 2003). Limestone soils on slight slopes are the most appropriate substrates for cultivating vanilla because they tend to be less acidic and are well-drained (Childers et al., 1948; Bouriquet, 1954; Dequaire, 1980; Ranadive, 2005); acidic soils are the least recommended because they favor the establishment of fungal pathogens (Soto, 2003).

The areas where vanilla cultivation has been practiced in the recent past are less desirable because they are likely to contain remnant populations of *Fusarium oxysporum*.

Land Preparation

The most prevalent form of vanilla cultivation in Mexico is the *vainillal*: a secondary forest regrowth the managed to cultivate vanilla. The first step in its preparation is to thin the forest to increase luminosity and space, leaving only those trees or shrubs that would serve as supports. Where deforestation has occurred, weedy herbs are cut and left to compost on-site in lieu of being burned. On level ground, where the soil may become waterlogged, drainages are dug to eliminate excess water. On steep terrains, terracing is done to control erosion, mitigate runoff of organic material, and to conserve soil humidity.

Support Trees

The primary function of the support tree is to provide an appropriate framework and microenvironment for the growth and management of the clambering vanilla plant. In addition to providing physical aid, the support trees give shade and, in deciduous or semideciduous species, apportion organic material. The choice of the support tree should be the species that is most adapted to the area, and feature the following characteristics: (1) maintain a canopy throughout the year, (2) be easy to propagate and able to reestablish readily, (3) respond positively to periodic pruning for form/shade management, (4) lack spines and leaves near its base, (5) be strong enough to support the weight of the vanilla plant, and strong winds, (6) possess bark that does not shed, (7) be fairly resistant to pests/diseases, and (8) have a deep root system that does not compete in the shallow soil layers for nutrients and water with the vanilla plants.

In the low-intensity *vainillales* of Mexico, where other crops and species of economic importance are managed, the species chosen for supports are the shrubs and trees that grow in that area. Fruit trees are also chosen: orange (*Citrus sinensis*), grape-fruit (*Citrus grandis*), and mandarin (*Citrus reticulata*). In intensively managed monoculture of vanilla (Figure 6.1), support trees must first be planted. The following are the most commonly planted vanilla support trees in the world:

Erythrina sp. (Mexico, Costa Rica, India, and Indonesia)
Gliricidia sepium (syn. *G. maculata*) (Madagascar, Tonga, India, Indonesia, Reunion Island, and French Polynesia)
Leucaena leucocephala (Tonga and Indonesia)

FIGURE 6.1 Intensive cultivation of vanilla on living supports (*Erythrina* sp.).

Casuarina equisetifolia (Madagascar, Tonga, Reunion Island, and India)
Jatropha curcas (Madagascar, Uganda, Tonga, Reunion Island, and French Polynesia)
Plumeria alba (India)

In the most recently established system of vanilla cultivation, artificial or "dead" supports are used (Figure 6.2). In this case, the function of the support is purely physical,

FIGURE 6.2 Vanilla intensively cultivated on concrete support under shade houses.

while the shade is provided by 50% shading cloth (green, red, or black), which is stretched above all the plants at ca. 3–5 m high, on the four sides of the planted area. These systems are referred to as "shade houses." In size, they are usually on the order of 25 m × 40 m (1000 m^2), and some are up to 1 ha (Hernández Hernández 2007a).

Establishing Living Support Trees

Supports such as *Erythrina* sp. and *Gliricidia* sp. are propagated mainly through stem cuttings, although some species such as *L. leucocephala* are propagated by seed.

In Mexico, transplanting and establishing the supports can be done at any time of the year, but is most successful when done at the onset of the rainy season. Cuttings are normally 1.5–2 m long and at least 5 cm in diameter. Cuttings are planted at about 30–40 cm depth, and six months to one year in advance of establishing the vanilla cuttings.

The most common planting distances between supports are: 1.2 × 1.50 m, 1.5 × 2.5 m, 2.0 × 2.5 m, 1.8 × 2.5 m, and 2.0 × 3.0 m, or a density of approximately 1600–5000 supports/ha (Childers et al., 1948; Dequaire, 1980; Tiollier, 1980; Sasikumar et al., 1992; Soto, 2003; Anandaraj et al., 2005; Ranadive, 2005). Smaller planting areas are problematic due to insufficient air circulation, which lead to diseases (Childers et al., 1948; Soto, 2003; Ranadive, 2005). For each support, 1–2 vanilla cuttings are planted.

Planting distance between fruit tree supports varies with the species. For orange, the planting distances are normally 4 × 4 m, 5 × 5 m, 6 × 6 m, and 7 × 7 m (roughly equivalent to 204–625 supports/ha). For each orange tree, 3–6 vanilla cuttings are planted, producing an estimated 1224–1875 vanilla plants/ha.

In polycrop systems that include vanilla, usually associated with coffee, banana, and coconut palm cultivation, it is common to use *Erythrina* sp. or *Gliricidia* sp. as vanilla supports, planting them at distances of 1.5 × 2 m. These are smaller densities than are encountered in monocrop systems, but have the advantage of allowing for at least two products to be exploited.

When used, the native tree species are planted at variable distances, depending on the standing distribution of those that are selected, and can be 500–1500 supports/ha. In shade houses, supports are spaced 1–1.5 m between each tutor and 2–2.5 m apart between each row. In 1 ha, 4000–5000 supports are set up, with 1–2 vanilla plants/support. Vanilla generally thrives in shade house cultivation. However, at the reproductively mature adult stage, their management becomes difficult and tedious, air circulation is reduced, and the spread of pathogens can be very rapid and devastating (Ranadive, 2005).

PROPAGATION

Vanilla is propagated almost entirely by stem cutting. The cuttings are procured from another grower or from a government agricultural entity. Cuttings are made from highly productive and vigorous individuals, which are selected and marked before the harvest. The cutting itself should not be a flowering shoot and should have at least three nodes with viable axillary buds for producing new shoots from which the plant grows. Cuttings should be free of damage or of symptoms of pests/diseases so as to avoid future proliferation of the disease.

Cuttings are normally 6–8 nodes (80–120 cm long, 1 cm in diameter) in length. Longer or thicker cuttings form new vegetative and reproductive shoots more rapidly (Ranadive, 2005), but are more difficult to deal with during planting and are also more expensive. Cuttings that are less than 60 cm long are best managed as nursery plants before they are transplanted permanently (Anandaraj et al., 2005; Ranadive, 2005).

In Vitro Plants

Another option for sourcing and reproducing commercial vanilla stock is *in vitro* propagation. *In vitro* propagules are accustomed in having their nourishment provided, and so before they are transplanted directly into a cultivated area, they must first be placed in climate-controlled greenhouses, where they can adapt in providing their own food via photosynthesis. These propagules are transplanted as soon as they reach a height of about 30 cm.

In Vitro Propagation

Vegetative growth proceeds slowly following establishment, usually taking at least one year before the plants reach the flowering/fruiting stage. Plants established from *in vitro* propagules are more phenotypically uniform and healthier, and consequently yield higher than vegetative cuttings. The cost and care of *in vitro* propagation is prohibitively expensive for many vanilla farmers.

Preparation and Disinfection of Cuttings

Cuttings are prepared prior to planting. The three most basal leaves are removed by hand by twisting at the petiole and taking care not to tear into the stem where open wounds can facilitate the spread of pathogens.

In order to prevent stem rot, caused primarily by *F. oxysporum*, stem cuttings are disinfected prior to planting. The basal portion of the cutting is submerged for 2–5 min in a fungicidal solution. The solution may consist either of carbendazim (2 g/L) or Bordeaux mixture (1 kg lime + 1 kg copper sulfate in 100 L of water), the latter being less effective but authorized for the production of organic crops. Fungicidal solutions are handled with rubber gloves to avoid harmful exposure to the human body.

After disinfection, cuttings are hung separately on a structure 1–1.5 m tall, in a shaded and well-ventilated area for a period of 7–15 days. The cuttings slightly dehydrate allowing for more flexible material for planting. Calluses form over areas of the cuttings that were damaged during leaf removal.

Establishing Cuttings—Timing

Cuttings are planted when support trees have developed sufficient foliage to prevent the young vanilla plants from sunburns. With shade cloth, cuttings are planted immediately after the establishment of support trees.

Cuttings can be planted practically at any point during the year given a sufficient availability of water. In the winter, new vegetative shoots are slower to develop and may be "burned" by low temperatures, which slow the overall development of the plant. In the rainy season, excess humidity can cause up to 50% of cuttings to suffer some degree of rotting.

The best conditions for planting cuttings are in humid substrates during warm, dry months preceding the onset of the rainy season (Ranadive, 2005). This timing favors a high percentage (>90%) of successful establishment of cuttings since high temperatures are conducive to the emergence of new shoots and roots.

Establishing Cuttings—Planting

Cuttings are planted in the following manner: adjacent to the support, a shallow ditch is dug 5–10 cm deep, into which the cutting is placed horizontally (but only the part that has had the leaves removed). The cutting is then buried with 3–5 cm of organic material and/or fertile soil or leaves, which will serve as mulch and as a source of nutrients. The extreme basal end of the cutting (2–3 cm) is left uncovered to prevent rot (Lepierre, 1988; Wong et al., 2003; Ranadive, 2005), especially when the substrate is humid. Some cuttings are established without making ditches, and are placed on top of a humid substrate.

Once planted, the rest of the cutting (with leaves, ca. 4–5 nodes) is positioned vertically on the support and fastened with biodegradable material such as banana leaves, tree bark, or henequen fiber.

Under optimal conditions of humidity and temperature, and with vigorous, healthy cuttings, the first roots begin to emerge the first week after planting and during the first shoots in about one month. This relatively early rooting negates the need to purchase rooting hormones or products.

TECHNIQUES/PRACTICES FOR VANILLA CULTIVATION

After cuttings are successfully established, several activities and practices are necessary to ensure the development and optimal production of the vanilla crop.

Irrigation

Water is the main factor in the growth and development of vanilla. Sufficient availability of water is most critical during flowering/pollination. Irrigation has three benefits: (1) encouraging growth and development of the plant, (2) increasing fruit yield and quality, and (3) prolonging the productive life of the plant.

Irrigation Systems

The most frequent form of irrigation in Mexico is the use of microemitters to moisten the mulch layer of the *vainillal*.

Irrigation Criteria

One criterion for irrigation is to maintain at all times a moist layer of mulch without reaching saturation levels. The frequency and amounts of irrigation therefore depend on the type of mulch, phenological stage of the vanilla plants at the time of watering, and the prevailing climatic conditions such as rain and solar radiation as well as the amount of shade covering the *vainillal*. Generally, in the dry season, watering is performed once to twice per week.

Nutrition

The primary source of nutrition for vanilla in cultivation is organic material (humus) that results from the natural decomposition of vegetable/animal residues (mulch), composting (via microorganisms), or vermiculture (worm-mediated breakdown of organic material).

Mulch

In addition to apportioning nutrients, mulch has the following benefits: (1) it helps maintain soil humidity, (2) serves as a porous substrate, aiding in soil aeration and permitting the unrestrained development of roots, (3) maintains an adequate temperature, and (4) decreases the incidence of weeds.

Mulches increase the biological activity of microorganisms and promote the development of mycorrhizal associations (Porras Afaro and Bayman, 2007). For example, the fungus *Rhizoctonia* sp. is known to establish a mycorrhizal symbiosis with vanilla roots, from which the vanilla derives a better absorption of nutrients and water while the fungus gains carbohydrates produced by the plant (Wong et al., 2003). Other fungal species belonging to the genera *Ceratobasidium*, *Thanatephorus*, and *Tulasnella* also positively effect seed germination and plant growth (Porras-Alfaro and Bayman, 2007).

Types of Mulch

The most common mulch for vanilla is from decaying leaf litter deriving from leaf fall from trees, pruning, and from herbaceous plants in the *vainillal*. Decomposed and rotting tree bark is also used in the form of sawdust.

When there is not sufficient organic plant material found in the *vainillal*, mulch can be made from stubble of kudzu (*Pueraria phaseolides*), elephant grass (*Pennisetum* sp.), or Guinea grass (*Panicum maximum*) (Bouriquet, 1954; Ranadive, 2005).

Coconut fiber is also a popular mulch because it is porous, lightweight, and has an excellent capacity for retention of humidity and for conserving an appropriate microclimate for the promotion of root growth.

Sawdust in a fresh state may contain toxic substances to plants such as phenols, resins, terpenes, and tannins. Fresh manure or manure that is not well decomposed can also burn or cause root rot and mortality. It is important that these materials are composted before they are applied to the vanilla plants.

Mulch—Thickness

The mulch should be 10–20 cm deep and approximately 50–100 cm wide, depending on the extent of root growth. In areas of high humidity and deficient drainage, the depth is less, in order to mitigate the development of fungal pathogens, which can favor root rot.

The mulch is laid down on either side of the support where the vanilla roots will grow. To prevent the loss of mulch from runoff from heavy rains, most prevalent in *vainillales* managed on slopes, borders are constructed out of trunks of wood, bamboo canes, rocks, or other materials.

New applications of mulch are made when roots are observed growing out of the surface of the mulch, generally 2–3 times/year, and mostly in the hot/dry months, when mulch is carefully managed to prevent dehydration.

Nutrients

The most important nutrients for vanilla are calcium, potassium, nitrogen, phosphorous, iron, and copper (Cibes et al., 1947; Childers et al., 1959; Domínguez, 2005; Ranadive, 2005). The normal or optimal levels of nutrition required by vanilla has not been studied in detail, and in practice, vanilla nutritional requirements are inferred from horticultural species of other members of the Orchidaceae (A.S. Anderson, pers. comm.).

Chemical Fertilizers

Generally, vanilla is not fertilized beyond provisioning of mulch. In India, growers are encouraged to apply 40–60 g of nitrogen, 20–30 g of phosphorous (P_2O_5), and 60–100 g of potassium (K_2O) per plant per year. Foliar sprays are also recommended such as 1% applications of Triple 17 fertilizer (17:17:17) once per month to stimulate growth and flowering (Anandaraj et al., 2005).

Management Activities

Weed Control

Weeding is done by hand. Between rows, tools such as a hoe or machete are used, but at the base of the plants themselves, weeds are carefully pulled out by hand so as not to disturb the shallow rooting structure of the vanilla plants. Once deracinated, weeds that are annual herbs can be added to the mulch or composted and later added. Perennial weeds such as *Commelina diffusa* and *Syngonium podophyllum* are removed from the *vainillal* because they do not readily decompose. Weeds should be dealt with whenever they impede access to the vanilla plants and/or when support trees defoliate in a disproportionate amount. In general, weeding is performed 3–4 times/year.

Regulation of Light and Shade

In *vainillales* with living support trees such as *Erythrina* sp. or *Gliricidia* sp., shade is controlled by periodic pruning, usually two or three times/year.

Pruning should be timed to take place in the rainy season to avoid the development of diseases in vanilla due to inadequate sunlight. Shade levels are between 30% and 50% during the rainy season. In dry and hot times of the year, which coincides with flowering/pollination and fruit development, support trees should have a denser canopy to provide 70–80% shade, which conserves humidity, prevents burning from intense sunlight, and decreases the incidence of young fruitdrop.

Pruning is accomplished by removing the thicker central branches and leaving the laterals in order to achieve a canopy in the shape of a parasol that also maximizes the equitable distribution of vanilla shoots. Branches are pruned with either saws or machetes, down to about 40 cm from where they diverge from the trunk. The thinnest of the cut branches are broken into longitudinal pieces and placed at the base of the support as an additional source of organic material. Thicker branches are removed from the *vainillal* entirely. Overpruning results in sunburns to the vanilla plants, and should be avoided.

In cultivation systems with artificial or "dead" supports, 50% shade is provided year-round by shade cloth. This system allows for uniformity of shade and negates the cost of having to periodically prune, but in hot months the 50% shade has been observed to be inadequate.

Shoot Management—Looping

The most common practice involving shoot management is "looping," that is, redirecting a growing shoot over a branch and toward the ground once it reaches the height of the first branches of the support tree. This practice maintains the height of the vanilla at roughly 2 m, facilitating hand pollination and harvesting. Another consequence of looping is hormonal induction promoting flowering and new shoot formation (usually just below the height of the fork in the tree where the shoot is bent). Shoots are managed so that they are equally distributed among the branches of the support tree such that no shoot shades out another.

Shoot Management—Rooting

Once a shoot has been looped and has reached the level of the ground, a portion of it, usually 2–3 internodes long, is buried, leaving the growing apical meristem uncovered. This practice promotes root formation at the buried nodes. The shoot apex is fastened back to the support tree to continue growth. Rooting of shoots is performed every instance a new shoot has reached ground level, which helps maintain the vigorous growth of the plant that obtains more nutrients and is more resistant to *F. oxysporum*. In this way, rooting helps counteract the mortality of plants due to pathogens (Hernández Hernández 2005). Alternatively, for plants that are at least three-years old and are near or at ground level, a grower may elect to trim off the apical growing tip (ca. 20 cm) to induce flowering in lieu of rooting (Anonymous, 1998, 2004).

Another technique for shoot management is to let the shoots attain a height of 1.5–2 m, at which point the shoots are trained onto horizontal supports such as bamboo canes, hoses, or plastic PVC pipes.

Disease/Pathogen Prevention

Keeping vanilla plants free of pests/disease requires frequent oversight and consists of removing any parts of the stem, leaves, or roots that manifest disease. Sometimes entire plants require removal to terminate the spread of disease from plant to plant. Diseased material is burned or buried outside the *vainillal* to eliminate sources of inoculum. Leaves attacked by pests are also removed.

Flowering and Pollination

In general, the first flowering happens three years following planting. When *Citrus* spp. are used as supports, or when vanilla is cultivated in shade houses, flowering initiates in the second year, even when smaller cuttings (80–100 cm long) are used, since the plants tend to grow more vigorously as a result of more consistent shade and management. Vanilla normally flowers once a year.

Factors Promoting Flowering

The physiological cue to flower is promoted by climatic or mechanical stress. The following are some examples:

a. Drought: Water stress induces reproductively mature individuals of vanilla and many other plants to flower. In Uganda, vanilla flowers twice a year, a consequence of the country having two distinct dry seasons (Anonymous, 2000, 2005).
b. Cool temperatures: The principal stress in Mexico that induces flowering are the low temperatures of autumn–winter, when cool air masses known as "nortes" blow down more or less unimpeded from the Arctic Circle, dropping temperatures to below 10°C; the lower the temperature, the greater the expectation of a good flowering year. The cool temperatures "burn" the apical tip, killing it, and break the apical dominance of the plant while stimulating lateral floral buds to develop.
c. Plant management: In India, growers are recommended to perform a series of tasks two months before flowering: (1) pruning of the apical tip, at least 10–15 cm, (2) temporary suspension of irrigation, (3) abundant use of irrigation once 10% of blooms have opened, (4) pruning of the support tree canopy up to 75% luminosity (Ranadive, 2005), but only when the intensity of sunlight is not high to cause burning (light is an important factor in activating floral buds; Hernández Apolinar, 1997), and (5) allowing the shoots to reach a height of 1.5 m and training them on a horizontal support to increase flower quantity and the number of aerial roots.

Inflorescence

The inflorescence of vanilla is a raceme, meaning individual flowers are borne on stalks arranged around a central axis. An individual plant normally produces 8–15 racemes or more (Anonymous, 1998, 2004; Ranadive, 2005), depending on the age, the quantity of flowering and rooting shoots, and the prevailing environmental conditions. From the initial formation of the inflorescence to the time the first flower opens is a period of 45–60 days.

Each raceme develops 10–20 floral buds, which open sequentially, starting from the base of the raceme. Normally, only one flower per raceme opens per day. Sometimes two flowers open simultaneously and sometimes no flowers open when there is rain or when temperatures are low. The flowers are fully open in the early morning and last 6–8 h before closing in the heat of the afternoon.

Flowering Period

Vanilla-producing countries in the northern hemisphere (e.g., Mexico, India) experience flowering in March–May, with a peak in April. Southern hemisphere countries (e.g., Madagascar, Indonesia) have a flowering season from September–December.

Natural Pollination

Mexico is one of the few countries where it is possible to obtain vanilla beans through natural pollination, although it happens rarely, accounting for only about 1% of all fruits. The identity of the natural pollinator(s) of vanilla is unclear, and for a long time it has been said that bees (*Melipona beechii*), hummingbirds (*Cynniris* sp.), and bats pollinate vanilla. The preponderance of evidence favors the hypothesis that the most common pollinator is the shiny green orchid bee *Euglossa viridissima* (Soto, 1999a, 2003; Lubinsky et al., 2006). These bees have been documented visiting vanilla flowers but their visits are irregular and their potential for effecting pollination even smaller, perhaps only just 1 fruit per 100 or 1000 flowers (Soto, 1999a, 1999b; Hagsater et al., 2005).

With natural pollination, only one fruit is observed per raceme. Sometimes up to four are observed, all of which are usually larger and heavier than vanilla beans obtained through hand pollination because more resources are allocated to their development.

Other orchid bees, namely, individuals of *Eulaema* sp. (*jicotes*) frequently visit the flowers of *Vanilla pompona* (Figure 6.3) in northern Veracruz, Mexico. On rare occasions, they also effect pollination of the flowers (5%) while looking for nectar inside and at the base of the labellum.

FIGURE 6.3 (See color insert following page 136.) *Eulaema* sp. (*jicote*) bees, probable natural pollinator of *V. pompona*.

The mechanism by which the aforementioned bees actually pollinate flowers of vanilla is yet to be documented.

Hand Pollination

Inside the labellum of the vanilla flower, the part which attaches to and wraps around the column, is a tissue that flaps down from the column called the rostellum. The rostellum hangs exactly in between the stigma (female organ) and the anther sac (male organ), and is considered to be a product of evolution selected to prevent self-fertilization. In hand pollination, pollen is manually moved from the anther sac to the stigma, bypassing the rostellum.

Hand pollination is performed with a small, thin stick roughly the size and shape of a toothpick, but can be made from bamboo, bone, spines, or other materials. The relevant parts and reproductive organs of the vanilla flower are shown in Figure 6.4.

Hand pollination was discovered by Charles Morren in 1836 and the first to put into practice on the island of Reunion was by Edmond Albius in 1841 (Lecomte, 1901; Childers and Cibes, 1948). This method of hand pollination is the same one that is in use today (Figure 6.5) and consists of the following steps:

1. Use a toothpick or similar tool to make a longitudinal slit in the labellum on the side opposite of the column to reveal the reproductive structures.
2. With the same end of the toothpick, lift underneath the rostellum and flip vertical so that the anther sac can hang down unimpeded over the stigma lobes.
3. Gently press the anther to the stigma until the two stick together and remove the toothpick.

Hand Pollination—Timing

Pollen of *V. planifolia* has been found to be viable for a period that begins 23 h before anthesis and ending 16 h after the flower closes. Likewise, the stigma is receptive

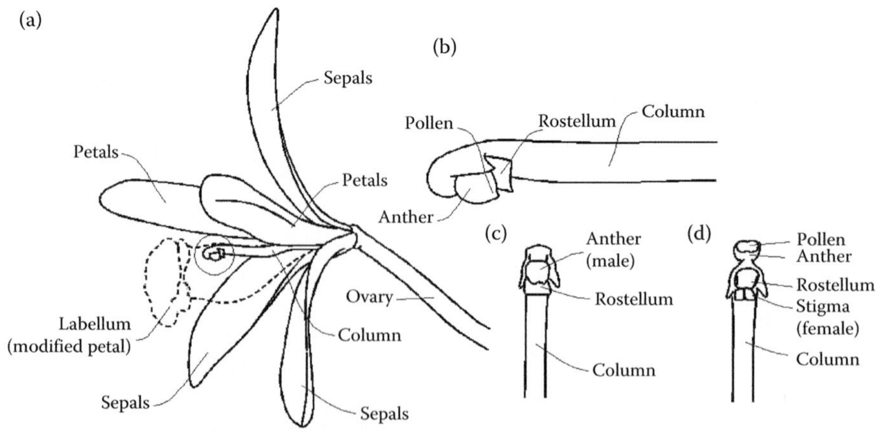

FIGURE 6.4 Floral structures of *V. planifolia*. (a) Complete flower, (b) column (side view), (c) column (front view), and (d) rostellum and anther sac lifted revealing the stigma.

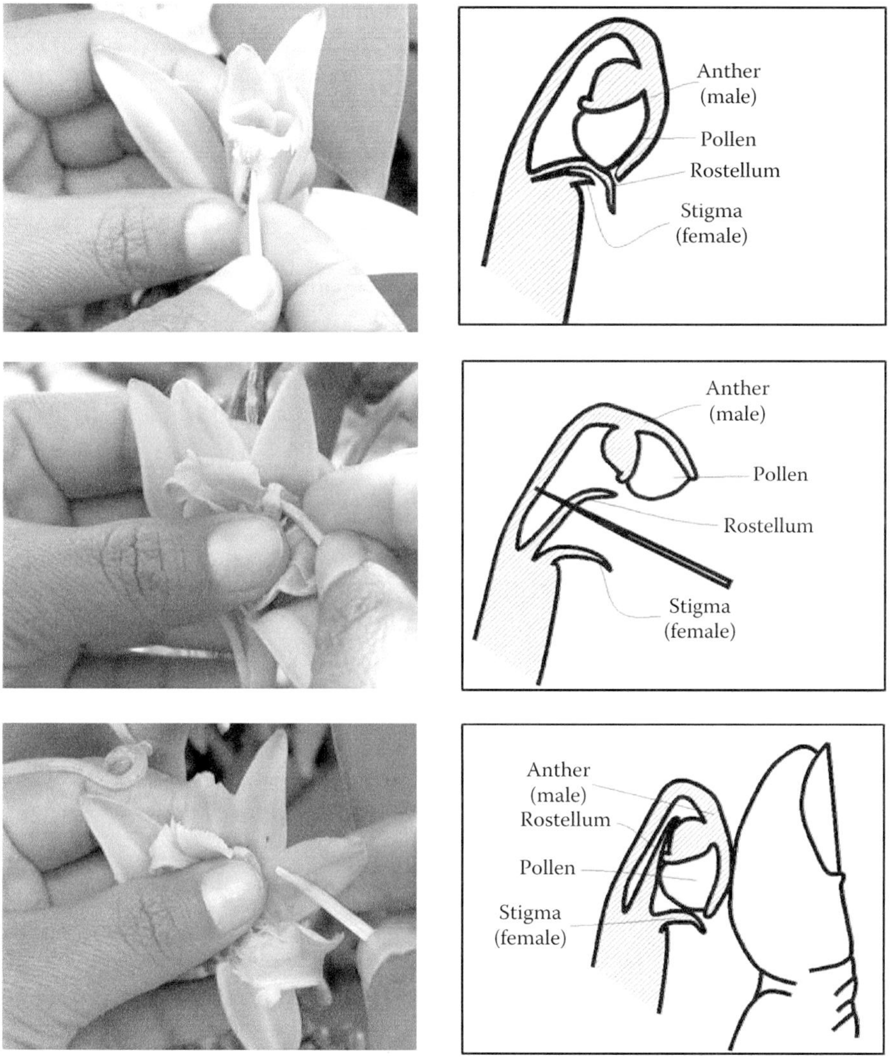

FIGURE 6.5 Method for hand pollination of *V. planifolia*.

41 h before flower opening and remains so until 17 h after the flower has closed (Shadakshari et al., 2003). For practical reasons, hand pollination is performed from 7 a.m.—noon, or a little bit later when it is overcast, but never when the flowers have already closed or withered.

Hand pollination should be conducted by able and experienced people. Women are more commonly involved in the task. An experienced person pollinates 1000–1500 flowers per 5–7 h period (ca. 4 flowers/min), assuming that the plants are more or less in the same area. The first flowers in the raceme that are pollinated yield longer and straighter fruits while the last flowers to open characteristically produce smaller and curved fruits that have less value.

Hand pollination is a daily task for a period of three months. Per hectare, 300–600 days of work is required to carry out pollination, depending on the abundance of flowers, their location, efficacy of the pollinator, and distance between plants.

Indicators of Success or Failure in Hand Pollination

Fertilized/pollinated flowers do not separate and fall from their pedicel (in the case of vanilla, the pedicel is also the inferior ovary). On occasion, the column may abscise from the fertilized ovary, but the ideal is for the column and petals to stay attached to the developing fruit as they serve to keep the fruit hydrated and diminish the colonization of pests or fungal disease.

Flowers that are not pollinated, pollinated incorrectly, or have been damaged by rain or high temperatures abscise or abort 2–3 days following pollination. On rainy days, pollinated flowers abort up to 50% of the time because the pollen humidifies, loses its adhesive quality, and falls away from the stigma.

Number of Fruits per Raceme

In general, 6–8 flowers per raceme are pollinated to ensure a minimum yield of 4–5 fruits per raceme of acceptable quality. Obtaining 100–120 fruits per plant requires 8–15 flowers per raceme to be pollinated (Anonymous, 1998). These approximations are rough since much depends on environmental conditions, the position and vigor of the plants, as well as the biological characteristics of the clone or cultivar. Vanilla growers determine the amount of flowers to be pollinated by considering pricing as well. Overpollination leads to an abundance of many smaller fruits of lesser value that increase the cost of pollination and exert a heavy cost on the plants. Overpollination is also associated with major fluctuations in production volume from year to year.

Fruit Development

Immediately following hand pollination, pollen tubes begin their germination and growth and eventual fertilization of the ovules. The ovary quickly begins to enlarge and assume a strong, dark green aspect as it orients itself downward. The maximum length and diameter of the fruit is achieved 45 days after hand pollination (Figure 6.6). Thereafter, growth ceases, and the fruit enters into a period of maturation lasting roughly 7–8 months.

Harvest

Indicators of When to Harvest

Vanilla is harvested when it reaches its commercial maturity, when the distal point of the fruit changes from green to yellow (Figure 6.7), generally 8–9 months after pollination.

Early harvested fruits weigh less, are more susceptible to fungal attack and, when cured, yield smaller quantities of vanillin (Tiollier, 1983). Fruits that are harvested

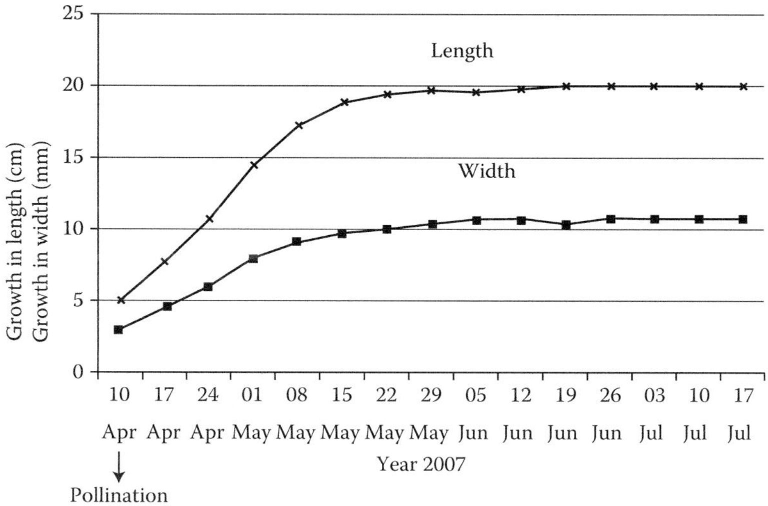

FIGURE 6.6 Time-series data on the development of a fruit from *V. planifolia* cv. *mansa* (length and diameter) observed at the Ixtacuaco experimental station (INIFAP, Mexico).

FIGURE 6.7 Commercial maturity of green beans (8–9 months after pollination).

past nine months begin to dehisce, and change from green to yellow to dark brown or black. These fruits are called "splits" and have less value.

Harvest—Timing

Northern hemisphere countries have harvest seasons from December to February while Southern hemisphere countries harvest from June to September. Most vanilla growers harvest at only one time, and collect all fruits, that is, those that are mature and those that are still developing. The still developing fruits dehydrate faster than mature fruits during curing. Some growers, for example, in Uganda and French Polynesia, harvest only when fruits are mature, generally once a week, prolonging the harvest to 2–3 months.

Units of Harvest

In Mexico, the whole bundle or raceme of fruits is harvested with the central stalk, or rachis, still attached. A more ideal practice is to harvest each mature fruit individually. Harvested fruits are placed in baskets or plastic crates to prevent mechanical damage, which can lead to pathogen infection. The fruits are also kept in well ventilated and shady areas.

Pruning of Flowering Shoots

After harvesting, it is customary to prune shoots that have already flowered. These shoots will not produce again (or as much) unless they retain buds. The pruning is performed with a knife or blade that is disinfected prior to use in a solution of 1 part bleach to 6 parts water.

The removal of these "spent" shoots serves to eliminate unproductive parts of the plant that occupy space and take away the energy resources from the plant. Their removal facilitates the maintenance of adequate ventilation and light conditions for the plant. Some of these spent shoots may serve as cuttings to start new plants if they retain meristematic tissue.

Yields—Green Vanilla

Yields from *vainillales* are extremely variable, and depend on the ages of the plants, density of the plantings, method of cultivation ("traditional" or modern), source of water (rain-fed or irrigated), the soil and climatic conditions of each site, and so on.

Worldwide, average yields of green vanilla are less than 500 kg/ha, since most growers employ few technologies. Those who use technologies for irrigation or fertilization or use shade houses may obtain 3–4 tons of green vanilla/ha, or approximately 500–800 kg cured vanilla/ha (Ranadive, 2005).

The average weight of fruits per plant is 1–2 kilos and up to 5.5 kilos for plants that have been grown on orange tree supports (Lopez Méndez and Mata García, 2006).

In whatever system of vanilla cultivation, the maximum yields occur in the fourth or fifth year following planting (second or third harvest). After this time, production volume can be lower or higher, but after nine years yields steadily decline until productivity ceases almost completely by the twelfth year.

Useful Life Span of Vainillales

Experience has demonstrated that *vainillales* that are densely planted can produce high yields, but generally only once. Later, yields decline drastically and abruptly terminate usually six years after planting because of major problems with disease. In *vainillales* that are less densely planted, a productive life span of 12 years is expected because adequate parameters exist for ventilation and disease/pest prevention. The best model for this is vanilla cultivated on orange supports.

Labor Requirements

Vanilla is a labor-demanding crop, requiring 172–575 workdays/year ha depending on the system of production, and stage of the plants (i.e., vegetative growth versus productive). The fourth and fifth years are normally the most demanding since production is greatest then. In Mexico, in the semi-intensive system using *Erythrina* sp. as a support tree, growers have needed a maximum of 575 workdays over the year (on an average, it is 316.2 workdays).

The majority of workdays, more than 50%, are needed for hand pollination, followed by shoot management (looping and rooting, 10.4%), pruning (7%), and weed control, accommodation of organic materials/compost, and phytosanitary maintenance (7%) (Table 6.1).

Labor requirements are similar in other systems of vanilla cultivation. One family can cultivate a maximum of 1 ha of vanilla, but needs to contract wage labor for help with pollination in a larger area.

TABLE 6.1
Number of Workdays per Activity for the Maintenance of 1 ha of Vanilla in the Semi-Intensive System using *Erythrina* sp. as Support Trees (5th Year), in the Region of Martínez de la Torre, Veracruz, México

Activity	No. of Workdays	Percent of Total
1. Pruning	40	7.0
2. Composting	20	3.5
3. Mulch acquisition	20	3.5
4. Irrigation	10	1.7
5. Weed control/ mulch maintenance	40	7.0
6. Shoot management	60	10.4
7. Pest/disease control	40	7.0
8. Hand pollination	300	52.1
9. Harvesting	30	5.2
10. Inspection/removal of diseased plants	15	2.6
Total	575	100

Perspectives on Crop Improvement

The fungus *F. oxysporum* is the major cause of economic damage to vanilla cultivation worldwide. Management practices and chemical control have so far not proved effective at controlling outbreaks of *Fusarium*. Strategies should be developed to identify a genetic basis for resistance through selection and breeding. Genomic/genetic technologies such as marker-assisted selection and gene chips can help in this regard.

Knowledge of the nutritional requirements of vanilla is still imperfect. The quantity and types of nutrients needed by vanilla at each phenological stage should be studied in more detail. Such information can be used to generate mulches and fertilizer mixtures specific to vanilla. This in turn would have a beneficial effect on yields. Similarly, water is an indispensable element in the cultivation of vanilla, but the precise water needs of vanilla—how much, and at what intervals—are yet to be studied.

Vanilla pests are readily managed with organic products, when such chemicals are used efficiently and opportunistically. In addition to learning more about the effects of specific management practices, vanilla cultivation stands to improve from insights into how flowering can be induced using chemical or hormonal applications.

REFERENCES

Anonymous, 1998. *Manuel de vulgarisation de la culture semi-intensive de la vanille*. UE/GES/GNEV. Antanarivo Madagascar, 80pp. (audiovisual material).

Anonymous, 2000. Agribusiness Development Centre (ADC) of the USAID-funded Uganda is Investment in Developing Export Agriculture (IDEA) project. *ADC Commercialization Bulletin* No. 1—*Vanilla*. 2000, 11pp. www.foodnet.cgiar.org/market/Uganda/report/vanilla.pdf. March, 2009.

Anonymous, 2004. Dossier: La vanille aux Loyauté. Les Cahiers de l'agriculture et de l'environnement. No. 9/septembre–octobre 2004. *Revue d´ information editée par la Province des îles Loyauté*, 11–22. www.symbiose.asso.nc/cahiers_agri/. Octubre 2004.

Anonymous, 2005. Vanilla. Product Profile No. 9. Uganda Export Promotion Board. www.data.mtti.go.ug/docs/vanilla.pdf. Marzo, 2009.

Anandaraj, M., J. Rema, B. Sasikumar, and R. Suseela Bhai, 2005. *Vanilla (extension pamphlet)*. Indian Institute of Spices Research. Calicut, Kerala, India, 11pp.

Bouriquet, G. 1954. Culture. In: *Le vanillier et la vanilla dans le monde*. Editions Paul Lechevalier, Paris, 429–458.

Cibes, H.R., N.F. Childers, and A.J. Laoustalot. 1947. Influence of mineral deficiencies on growth and composition of vanilla vines. *Plant Physiology* 22:291–299.

Curti D. E. 1995. Cultivo y beneficiado de la vainilla en México. Folleto técnico para productores. Organización Nacional de Vainilleros Indígenas, Papantla, Veracruz, 96pp.

Childers, N.F. and H.R. Cibes. 1948. Vanilla culture in Puerto Rico. Cir. No. 28. Federal Experiment Station in Puerto Rico (USDA). Mayaguez, Puerto Rico, 94pp.

Childers, N.F., H.R. Cibes, and E. Hernández Medina. 1959. Vanilla—The Orchid of commerce. In: C. L. Withner, ed. *The Orchids. A Scientific Survey. Chronica Botanica. New Series of Plant Science Books*. The Ronald Press Co., New York, 477–508.

Dequaire, J. 1980. La culture et la préparation de la vanille a la Reunion. Coopérative Agricole des Producteurs de Vanille de la reunión 1980/02. 43pp.

Domínguez G.R. 2005. Crecimiento y niveles nutrimentales en vainilla planifolia. Tesis de maestría, Colegio de Postgraduados, Montecillos, México, 59pp.

Hagsater, E., M.A. Soto-Arenas, G.A. Salazar Chávez, R. Jiménez Machorro, M.A. López Rosas, and R.L. Dressler. 2005. Las orquídeas de México: Orquídeas y gente. Instituto Chinoin, México D.F, 38–71.

Hernández Apolinar, M. 1997. Crecimiento y reproducción de *Vanilla planifolia* (Orchidaceae) en Usila, Oaxaca. Tesis de maestría en ciencias, UNAM, 105pp.

Hernández Hernández, J. 2005. Comparación de dos Sistemas de Producción Intensiva de Vainilla. Resultados Finales. En: Avances en la Investigación Agrícola, Pecuaria, Forestal y Acuícola en el Trópico Mexicano 2005. Libro Científico No. 2. Veracruz, México, 81–94.

Hernández Hernández, J. 2007a. Tecnología de "casa-sombra" Producción Intensiva de Vainilla, en la Región del Totonacapan. *Revista Agroentorno* No. 90/Año Octubre 10, 2007, 4pp.

Hernández Hernández, J. 2007b. Bad weather and hurricane dean: The beginning and of México's Vainilla production collapse of 2007. In: *Presentation (conference) from Vanilla 2007, International Congress*. East Brunswick, NJ, USA. November 6–8, 2007, 24pp.

Hernández Medina, E. 1943. Studies of the shade requirements of vanilla. *The Journal of Agriculture of the University of Puerto Rico* 27, 27–37.

John, T.V. 2005. Indian vanilla. Cultivation, curing, quality, and authenticity. *The Proceedings of the First International Congress*. Princeton, NJ, USA, November 11–12, 2003, 17–24.

Lecomte, H. 1901. *Le vaniller: Sa ulture, preparation et commerce de la vanille*. C. Naud Editeur, París, 228pp.

Lepierre, S. 1988. *Cultiver La Vainille en Nouvelle-Caledonie*. CIRAD-CA. (NCL): s.n., 23pp.

López Mendez, S. and B. Mata García. 2006. *La vainilla en el Totonacapan, símbolo de la sustentabilidad. Extensión al Campo*. Publicación trimestral de la Universidad Autónoma Chapingo. Num. 02, 21–28.

Lubinsky, P., M. Van Dam, and A. Van Dam, 2006. Pollination of Vanilla and evolution in Orchidaceae. *Lindleyana* 75:926–929.

Porras Afaro, A. and P. Bayman, 2007. Mycorrhizal fungi of Vanilla: Diversity, specificity and effects on seed germination and plant growth. *Mycologia* 99:510–525.

Purseglove, J.W., E.G. Brown, C.L. Green, and S.R.J. Robbins. 1981. *Spices*. Vol. 2, Longman, London, pp. 644–733.

Ranadive, A.S. 2005. Vanilla cultivation. *The Proceedings of the First International Congress*. Princeton, NJ, USA, November 11–12, 2003, 25–31.

Sánchez Morales, S. 1993. Manual de producción de vainilla en el estado de Veracruz. Folleto para productores No. 6. Campo Experimental Papantla. CIRGOC-INIFAP, 28pp.

Sasikumar, B., J. Rema, and P.N. Ravindran. 1992. Vanilla. *Indian Cocoa, Arecanut and Spices Journal* 16:6–10.

Shadakshari, Y.G., D. Madaiah, M. Dinesh Kumar, K.V. Shivakumar, and K.H. Bhagavantha Goudra. 2003. Pollen viability and stigma receptivity in vanilla (*Vanilla planifolia* ***Andrews***). *Journal of Spices and Aromatic Crops* 12(2):194–196.

Soto Arenas, M.A. 1999a. Conservation of the genetic resources of Vanilla. *Canadian Orchid Congress Newsletter* 11:1. www.chebucto.ns.ca/recreation/Orchidcongress/news. Marzo, 2008.

Soto Arenas, M.A. 1999b. Filogeografía y recursos genéticos de las vainillas de México. Informe final del proyecto J101. Instituto Chinoin A.C. Herbario de la Asociación Mexicana de Orquideología, A.C., 85pp.

Soto Arenas, M.A. 2003. Vanilla. In: Pridgeon, A. M., P. J. Cribb, M. W. Chase, and F. N. Rasmussen, eds. *Genera Orchidacearum Vol. 3. Orchidoideae (Part two). Vanilloideae*. Oxford University Press, Oxford, 321–334.

Tiollier, V. 1980. Vanilla cultivation in Tonga. *Technical Bulletin* 1:37.

Tiollier, V. 1983. Vanilla curing in Tonga. *Technical Bulletin* 5:73.

Wong, C., Wong, M., and Grisoni, M. 2003. Culture de la vanilla. Fiches techniques/technical leaflets, 37pp. (audiovisual material).

7 Virus Diseases of Vanilla

Michel Grisoni, Michael Pearson, and Karin Farreyrol

CONTENTS

Virus diseases became a major concern for vanilla production over the last few decades probably as a consequence of the diversification of the cultivation areas and intensification of vanilla growing. This chapter reviews the data accumulated on the viruses affecting vanilla plantation throughout the world, particularly Cymbidium mosaic virus (CymMV), potyviruses, and Cucumber mosaic virus (CMV). It discusses how the environmental changes induced by the intensification of vanilla cultivation have favored the emergence of viral epidemics, the possible control strategies available at present, and the research perspectives to improve them.

INTRODUCTION

Viruses are obligatory cellular parasites that can infect bacteria, fungi, plants, and animals. They have probably emerged and evolved in their hosts at the beginning of the tree of life (Forterre, 2006). Virus particles are basically made of one or few genomic ribonucleic acid (RNA) or deoxyribonucleic acid (DNA) molecules encapsulated in a protein shell, the capsid. More than 5000 virus species are currently recognized (Fauquet et al., 2005), of which ~1000 infect plants and have been classified into more than 60 genera (Wren et al., 2006). Most of them are disease causing since virus development induces symptoms in host tissues that may generate severe crop losses. Plant viruses are only infrequently transmitted by seeds but somatic plant parts such as cuttings, bulbs, and buds ensure the propagation of many plant viruses. Horizontal transmission of viruses often involves animal vectors (mostly insects), but some viruses have no vector and only infect new hosts by contact with epidermal wounds.

More than 20 virus species have been reported to infect orchids (Gibbs and Mackenzie, 1997). The first demonstration of vanilla infection by a virus was by Wisler et al. (1987), in French Polynesia (FP). To date 10 virus species belonging to four families of positive-sense RNA genome viruses have been described from vanilla.

CYMBIDIUM MOSAIC VIRUS (CYMMV)

CymMV is the most prevalent and economically damaging virus infecting orchids (Zettler et al., 1990). Since its first description by Jensen in California (Jensen, 1950; Jensen and Gold, 1955), CymMV has been reported in most cultivated orchid species in many countries and is now considered to be present worldwide. CymMV was first detected in vanilla during a survey in FP (Wisler et al., 1987). It was subsequently found in vanilla plots of many countries: Cook Islands, Fiji, Niue, and Tonga (Pearson et al., 1993), Madagascar (Grisoni et al., 1997), Reunion Island (Pearson, 1997), Mauritius (Rassaby, 2003), India (Bhat et al., 2006) as well as in material conserved in botanical gardens (Grisoni et al., 2007).

Virus Structure and Genetic Diversity

As a member of the genera *Potexvirus*, family *Flexiviridae* (Adams et al., 2004), CymMV has flexuous particles (~13 nm × 480 nm) containing a single-stranded RNA (+) genome of about 6.3 kb with five open reading frames (ORFs) flanked by 5′ and 3′ noncoding regions plus a 3′ polyA tail (Francki, 1970; Wong et al., 1997).

The GenBank database contained (in October 2009) the complete genome sequence for nine isolates of CymMV and over 100 sequences for the coat protein (CP) gene, most of which are of Asian origin. The overall diversity between isolates is low at the amino acid level, with less than 14% divergence for CP (Ajjikuttira et al., 2002; Bhat et al., 2006; Gourdel and Leclercq-Le Quillec, 2001; Moles et al., 2007; Sherpa et al., 2006) and less than 3% divergence for RNA-dependent RNA polymerase (Moles et al., 2007). Nucleotide sequence analyses revealed, however, that the CymMV population splits into two diverging haplogroups, which may reflect a dual origin for the isolates found worldwide. However, no biological difference has been observed between the two clusters of strains, which can coexist and recombine within the same host (Sherpa et al., 2007; Vaughan et al., 2008). Strains belonging to both CymMV subgroups have been found in vanilla (Moles et al., 2007).

Symptoms and Diagnosis

In ornamental orchids, CymMV causes chlorotic or necrotic spots and streaks on leaves and flowers (Albouy and Devergne, 1998; Gibbs et al., 2000; Yamane et al., 2008) and reduces plant growth (Izaguirre-Mayoral et al., 1993; Pearson and Cole, 1991; Wannakrairoj, 2008). In vanilla, CymMV infection is generally symptomless but has occasionally been associated with flecking on leaves (on *Vanilla planifolia* and *Vanilla tahitensis*) and necrotic spots on the stem and leaves (on *V. planifolia*) (Grisoni et al., 1997, 2004; Leclercq Le Quillec et al., 2001) (Figure 7.1). It has been suggested from field observations that CymMV weakens the vine and increases decline due to stresses, resulting from overcropping or infection with another pathogen (Benezet et al., 2000), but this remains to be demonstrated. Even in the absence of symptoms, CymMV infection can be responsible for up to 40% reduction of stem growth (Bartet, 2005). The impact of CymMV on aromatic content of the pods has not been investigated so far.

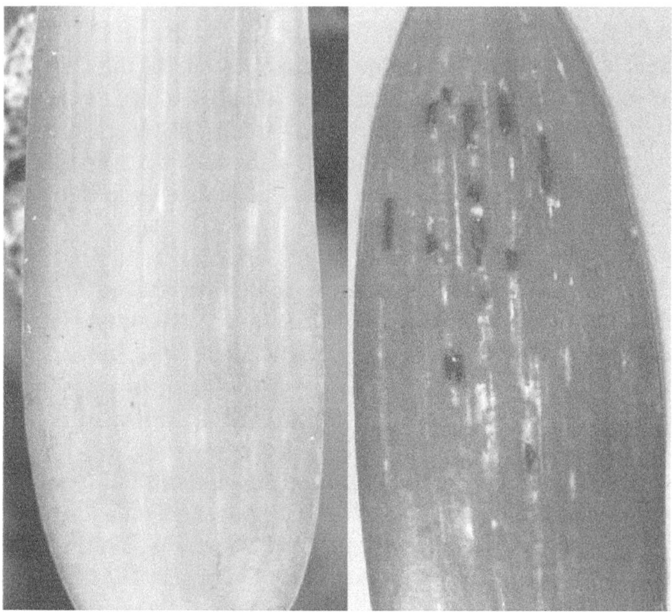

FIGURE 7.1 **(See color insert following page 136.)** Chlorotic (left) and necrotic (right) flecks induced by CymMV on vanilla leaves.

Observations in field surveys indicated that CymMV symptoms were more severe on *V. planifolia* (in Indian Ocean area) than on *V. tahitensis* (in the Pacific). The evaluation of disease severity in comparable greenhouse conditions using a local subgroup A strain in Reunion Island did not show any difference in virus disease between these two *Vanilla* species. Surprisingly however, a *Vanilla pompona* accession included in the trial exhibited partial resistance to this virus. This resistance consisted of two components: (1) resistance to inoculation (lower rate of infected plants in mechanical inoculation tests compared to *V. tahitensis* and *V. planifolia* accessions, and (2) reduction of virus titer in the infected leaves, resulting several months after inoculation with apparent elimination of the virus (Table 7.1). This resistance evokes transient inhibition of posttranscriptional gene silencing by CymMV in *V. pompona* rather than a hypersensitive response (Baures et al., 2008), but the mechanism underlying this resistance is unclear and needs further investigation.

CymMV cannot be reliably detected in vanilla on the basis of symptoms and therefore diagnosis has to be based on detection of the virus. Numerous and innovative molecular techniques have been described for CymMV detection, such as: quartz crystal microbalance-based DNA biosensors (Eun et al., 2002), immunocapillary zone electrophoresis (Eun and Wong, 1999), DIG-labeled cRNA probes (Hu and Wong, 1998), antisera produced to recombinant capsid proteins (Lee and Chang, 2008), super paramagnetic beads (Ooi et al., 2006), TD/RT-PCR (Seoh et al., 1998), rapid immunofilter paper assay (Tanaka et al., 1997), and wash-free antibody-assisted magnetoreduction assays (Yang et al., 2008).

Since CymMV particles are highly immunogenic (Francki, 1970) and abundantly and evenly distributed in host tissues (Lawson and Hearon, 1974; Leclercq-Le Quillec and Servé, 2001), serological techniques, such as ELISA or DIBA, which are cheap and robust, are particularly suitable for routine diagnosis of the virus. As would be expected from the low amino acid divergence of the CP, monoclonal as well as polyclonal antibodies have detected a wide range of isolates (Lee and Chang, 2008; Vejaratpimol et al., 1998; Wisler et al., 1982). However, Read et al. (2007) reported a strain (Thailand 306) that did not react with a commercial monoclonal antibody. This strain was characterized by aspartic acid instead of asparagine in position 123 of the CP, unique to this isolate.

More specific and sensitive detection of the virus can be achieved using nucleic acids-based methods, and a number of primer pairs and various formats have been designed for RT-PCR detection of CymMV (Barry et al., 1996; Bhat et al., 2006; Lim et al., 1993b; Martos, 2005; Ryu et al., 1995; Seoh et al., 1998; Yamane et al., 2008).

Epidemiology

CymMV incidence in vanilla plots was assessed in several surveys conducted in various agronomic conditions and revealed very disparate situations, not explicable by sampling bias alone. In Comores, for instance the entire planting material seems to be healthy since a recent and extended survey did not find a single virus infected vine (Grisoni and Abdoul-Karime, 2007). In Madagascar, the situation was similar at the beginning of 2000s with the large majority of traditional farmer plantations being CymMV free. However, most of the cuttings from new cultivars propagated in

TABLE 7.1
Partial Resistance of *V. pompona* to Infection and Replication of CymMV

(A) Rate of Infection Determined by ELISA 6 Weeks after Mechanical Inoculation

Species-Accession	Proportion of Infected Plants			
	Exp 1	Exp 2	Exp 3	Average (%)
V. pompona–CR0018	5/12	4/10	0/20	21
V. pompona–CR0031	Nd	4/10	6/28	26
V. planifolia–CR0044	13/13	10/10	42/48	92
V. planifolia–CR0036	12/12	10/10	36/40	93
V. tahitensis–CR0017	10/10	10/10	Nd	100
V. bahiana–CR0087	Nd	7/10	Nd	70
V. crenulata–CR0091	Nd	10/10	Nd	100

(B) Virus Titer in Infected Plants Determined by SyBr Green Q-RT-PCR (log of Number of Copies per μL RNA extract ± Confidence Interval at 5%)

	Exp 1	Exp 2			
Species/Accession	8 Months pi	3 Months pi	8 Months pi	20 Months pi	34 Months pi
V. pompona–CR0018	4.40 ± 0.52	8.51 ± 0,82	6.26 ± 0.36	6.61 ± 0.98	3.28 ± 0.12
V. pompona–CR0031	4.35 ± 0.49	Nd	Nd	Nd	Nd
V. planifolia–CR0044	6.53 ± 0.04	8.17 ± 0.46	8.13 ± 0.64	10.4 ± 0.23	8.95 ± 0.46
V. planifolia–CR0036	Nd	8.58 ± 0.12	8.31 ± 0.32	9.82 ± 0.70	7.90 ± 0.54
V. tahitensis–CR0017	Nd	8.67 ± 0.14	8.43 ± 0.04	10.3 ± 0.31	Nd
V. bahiana–CR0087	6.55 ± 0.08	Nd	Nd	Nd	Nd
V. crenulata–CR0091	6.37 ± 0.15	Nd	Nd	Nd	Nd

Nd, Not determined.

the germplasm repository at Ambohitsara (Antalaha district) were infected (Grisoni et al., 1997; Leclercq-Le Quillec and Nany, 1999, 2000). In Reunion and Society Islands relatively high CymMV incidence was recorded after intensive cultivation programs were initiated in the 1990s (Benezet et al., 2000; Grisoni et al., 2004; Leclercq Le Quillec et al., 2001) and epidemiological analysis could trace back the propagation of the virus in vanilla plots.

Owing to the high stability of the virion and its elevated concentration in host tissues, CymMV is readily transmitted by mechanical means (Jensen and Gold, 1955). It can therefore spread efficiently in the field or in green houses in the absence of specific vectors, solely as a result of virus from infected sap entering micro injuries in epidermal cells of the host. This has been specifically demonstrated for vanilla in shade house programs in Reunion Island (Leclercq Le Quillec et al., 2001) and FP (Grisoni et al., 2004). CymMV monitoring in these shade houses showed a rapid spread in large clusters primarily along the rows. These results supported the assumption that CymMV is introduced into shade houses via infected cuttings collected in the field, and then spread from plant to plant mainly by cultural practices (artificial pollination, looping) and possibly by root anastomosis between adjacent vines.

It is most important for vanilla growers to appreciate that the absence of symptoms increases the probability of both primary and secondary dissemination of this virus, and therefore the importance of planting material that has been certified free of the virus.

Control

In the absence of curative means or resistant commercial vanilla varieties, prophylaxis is the only approach to avoid CymMV losses in vanilla crops. Since humans are the principal vector of the virus (by planting infected cutting or as a mechanical transmission agent) a simple strategy combining plant phytosanitary certification and good management practices is sufficient to prevent the disease. This was successfully implemented in Reunion Island and FP in new plantation programs at the beginning of the 2000s (Benezet et al., 2000; Richard et al., 2009).

A basic requirement of phytosanitary certification is the need for virus-free material, which should be multiplied according to a specific protocol. A healthy starting material can generally be found within the cultivated stock but where a particular variety is required for which only infected germplasm is available, plants need to be cured of the virus. Vanilla is a clonal crop with a high degree of heterozygosity (see Chapter 2), and although seedlings give rise to virus-free vanilla (Jensen and Gold, 1955; Yuen et al., 1979), they are not true to type. Virus elimination has been achieved for several orchid species by chemotherapy *in vitro* (Albouy et al., 1988; Lim et al., 1993a) and similar strategies should be transferable to vanilla.

Genetic resistance or tolerance is particularly desirable when prophylaxis fails to provide sufficient control of viral disease. Apart from the resistance described in *V. pompona*, no natural resistance is available in *V. planifolia*. Pathogen-derived acquired resistance to CymMV has been bio-engendered in some orchid species (Chang et al., 2005; Liao et al., 2004; Lim et al., 1999), and although this strategy has not yet been developed for vanilla, it is theoretically feasible, since regeneration of vanilla protoplasts has been recently achieved (Minoo et al., 2008).

POTYVIRUSES

The genus *Potyvirus* represents the second largest group of plant pathogenic viruses (about 25% of all known plant viruses species), making the potyviruses a worldwide agricultural concern (Fauquet et al., 2005). A potyvirus causing leaf distortion and mosaic in *V. tahitensis* vines was identified in 1986 in FP (Wisler et al., 1987). This virus, named Vanilla mosaic virus (VanMV), was shown to be serologically, then genetically, related to *Dasheen mosaic virus* (DsMV) (Farreyrol et al., 2006; Wang and Pearson, 1992; Wisler et al., 1987). Contemporaneously another potyvirus causing necrosis in *V. planifolia* was identified in 1986 in Tonga and tentatively named Vanilla necrosis virus (VNV) before being characterized as a strain of Watermelon mosaic virus (WMV-Tonga) (Pearson and Pone, 1988; Pearson et al., 1990). The biological properties of WMV-Tonga have been well characterized and its full CP gene sequence determined (Pearson et al., 1990; Wang et al., 1993). Wang and Pearson (1992) demonstrated that WMV-Tonga and DsMV-Vanilla were distinct viruses. DsMV-Vanilla was subsequently detected serologically in *V. tahitensis* in the Cook Islands, and in *V. planifolia* in Fiji and Vanuatu (Pearson et al., 1993), and WMV was detected in *V. planifolia* in FP (Grisoni et al., 2004). Several potyvirus isolates that did not react with either WMV or DsMV antisera were also reported (Grisoni et al., 2004; Pearson, 1997; Pearson et al., 1993), suggesting that other potyviruses were present in vanilla crops.

THE VIRUSES

The *Potyvirus* genus (family *Potyviridae*) encompasses 129 recognized species plus about 15 tentative species (Carstens and Ball, 2009; Fauquet et al., 2005). Potyviruses are found worldwide and infect more than 30 plant families, although individual viruses often have a restricted host range. They have flexuous, nonenveloped, rod-shaped particles (680–900 nm long, 12–15 nm in diameter) that contain a positive-sense single-stranded RNA molecule (Hull, 2002). The monopartite genome is about 10 kb in size and contains a single ORF that encodes a polyprotein of 3000–3300 amino acids. The polyprotein is cleaved co- and post-translationally by three virus-encoded proteinases, into 10 mature proteins, most of which are multifunctional (Adams et al., 2005a; Revers et al., 1999; Urcuqui-Inchima et al., 2001). An extensive study of gene diversity within the family *Potyviridae* (Adams et al., 2005b) concluded that the cytoplasmic inclusion (CI) protein gene was superior for potyvirus identification when using only a subportion of the genome. Although less informative than the CI, the CP gene can also efficiently separate the species, the demarcation criterion being defined as 76–77% nt identity by these authors.

The CP gene analysis from 36 vanilla samples collected in the Indian Ocean and Pacific regions between 1997 and 2005 added five potyvirus species to the two previously described infecting vanilla, namely: Bean common mosaic virus (BCMV), Cowpea aphid-borne mosaic virus (CABMV), Wisteria vein mosaic virus (WVMV),

Bean yellow mosaic virus (BYMV), and Ornithogalum mosaic virus (OrMV) (Grisoni et al., 2006).

Remarkably, among the seven species that infect vanilla, five (BCMV, CABMV, DsMV, WMV, and WVMV) belong to the BCMV subgroup, which preferentially infect leguminous crops and weeds. They were detected in Reunion Island (BCMV and CABMV), in Madagascar (BCMV), in Mauritius (CABMV), in FP (BCMV, DsMV, and WMV), in Tonga (WMV), and in Samoa (WVMV). The two other species, BYMV and OrMV, were only detected in Reunion Island. The mosaic inducing potyvirus reported in India (Bhai et al., 2003; Bhat et al., 2004; Thomas et al., 2002) were not identified but may involve BYMV as suggested by the two GenBank accessions (AY845011 and AY845012) originating from vanilla material collected in Karnataka (India). Potyvirus-like symptoms have also been reported in Papua New Guinea (Kokoa, 2000) and in Puerto Rico (Childers and Cibes, 1948).

The Case of DsMV-Vanilla

The DsMV strains infecting vanilla were at first tentatively referred to as VanMV because they showed only distant serological relationship to dasheen-infecting strains of DsMV and because of the failure of cross-inoculation between the strains originating from the two hosts. Subsequently, the nucleotide analysis of the 3′ end of the genome of two VanMV isolates (from Cook Islands [CI] and FP [FP]) confirmed that they should be classified as DsMV strains, since they shared more than 78% identities, notably in the core CP (Farreyrol et al., 2006). Interestingly, the DsMV-Vanilla [CI] and DsMV-Vanilla [FP] were as divergent from each other as from published DsMV-dasheen (*Colocasia esculenta*) strains (Farreyrol et al., 2006). However, recently obtained sequences for dasheen isolates from Cook Islands and FP were very similar to vanilla isolates from the same country (M. Pearson et al., unpubl. data). Since dasheen has been grown on these islands for much longer than vanilla, this suggests that "virus jump" from dasheen to vanilla may have occurred independently in the two Pacific Islands. In addition, the DsMV-Vanilla [FP] showed unusual features which makes it a particular virus among potyviruses (Farreyrol et al., 2006): a DVG aphid transmission motif (instead of the more common DAG motif) upstream of an unusual stretch of amino acid repeats (GTN) typical of natively unfolded proteins and an uncommon NIb/CP proteolytic cleavage site (Q//V).

Symptoms and Diagnosis

In vanilla, potyviruses cause a more or less pronounced mosaic and deformation of leaf blades sometimes associated with necrotic lesions (Figure 7.2). The mosaic and necrosis can also be visible on the stem. These symptoms are particularly spectacular on the young leaves and are much attenuated on older leaves. In Reunion Island, potyvirus symptoms were most obvious during the cool months with a remission of symptoms observed during the hot season (Benezet et al., 2000; Leclercq Le Quillec et al., 2001). The impact of the viruses on flowering and fruiting of the vanilla vines has not been established, although some reduction in growth can be inferred from the severe mosaic and deformation, affecting the shoots. Field observations revealed that the virus particles were unevenly distributed in the plant and symptomatic vines may test negative in

FIGURE 7.2 (See color insert following page 136.) Potyvirus symptoms on leaves of *V. planifolia* (a = BYMV-Réunion, b = BCMV-Madagascar, c = necrotic strain of WMV-Tonga) and *V. tahitensis* (d = WMV-FP, e = DsMV-FP) and on stem (f = WMV-FP).

ELISA (Leclercq Le Quillec et al., 2001). As a consequence, visual observation of mosaic on leaves and ELISA tests are complementary to assess potyvirus in vanilla plots, none of the technique used alone being sufficiently reliable for diagnosis.

Several molecular techniques are available to diagnose potyvirus infection of vanilla, at the generic or specific level. The generic anti-potyvirus monoclonal antibodies (Jordan and Hammond, 1991), commercialized by Agdia (USA), proved to be efficient in detecting all the potyvirus species identified in vanilla by Indirect Simple Sandwich ELISA. Likewise, several motifs conserved in the genome of potyviruses have been used to design a number of degenerate primers for RT-PCR amplification of the corresponding sequence in the potyviruses genome (Chen et al., 2001; Colinet et al., 1998; Gibbs and Mackenzie, 1997; Ha et al., 2008; Langeveld et al., 1991; Marie-Jeanne et al., 2000; Pappu et al., 1993). Both techniques can be used to assess the potyvirus status of material but do not specifically identify the potyvirus species detected.

For potyviruses, specific identification by serology is unreliable because of the numerous cross reactions between virus species (Shukla et al., 1994). To overcome this problem a simple one tube, one-step, RT-PCR assay using degenerate primers followed by direct sequencing of a short fragment can be used (Grisoni et al., 2006). As RT-PCR and sequencing are getting easier and cheaper technologies, this method can be used in epidemiological surveys requiring high throughput. Microarray technologies, which are a promising way to get broad spectrum, specific, and sensitive detection tool for potyviruses, are under development (Boonham et al., 2007; Wei et al., 2009).

Epidemiology

Potyviruses are responsible for major losses of many economically important crops because they spread readily in the fields. Several potyviruses are seed transmitted, most are mechanically transmitted (albeit inefficiently), but all are efficiently transmitted by aphids in a nonpersistent manner (Brunt, 1992).

VanMV was readily transmitted from *V. tahitensis* to *V. pompona* using the peach aphid *Myzus persicae* (Wisler et al., 1987). Experimental transmission tests of WMV-Tonga from infected to healthy *Nicotiana clevelandii* using *Aphis gossypii* was successful but transmission tests involving *V. planifolia* as a donor plant with the same aphid had failed (Pearson and Pone, 1988; Pearson et al., 1990; Wang and Pearson, 1992; Wang et al., 1993, 1997). However, several field data on potyvirus infections support the role of aphids in the spread of potyviruses in vanilla plots (Richard et al., 2009). Successful aphid transmission trials were consistent with a nonpersistent mode of transmission. In nonpersistent transmission, a few seconds or minutes are sufficient for virus acquisition or virus inoculation by the vector in feeding probes. It is therefore not surprising to observe heavy potyvirus infection in the absence of established aphid had colonies on vanilla (excepting *Cerataphis orchidarum*, for which only wingless forms have been described and which is not considered an efficient virus vector).

A few potyviruses are transmitted by seeds, notably CABMV in several leguminous species (Gillaspie, 2001), BCMV in some *Phaseolus* and *Vigna* cultivars, and BYMV in several leguminous species (Aftab and Freeman, 2006). In addition, BCMV is transmitted by bean pollen (Card et al., 2007). DsMV, WMMV, and WVMV are not known to be seed transmissible, but it has been suggested that seed transmission might be more common than currently recognized particularly within the BCMV subgroup (Gibbs et al., 2008a).

Seed transmission of viruses has epidemiological implications through intercropping and weed management. In the field, primary inoculum of potyvirus may come from infected vanilla cuttings or from surrounding plants, either perennial plants or annual weeds, in which the virus can overwinter when it is transmitted by seeds. From this primary inoculum, the potyvirus is then disseminated in the plot by aphids. Molecular similarity between isolates infecting vanilla and weeds provides support to the passage of viruses from one species to another. For instances, in Madagascar, BCMV isolates were identified in vanilla and in the bordering weed *Senna* sp. (*Leguminosae*), which were 100% identical in CP gene sequences (Grisoni et al., 2006). In Reunion Island, a rapid outbreak of CABMV in a recently planted vanilla plot using virus-free cutting was correlated to the intercropping with *Vigna unguiculata* in which CABMV is seed transmitted. On the contrary, the use of *Gliricidia sepium*, a legume tree frequently used as support in vanilla plots and which is, because of frequent pruning, heavily infested by *Aphis craccivora* aphids (an efficient vector of BCMV subgroup of viruses) does not seem to increase potyvirus incidence in vanilla plots compared to when aphid-free support trees are used, such as *Leucena glauca* (Grisoni et al., 2004). However, in the presence of other virus sources the high aphid populations could become an issue.

Control

Epidemiological data showed that severe potyvirus outbreaks occur primarily in intensive vanilla farming system and may originate from a diversity of virus species involving a diversity of host species, virus reservoirs, and probably aphid species.

Thus, apart from the universal recommendation of planting virus-free material, the control strategy needs to be adapted to each local circumstance. As a consequence, it is advisable to assess potyvirus risk prior to any development program in a new area or when changing the cultivation system. In addition, molecular tools that enable rapid identification of potyviruses help provide insight into the appropriate control strategy to use in case of unexpected potyvirus outbreaks.

CUCUMBER MOSAIC VIRUS (CMV)

CMV is a widespread virus infecting a very broad range of plants. It was identified in vanilla in the early 2000s, associated with severe distortions of vines in FP (Farreyrol et al., 2001). The virus was later found in vanilla in Reunion Island and in India (Madhubala et al., 2005).

Virus and Genetic Diversity

CMV is the type member of the *Cucumovirus* genus in the family *Bromoviridae*. It has a tripartite positive-sense RNA genome encoding five proteins (two polymerase components on RNA1 and RNA2, the 2b protein involved in the breaking of plant defense and virus movement in the plant, on RNA2, and a 30 K protein and the CP on RNA3). Each RNA is encapsulated in a distinct polyhedral particle (28 nm diameter) hence the infection of a plant requires the inoculation of the three types of virions. Some isolates also encapsulate a subgenomic RNA or a satellite (sat) RNA. The biology and ecology of this virus have been extensively reviewed by Palukaitis et al. (1992) and Gallitelli (2000).

The CMV strains are divided into two serologically distinct subgroups diverging in about 25% of their nucleotides and also differing in biological traits. The thermotolerant strains of subgroup 1 are further divided into two clades (1A and 1B) on the basis of genomic sequences of their RNAs (Roossinck et al., 1999). The subgroup 1A emerged from the subgroup 1B and (until recently) the 1A strains were only found in Asia. The thermosensitive strains of subgroup 2 are restricted to cool climate areas. Three mechanisms have been associated with the evolution of CMV: RNA reassortments between the RNA components, recombination in the noncoding regions and mutations in the coding regions. Several amino acid positions on the CP were shown to be positively selected in relation to virus transmission (Moury, 2004).

To our knowledge, 65 CMV–CP sequences have been obtained for vanilla isolates, originating from FP (58), Reunion Island (6), and India (1) (Farreyrol, 2005; Madhubala et al., 2005; Mongredien, 2002), 53 of which are deposited in Genbank. The majority of the sequences were most closely related to subgroup 1B isolates and clustered in a few clades (Farreyrol et al., 2009). Trees generated from 3′ UTR of RNA3 (Farreyrol, 2005) and ORF2b of RNA2 (Mongredien, 2002) sequences were congruent with CP analysis. Interestingly, an isolate belonging to group 2 (NZ100, AY861389) was mechanically inoculated in vanilla plants, which induced leaf mosaic and deformation (Farreyrol et al., 2009), suggesting that a wide range of CMV isolates can infect vanilla. Attempts to detect CMV sat-RNA associated to vanilla isolates of CMV from FP by RT-PCR using degenerate primers (Escriu et al., 2000; Varveri and Boutsika, 1999) have failed (M. Grisoni, unpubl. data).

Phylogenetic analysis of CMV isolates collected from vanilla plus several associated crops or weeds revealed a clustering according to the geographic origin rather than to the host plant. For instance in the Leeward Islands (FP), the CMV isolates from the islands of Raiatea and Huahine were in distinct clusters, while high similarities were observed between isolates from vanilla and from *Commelina diffusa* (Farreyrol et al., 2009). These findings supported the hypothesis of transmission of the virus from one plant to another, particularly from *C. diffusa* to *V. tahitensis*. Similarly, the CP gene of CMV from vanilla sequenced in Kerala (India) showed the highest identities (99%) with that of another Indian isolate infecting black pepper in Karnataka (Madhubala et al., 2005).

Symptoms and Diagnosis

A variety of symptoms has been associated with CMV on its numerous hosts; most common are mosaics and stunting, but symptoms can be as severe as complete systemic necrosis. Some strains are symptomless on certain hosts or at elevated temperature (subgroup two strains) and symptom expression may also vary over time due to temporary remission of the virus infection or be attenuated or aggravated by the coinfection with a sat-RNA.

In vanilla, the CMV isolates described in FP induced severe leaf and shoot deformation and the infected plants are often sterile, either because the flowers do not fully develop or because they are abnormally formed and contain no viable pollen. On *V. tahitensis* vines mechanically inoculated with a field isolate of CMV the first foliar symptoms appeared two months after inoculation with an embossing of the young leaves, which usually had a slender and asymmetrical (comma-like) shape (Figure 7.3a). The vines subsequently developed an abnormal phyllotaxy and irregular growth of the apices and flowers (Figure 7.3b and c), typical of the disease. Similar leaf symptoms were described in India and the Reunion Island on *V. planifolia*. In a few instances, shoot deformations on CMV infected vines have been observed (Farreyrol, 2005) resembling the *abnormal tubular leaves* described by Jacob de Cordemoy (1899) (Figure 7.4), which are possibly historic evidence of remote infections by CMV in the Reunion Island.

CMV symptoms are spectacular on vanilla and rather specific and can be good indicators of virus infection. However, these symptoms can be confused, in the early stages, with potyvirus infection, and in the later stages, with physiological disorders. In addition, temporary symptom remission may occur. Therefore, tests based on serology (Devergne et al., 1981; Hu et al., 1995; Maeda and Inouye, 1991; Yu et al., 2005) or nucleic acid amplification (Hu et al., 1995; Wylie et al., 1993; Yu et al., 2005) are recommended to diagnose the disease, or to select virus-free cuttings.

Epidemiology

CMV is a plant virus having the largest known host range of more than 1000 species belonging to more than 30 families (Douine et al., 1979). The virus is transmitted by

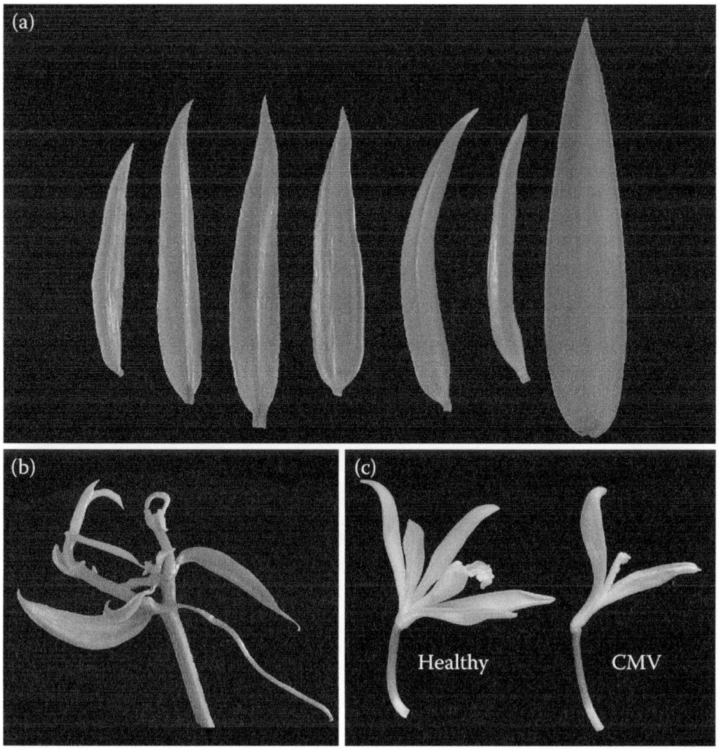

FIGURE 7.3 (See color insert following page 136.) Symptoms caused by CMV infection in *V. tahitensis*: (a) young leaves embossed and deformed (healthy leaf on the right); (b) proliferation on infected young shoot; and (c) abnormal flower (right) on CMV-infected vine.

more than 80 aphid species in a nonpersistent manner (Gallitelli, 2000). In several hosts, CMV is also transmitted at high frequency through the seeds (Palukaitis et al., 1992). Several studies have pointed out the key role of the co-occurrence of reservoir plants (that overwinter the primary inoculum) and aphids (that ensure secondary spread of the virus) in the epidemics of CMV in different crops and countries (Hobbs et al., 2000; Kiranmai et al., 1998; Lavina et al., 1996; McKirdy, 1994; Rist and Lorbeer, 1991; Skoric et al., 2000).

In vanilla plots of the Society Islands (FP), high prevalence of CMV was recorded in the early 2000s, with one out of three plots infected, and an incidence frequently exceeding 25% of vines (Farreyrol et al., 2009). Much lower virus prevalence was observed in the Reunion Island with only 2 infected plots out of 25 with sporadic occurrence of the virus (Farreyrol, 2005) as well as in India where mosaic viruses were observed in <5% of vines (Bhat et al., 2004; Madhubala et al., 2005). In Marquisas and Australes (FP), Comoros archipelagos, and Madagascar no CMV was found in vanilla despite intensive surveys (Grisoni, 2003; Grisoni, 2009; Grisoni and Abdoul-Karime, 2007; Leclercq-Lequillec and Nany, 2000).

The spread of CMV was monitored in several shade houses planted with virus-free cuttings and conducted in various locations in the Leeward Islands (Richard et al.,

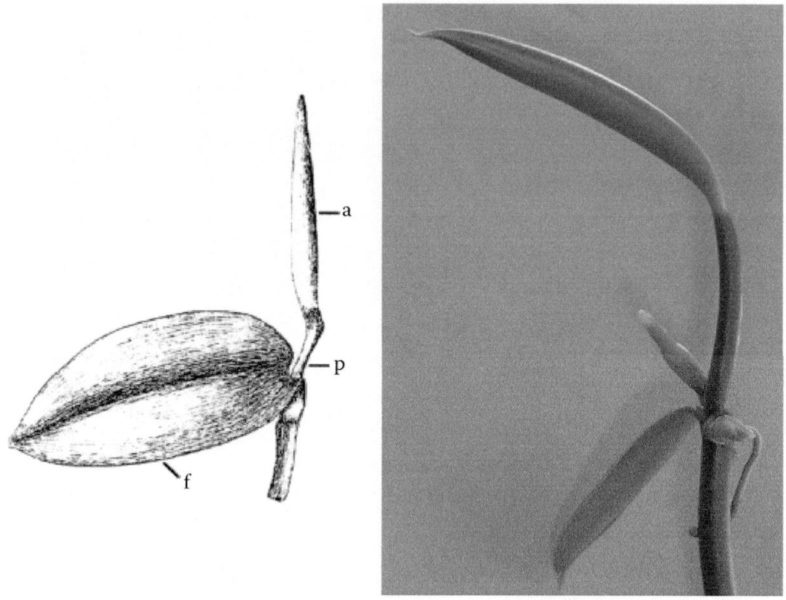

FIGURE 7.4 (Left) Abnormal tubular leaf a, observed on *V. planifolia* in 1899 by M.H. Jacob de Cordemoy (Jacob de Cordemoy, 1899); f, normal leaf; p, peduncle. (Right) Tubular leaf symptom observed during a survey in FP in 2005.

2009). The results showed that in the absence of insect-proofing and in favorable locations CMV can spread very quickly, with 50% of vines infected only two years after planting. In contrast, with insect proofing or in another location, a very low incidence of CMV was recorded. These results demonstrated the role of aphids in spreading the disease and also the importance of the environment possibly through the presence or absence of CMV reservoir among weeds.

C. diffusa (common name *climbing* or *spreading dayflower, ma'apape* in Tahitian or *herbe d'eau* in French, Figure 7.5) was shown to play a predominant role in CMV epidemics in FP (Richard et al., 2009): (1) *C. diffusa* was the only species that exhibited simultaneously high frequencies of CMV infection and aphid (*A. gossypii*) infestation, (2) the occurrence of CMV in vanilla vines was highly correlated with the proximity of *C. diffusa*, (3) the incidence of CMV in the plots surveyed was significantly correlated with the occurrence of infected *C. diffusa* within the plot but not with the other CMV-infected species, and (4) the ability of *A. gossypii* and *A. craccivora* to transmit CMV from *C. diffusa* to young plants of *V. tahitensis* was demonstrated in laboratory tests. All these data, along with the high sequence similarities between vanilla and *C. diffusa* CMV isolates (Farreyrol et al., 2009) strongly support the view that *C. diffusa* is a major active reservoir of CMV and contributes greatly to the spread of this virus in vanilla plots. This epidemiological scenario resembles that described for CMV in banana plantations (Eiras et al., 2004; Lockard, 2000; Magnaye and Valmayor, 1995).

FIGURE 7.5 *C. diffusa* Burm. f., the main aphid and virus reservoir involved in CMV epidemics in FP: (a) virus-infected plants; (b) flower; and (c) aphid-infested leaf.

Control

In view of the very severe symptoms induced on leaves and flowers and its potentially high incidence in the field, CMV is a major threat to vanilla plantations and should be a priority for risk assessment and control when implementing intensive cultivation systems.

In FP, an efficient control strategy was set up in 2003 relying on (1) a supply of virus-free cuttings, (2) roofing the shade houses with insect-proof netting, (3) roguing the diseased plants as soon as they are detected, (4) decontaminating tools, and (5) avoiding weeds (particularly *C. diffusa*) inside and at the proximity of the plantation. A postimplementation survey in 2007 showed that the viral impact (CMV and other viruses) was drastically reduced in the new plantations that followed the recommendations (Richard et al., 2009).

Epidemiological data indicate that CMV prevalence may vary greatly from one area to another, notably in relation to the presence of virus reservoirs. Therefore, control strategies have to be adapted according to each specific cultivation context.

OTHER VIRUSES

Besides the major viruses already detailed other viruses with minor economic importance such as Odontoglossum ringspot virus (ORSV) or barely characterized such as a Rhabdolike-virus and a Clostero-like viruses (Bhat et al., 2004; Pearson et al., 1993) have also been reported in vanilla.

Odontoglossum Ringspot Virus (ORSV)

ORSV is a member of the *Tobamovirus* genus of which the type member is Tobacco mosaic virus (Jensen and Gold, 1951). The rod-shaped particles of ORSV (300 × 18 nm) encapsulate a single-stranded positive RNA molecule of approximately 6.6 kb coding

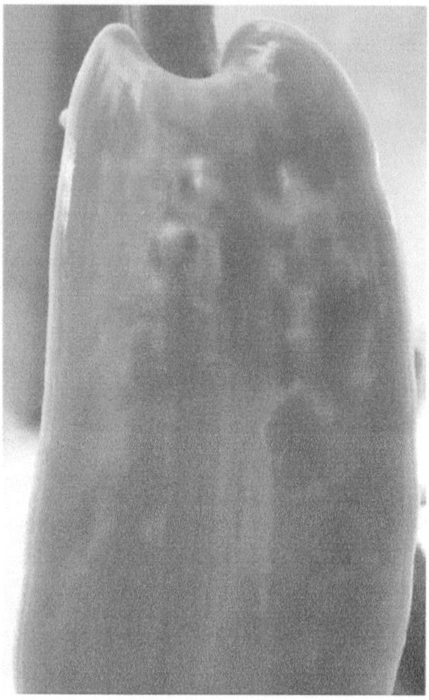

FIGURE 7.6 **(See color insert following page 136.)** Mosaic on leaf of *V. pompona* infected by ORSV.

for five proteins (Ryu et al., 1995). ORSV's natural host range is limited to orchids in which it may cause severe symptoms such as mosaic, ringspots, necrosis on leaves and flowers (Albouy and Devergne, 1998; Gibbs et al., 2000). As a tobamovirus, ORSV has a high stability in sap (temperature inactivation of 90°C and a dilution endpoint of 10^{-5}–10^{-6}) and is readily transmitted by mechanical means. It is frequently found in cultivated ornamental species worldwide (Freitas et al., 1999; Khentry et al., 2006; Zettler et al., 1990).

On vanilla, however, the virus has a very limited incidence and prevalence. It has been serologically detected in few vanilla vines in Tonga, Fiji, Cook Islands, Niue, FP, and the Reunion Island (Farreyrol et al., 2001; Pearson and Pone, 1988; Pearson et al., 1993; Wisler et al., 1987), but no symptoms were consistently associated with the virus. However, in plant quarantine in the Reunion Island, we recently observed a *V. pompona* vine originating from a botanical garden that exhibited mosaic on young leaves (Figure 7.6). This plant tested positive for ORSV and negative for CymMV, potyviruses, and CMV, by ELISA and RT-PCR.

Since it is relatively infrequent in vanilla, ORSV has so far been of little concern in production plots. However, due to its possible pathogenicity and absence of a natural vector it is worth testing the planting material for this virus in order to avoid its propagation. As with CymMV, a number of innovative methods have been developed to detect ORSV.

Unidentified Rhabdo-Like Virus

During a survey conducted in Pacific Islands (Pearson et al., 1993), unusual symptoms on leaves, consisting of necrotic spots were observed in vanilla plots (Figure 7.7a) in

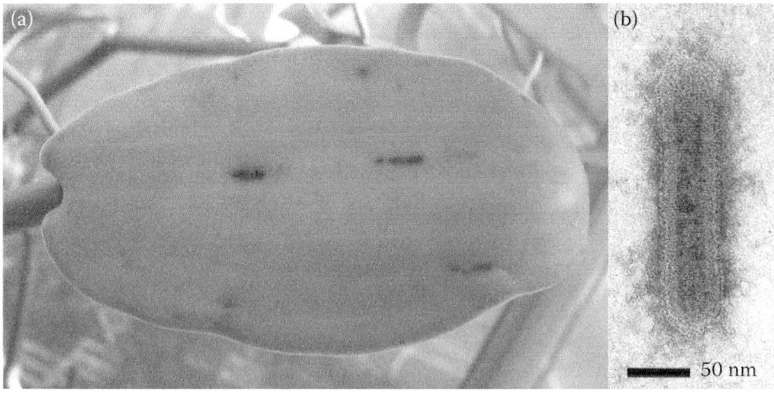

FIGURE 7.7 (See color insert following page 136.) (a) Enations and necrosis on *V. planifolia* leaves infected with (b) enveloped virus-like particles visible under electron microscope.

Vanuatu and Fiji. Leaf dip preparations from symptomatic leaves revealed the presence of enveloped bacilliform particles suggesting a possible *Rhabdovirus*.

The *Rhabdoviridae* family (order *Mononegavirales*) contains viruses infecting animals and plants, which are transmitted by arthropods and may multiply in the vector (Fauquet et al., 2005). The plant infecting rhabdoviruses are assigned to two genera (*Cytorhabdovirus* and *Nucleorhabdovirus*) but a number is still unassigned, including the putative rhabdovirus Orchid fleck virus (OFV), which has become more prevalent in orchids over the last decade (Kitajima et al., 2001; Kondo et al., 2006). OFV has been tentatively classified in a new *Dichorhabdovirus* genus because its genome, although showing sequence similarities with nucleorhabdoviruses, is bipartite and the particles are not enveloped (Kondo et al., 2009). This last feature contrasts with the putative rhabdovirus particles seen in vanilla which are clearly enveloped (Figure 7.7b). In addition, repeated attempts to amplify OFV sequences from vanilla symptomatic leaves using specific primers (Blanchefield et al., 2001) have failed (Mongredien, 2002). Since the statement from 1993, no further damage has been reported in vanilla that is reminiscent of this still uncharacterized putative rhabdovirus, and this virus should be considered as anecdotal in vanilla plots.

VIRUS INCIDENCE AND CULTIVATION SYSTEM

Converging studies, using distinct approaches and pathosystems, have linked the early emergence and radiation of plant pathogens to the modification of their environment (Stukenbrock and McDonald, 2008; Webster et al., 2007). In particular, the development of agriculture from the Eocene coincides with the dates of the radiation of RNA potyviruses, 6600 years ago (Gibbs et al., 2008b, 2008c) and DNA sobemoviruses 3000 years ago (Fargette et al., 2008), as estimated by phylogenetic analyses. The changes induced by agriculture and human migration favored "new encounters between plants and viruses" leading to the increasing number of virus species. Similarly, the recent history of potyvirus diseases of vanilla typically exemplifies the close relationship between agro-systems and virus pressure (Table 7.2).

TABLE 7.2
Prevalence and Incidence of Potyviruses in *Vanilla* Plots Conducted without Specific Control Measures in Varied Cultivations Systems

Country	Year of the Survey	Proportion of Infected Plots	Proportion of Infected Vines	Cultivation System	*Potyvirus* Identified	Reference
Comoros	2007	0/41	0%	UF/SI	None	Grisoni and Abdoul Karime (2007)
Fiji	1989	20/36	1–49%	SI/UF	NA	Pearson et al. (1993)
India	2000/01	9/35	0.2–5%	SI	NA	Bhai et al. (2003)
India	2003	52/65	0.25–10%	SI	NA	Bhat et al. (2004)
Madagascar	1997	0/21	0%	UF/SI	None	Grisoni et al. (1997)
Madagascar	2000	0/29	0%	UF/SI	None	Leclerq–Le Quillec (2000a, 2000b)
Madagascar	2004	NA	5%	SH	BCMV	Grisoni (2004)
Madagascar	2009	45	0%	SH/SI		Grisoni (2009)
Marquesas (FP)	2003	5/29	0.5–90%	SI	WMV-DsMV	Grisoni (2003)
Mayotte	2007	0/	0%	SI	None	Grisoni and Abdoul Karime (2007)
Reunion	1998/99	14/17	0.3–14.4%	SH	NA	Benezet et al. (2000)
Reunion	2000	9/25	NA	SI/SH	NA	Farreyrol (2005)
Samoa	2005	NA	6%	SI	WVMV	Grisoni et al. (2006)
Society Islands (FP	2000	18/32	NA	SI/SH	NA	Farreyrol (2005)
Society Islands (FP)	1998/99	27/49	0.6–97%	SI/SH	DsMV, WMV, BCMV	Grisoni et al. (2004)
Society Islands (FP)	1986	9/30	0.3–47.9%	SI/UF	DsMV	Wisler et al. (1987)
Tonga	1989	19/25	1–21%	SI/UF	NA	Pearson et al. (1993)

NA, not available; SH, intensive under shade house; SI, semi-intensive; UF, extensive under forest.

FIGURE 7.8 Simplification of *Vanilla* agrosystem consecutive to intensification: (a) Underforest counting more than 30 adventive species (shade and support trees, weeds, and associated crops); (b) Semi-intensive system (with 10–20 adventive species); and (c) intensive system under shade house (vanilla plus sometimes a few weeds).

Mosaic and leaf deformation on vanilla shoots were reported in Puerto Rico as early as 1948 (Childers and Cibes, 1948) and might represent the first record of potyviruses in vanilla. However, the first confirmed outbreak of potyviruses in vanilla was reported in FP in 1986 (Wisler et al., 1987). It coincided with the implementation of a vanilla development program promoting the intensification of the cultivation practices, which entailed environmental changes in the vanilla agro-system (Anonymous, 1984). Likewise for the later reports of potyvirus infection of vanilla plots where analogous intensification programs were implemented such as in Tonga (Pearson and Pone, 1988), Vanuatu (Pearson et al., 1993), Reunion Island (Benezet et al., 2000), India (Bhat et al., 2004), Madagascar (Grisoni et al., 2006), Mauritius (Rassaby, 2003), and Samoa (Grisoni et al., 2006). Conversely no or only exceptional potyvirus infection has been recorded so far in the under-forest vanilla plots (Grisoni et al., 1997; Leclercq-Le Quillec and Nany, 2000).

We hypothesize that the multiplicity of potyviruses infecting vanilla (seven species recorded) results from the diversity of the epidemiological circumstances where vanilla is grown and from the conditions in intensive cultivation that enhance aphid transmission of viruses present in weeds or associated crops. The incidence of the potyviruses and CMV in the vanilla plots varies greatly from one plot to another, although the heaviest infections (more than 50% infected vines two years after planting) were recorded in shade houses that are characterized by a very low plant diversity compared to the under-forest cultivation (Figure 7.8). It is therefore probable that other opportunistic viruses will infect vanilla as its culture expands into new environments. This highlights the need for rapid, specific, and widely applicable virus identification tools to understand the outbreaks affecting vanilla plantations and implement adapted control strategies.

CONCLUSIONS AND PERSPECTIVES

Virus diseases have become a major concern for vanilla production over the last few decades probably as a consequence of the diversification of the cultivation sites and intensification of vanilla growing systems. In the absence of curative means, prophylaxis is the only way to avoid viral diseases. The data accumulated on the viruses affecting vanilla throughout the world, particularly CymMV, potyviruses, and CMV enabled the introduction of control measures that proved sufficiently effective to preserve and expand a profitable vanilla industry.

However, viruses have the ability of evolving extremely fast and the emergence of new pathogenic isolates in vanilla plots are likely to occur, particularly in the context of quick environmental mutations induced by global change (Canto et al., 2009; Garrett et al., 2006). Huge progress has been accomplished recently in the comprehension of plant and virus biology and interactions, leading to powerful biotechnologies for engineering virus-resistant varieties. The regeneration of vanilla plants from protoplasts that was recently achieved (Minoo et al., 2008) makes such biotechnologies accessible for the production of transgenic vanilla, expressing high levels of resistance to RNA viruses. However, this strategy has yet to be implemented, possibly because virus pressure has been maintained at a tolerable level by conventional methods. In addition, consumers consider vanilla as a natural product and genetic bioengineering is not compatible with that image.

High-performance tools for nucleic acid detection and identification, such as portable molecular tests, polyvalent PCR, microarrays, next generation DNA sequencing, will also contribute to improve virus management in the future. In particular they will assist in better understanding of vanilla virus epidemiology and subsequently in designing agro-systems by reducing virus impact on crop.

Globally, these technologies will help produce planting material of the highest health and genetic status for the vanilla industry. It is undoubtedly on this elite material, properly multiplied and cultivated with adequate prophylactic measures, that the future of a profitable and healthy vanilla industry relies.

REFERENCES

Adams, M.J., Antoniw, J.F., and Beaudoin, F. 2005a. Overview and analysis of the polyprotein cleavage sites in the family *Potyviridae. Molecular Plant Pathology* 6:471–487.

Adams, M.J., Antoniw, J.F., and Fauquet, C.M. 2005b. Molecular criteria for genus and species discrimination within the family *Potyviridae. Archives of Virology* 150:459–479.

Adams, M.J., Antoniw, J.F., Bar-Joseph, M., Brunt, A.A., Candresse, T., Foster, G.D., Martelli, G.P., Milne, R.G., Zavriev, S.K., and Fauquet, C.M. 2004. The new plant virus family *Flexiviridae* and assessment of molecular criteria for species demarcation. *Archives of Virology* 149:1045–1460.

Aftab, M. and Freeman, A. 2006. *Temperate Pulse Viruses: Bean Yellow Mosaic Virus (BYMV): Agriculture Notes*. Department Primary Industries, State of Victoria, Australia.

Ajjikuttira, P.A., Lim-Ho, C.L., Woon, M.H., Ryu, R.H., Chang, C.A., Loh, C.S., and Wong, S.M. 2002. Genetic variability in the coat protein genes of two orchid viruses: *Cymbidium mosaic virus* and *Odontoglossum ringspot virus. Archives of Virology* 147:1943–1954.

Albouy, J. and Devergne, J.-C. 1998. *Maladies à virus des plantes ornementales*. INRA, Paris.

Albouy, J., Flouzat, C., Kusiak, C., and Tronchet, M. 1988. Eradication of orchid viruses by chemotherapy from *in vitro* cultures of cymbidium. *Acta Horticulturae (Virus Diseases of Ornamental Plants)* 234:413–420.

Anonymous, 1984. *La Vanille. Service Economie Rurale.* Papeete, French Polynesia.

Barry, K., Hu, J.S., Kuehnle, A.R., and Sughii, N. 1996. Sequence analysis and detection using Immunocapture-PCR of *Cymbidium mosaic virus* and *Odontoglossum ringspot virus* in Hawaian Orchids. *Journal of Phytopathology* 144:179–186.

Bartet, L. 2005. *Recherche d'une source de tolérance au Cymbidium mosaic virus (CymMV) chez les vanilliers: Bioprotection et Biotechnologies pour l'Environnement*. Université de Pau et des Pays de l'Adour, Pau, 39pp.

Baures, I., Candresse, T., Leveau, A., Bendahmane, A., and Sturbois, B. 2008. The *Rx* gene confers resistance to a range of potexviruses in transgenic nicotiana plants. *Molecular Plant-Microbe Interactions* 21(9):1154–1164.

Benezet, H., Picard, E., Côme, B., Grisoni, M., Leclerq-Le Quiliec, F., Gambin, O., and Jeuffrault, E. 2000. Les virus du vanillier à la Réunion. *Phytoma, La Défense des Végétaux* 526:40–42.

Bhai, R.S., Thomas, J., Potty, S.N., Geetha, L., and Solomon, J.J. 2003. Mosaic disease of vanilla (*Vanilla planifolia* Andrews)—the first report from India. *Journal of Spices and Aromatic Crops* 12:80–82.

Bhat, A.I., Bhadramurthy, V., Siju, S., and Hareesh, P.S. 2006. Detection and identification of *Cymbidium mosaic virus* infecting vanilla (*Vanilla planifolia* Andrews) in India based on coat protein gene sequence relationships. *Journal of Plant Biochemistry and Biotechnology* 15:33–37.

Bhat, A.I., Venugopal, M.N., Pant, R.P., and Bhai, R.S. 2004. Occurrence and distribution of viral diseases on vanilla (*Vanilla planifolia* Andrews) in India. *Journal of Spices and Aromatic Crops* 13:143–148.

Blanchfield, A.L., Mackenzie, A.M., Gibbs, A., Kondo, H., Tamada, T., and Wilson, C.R. 2001. Identification of orchid fleck virus by reverse transcriptase-polymerase chain reaction and analysis of isolate relationships. *Journal of Phytopathology* 149, 713–718.

Boonham, N., Tomlinson, J., and Mumford, R. 2007. Microarrays for rapid identification of plant viruses. *Annual Review of Phytopathology* 45(1):307–328.

Brunt, A.A. 1992. The general properties of potyviruses. *Archives of Virology* 20:3–16.

Canto, T., Aranda, M.A., and Fereres, A. 2009. Climate change effects on physiology and population processes of hosts and vectors that influence the spread of hemipteran-borne plant viruses. *Global Change Biology* DOI: 10.1111/j.365–2486.008.01820.x.

Card, S.D., Pearson, M.N., and Clover, G.R.G. 2007. Plant pathogens transmitted by pollen. *Australasian Plant Pathology* 36:455–461.

Carstens, E. and Ball, L. 2009. Ratification vote on taxonomic proposals to the International Committee on Taxonomy of Viruses (2008). *Archives of Virology* 154:1181–1188.

Chang, C., Chen, Y.C., Hsu, Y.H., Wu, J.T., Hu, C.C., Chang, W.C., and Lin, N.S. 2005. Transgenic resistance to *Cymbidium mosaic virus* in *Dendrobium* with the capsid protein gene. *Transgenic Research* 14:41–46.

Chen, J., Chen, J., and Adams, M.J. 2001. A universal PCR primer to detect members of the *Potyviridae* and its use to examine the taxonomic status of several members of the family. *Archives of Virology* 146:757–766.

Childers, N.E. and Cibes, H.R. 1948. *Vanilla Culture in Puerto Rico*. U.S. Dept. Agr., Washington, DC, Circular 28, 94 pp.

Colinet, D., Nguyen, M., Kummert, J., Lepoivre, P., and Xia, F.Z. 1998. Differentiation among potyviruses infecting sweet potato based on genus- and virus-specific reverse transcription polymerase chain reaction. *Plant Disease* 82:223–229.

Devergne, J.C., Cardin, L., Burckard, J., and Van Reegenmortel, M.H.V. 1981. Comparison of direct and indirect ELISA for detecting antigenically related cucumoviruses. *Journal of Virological Methods* 3(4):193–200.

Douine, L., Quiot, J.B., Marchoux, G., and Archange, P. 1979. Recencement des espèces végétales sensibles au virus de la mosaique du concombre (CMV). Etude bibliographique. *Annales de Phytopathologie* 11:439–475.

Eiras, M., Boari, A.J., Colariccio, A., Chaves, A.L.R., Briones, M.R.S., Figueira A.R., and Harakava R. 2004. Characterization of isolates of the cucumovirus Cucumber mosaic virus present in Brazil. *Journal of Plant Pathology* 86(1):61–69.

Escriu, F., Fraile, A., and García-Arenal, F. 2000. Evolution of virulence in natural population of the satellite RNA of Cucumber mosaic virus. *Phytopathology* 90(5):480–485.

Eun, A.J.C. and Wong, S.M. 1999. Detection of *Cymbidium mosaic potexvirus* and *Odontoglossum ringspot tobamovirus* using immuno-capillary zone electrophoresis. *Phytopathology* 89:522–528.

Eun, A.J.C., Huang, L., Chew, F.T., Li, S.F.Y., and Wong, S.M. 2002. Detection of two orchid viruses using quartz crystal microbalance-based DNA biosensors. *Phytopathology* 92:654–658.

Fargette, D., Pinel-Galzi, A., Sérémé, D., Lacombe, S., Hébrard, E., Traoré, O., and Konaté, G. 2008. Diversification of rice yellow mottle virus and related viruses spans the history of agriculture from the neolithic to the present. *PLoS Pathogens* 4(8):e1000125.

Farreyrol, K. 2005. *Characterization and Detection of Viruses (Cucumovirus, Potyvirus) Infecting Vanilla in Réunion Island and Polynesian Islands*. The University of Auckland, Auckland, 170pp.

Farreyrol, K., Grisoni, M., Pearson, M.N., Richard, A., Cohen, D., and Beck, D.L. 2009. Genetic diversity of *Cucumber mosaic virus* infecting vanilla in French Polynesia and Réunion Island. *Autralasian Plant Pathology* 39(2):132–140.

Farreyrol, K., Pearson, M.N., Grisoni, M., and Leclercq Le Quillec, F. 2001. Severe stunting of *Vanilla tahitensis* in French Polynesia caused by *cucumber mosaic virus* (CMV), and the detection of the virus in *V. fragrans* in Reunion Island. *Plant Pathology* 50, 414.

Farreyrol, K., Pearson, M.N., Grisoni, M., Cohen, D., and Beck, D.L. 2006. Vanilla mosaic virus isolates from French Polynesia and the Cook Islands are *Dasheen mosaic virus* strains that exclusively infect vanilla. *Archives of Virology* 151:905–919.

Fauquet, C., Mayo, M.A., Maniloff, J., Desselberger, U., and Ball, L.A., 2005. *Virus Taxonomy: Eighth Report of the International Committee on Taxonomy of Viruses*. Elsevier, Academic Press, London, UK, 1259 pp.

Forterre, P. 2006. The origin of viruses and their possible roles in major evolutionary transitions. Virus research. *Comparative Genomics and Evolution of Complex Viruses* 117:5–16.

Francki, R.I.B. 1970. Cymbidium mosaic virus. C.M.I./A.A.B. Description of plant viruses (27):3p.

Freitas-Astua, J., Rezende, J.A.M., and Kitajima, E.W. 1999. Incidence of orchid viruses in the state of Sao Paulo, Brazil. *Fitopatologia Brasileira* 24(2):125–130.

Gallitelli, D. 2000. The ecology of *Cucumber mosaic virus* and sustainable agriculture. *Virus Research* 71:9–21.

Garrett, K.A., Dendy, S.P., Frank, E.E., Rouse, M.N., and Travers, S.E. 2006. Climate change effects on plant disease: Genomes to ecosystems. *Annual Review of Phytopathology* 44:489–509.

Gibbs, A. and Mackenzie, A. 1997. A primer pair for amplifying part of the genome of all potyvirids by RT-PCR. *Journal of Virological Methods* 63:9–16.

Gibbs, A., Mackenzie, A., Blanchfield, A., Cross, P., Wilson, C., Kitajima, E., Nightingale, M., and Clements, M. 2000. Viruses of orchids in Australia: Their identification, biology and control. *Australian Orchid Review* 65:10–21.

Gibbs, A., Mackenzie, A., Wei, K.-J., and Gibbs, M. 2008a. The potyviruses of Australia. *Archives of Virology* 153:1411–1420.

Gibbs, A., Trueman, J., and Gibbs, M. 2008c. The bean common mosaic virus lineage of potyviruses: Where did it arise and when? *Archives of Virology* 153:2177–2187.

Gibbs, A.J., Ohshima, K., Phillips, M.J., and Gibbs, M.J. 2008b. *The Prehistory of Potyviruses: Their Initial Radiation was during the Dawn of Agriculture. PLoS ONE* 3:e2523.

Gillaspie, G., Jr., 2001. RT-PCR detection of seedborne *Cowpea aphid-borne mosaic virus* in peanut. *Plant Disease* 85:1181–1182.

Gourdel, S. and Leclercq-Le Quillec, F. 2001. Coat protein gene of *Cymbidium mosaic virus* from *Vanilla fragrans* in Reunion Island. *Journal of Phytopathology* 149:561–564.

Grisoni M. 2003. Prospection sur les virus des vanilliers aux Iles Australes et Marquises (Polynésie française), Cirad / SDR, Uturoa, Polynésie Française, 9p.

Grisoni, M. 2004. Rapport de mission pour la société Biolandes à Ambanja (Madagascar). Cirad, Saint Pierre, La Réunion, 25p.

Grisoni, M. 2009. *Mission d'expertise en Pathologie du Vanillier Dans la SAVA*. CTHT/FED-Stabex, Toamasina, Madagascar.

Grisoni, M. and Abdoul-Karime, A.L. 2007. Epidémio surveillance des vanilleraies de Mayotte et des Comores, Cirad/DAF-PV, Mamoudzou, Mayotte, 34pp.

Grisoni, M., Côme, B., and Nany, F. 1997. Projet de relance de la vanilliculture dans la région du Sava. *Compte rendu de mission à Madagascar du 05 au 18 mai 1997*, CIRAD/FOFIFA, Saint Pierre, 11pp.

Grisoni, M., Davidson, F., Hyrondelle, C., Farreyrol, K., Caruana, M.L., and Pearson, M. 2004. Nature, incidence and symptomatology of viruses infecting *Vanilla tahitensis* in French Polynesia. *Plant Disease* 88:119–124.

Grisoni, M., Moles, M., Besse, P., Bory, S., Duval, M.-F., and Kahane, R. 2007. Towards an international plant collection to maintain and characterize the endangered genetic resources of vanilla. *Acta Horticulturae (ISHS)* 760:83–91.

Grisoni, M., Moles, M., Farreyrol, K., Rassaby, L., Davis, R.I., and Pearson, M.N. 2006. Identification of the potyviruses infecting vanilla by direct sequencing of a short RT-PCR amplicon. *Plant Pathology* 55:523–529.

Ha, C., Coombs, S., Revill, P.A., Harding, R.M., Vu, M., and Dale, J.L. 2008. Design and application of two novel degenerate primer pairs for the detection and complete genomic characterization of potyviruses. *Archives of Virology* 153(1):25–36.

Hobbs, H.A., Eastburn, D.M., D'Arcy, C.J., Kindhart, J.D., Masiunas, J.B., Voegtlin, D.J., Weinzierl, R.A., and McCoppin, N.K. 2000. Solanaceous weeds as possible sources of cucumber mosaic virus in Southern Illinois for aphid transmission to pepper. *Plant disease* 84(11):1221–1224.

Hu, J.S., Li, H.P., Barry, K., and Wang, M. 1995. Comparison of Dot Blot, ELISA, and RT-PCR assays for detection of two cucumber mosaic virus isolates infecting banana in Hawaii. *Plant Disease* 79(9):902–906.

Hu, W.W. and Wong, S.M. 1998. The use of DIG-labelled cRNA probes for the detection of *Cymbidium mosaic potexvirus* (CymMV) and *Odotonglossum ringspot tobamovirus* (ORSV) in orchids. *Journal of Virological Methods* 70:193–199.

Hull, R. 2002. *Plant Virology*, 4th edn. Academic Press, London.

Izaguirre-Mayoral, M.L., de Uzcategui, R.C., and Carballo, O. 1993. Crassulacean acid metabolism in two species of orchids infected by *Tobacco mosaic virus*-orchid strain and/or *Cymbidium mosaic virus*. *Journal of Phytopathology* 137:272–282.

Jacob de Cordemoy, H. 1899. Sur une anomalie de la vanille. *Revue Générale de Botanique* 11:258–267.

Jensen, D.D. 1950. Mosaic of Cymbidium orchids (Pacific Division Report). *Phytopathology* 40:966–967.

Jensen, D.D. and Gold, A.H. 1951. A virus ring spot of odontoglossum orchid: Symptoms, transmission, and electron microscopy. *Phytopathology* 41:648–653.

Jensen, D.D. and Gold, A.H. 1955. Hosts, transmission, and electron microscopy of *cymbidium mosaic virus* with special reference to Cattleya leaf necrosis. *Phytopathology* 45:327–334.

Jordan, R. and Hammond, J. 1991. Comparison and differentiation of potyvirus isolates and identification of strain-, virus-, subgroup-specific and potyvirus group-common epitopes using monoclonal antibodies. *Journal of General Virology* 72:25–36.

Khentry, Y., Paradornuwat, A., Tantiwiwat, S., Phansiri, S., and Thaveechai N. 2006. Incidence of Cymbidium mosaic virus and Odontoglossum ringspot virus in Dendrobium spp. in Thailand. *Crop Protection* 25(9):926–932.

Kiranmai, G., Sreenivasulu, P., and Nayudu M.V. 1998. Epidemiology of cucumber mosaic cucumovirus isolates naturally infecting three solanaceous vegetable crops around Tirupati. *Indian Phytopathology* 51(4):315–318.

Kitajima, E.W., Kondo, H., Mackenzie, A., Rezende, J.A.M., Gioria, R., Gibbs, A., and Tamada, T. 2001. Comparative cytopathology and immunocytochemistry of Japanese, Australian and Brazilian isolates of orchid fleck virus. *Journal of General Plant Pathology* 67(3):231–237.

Kokoa, P. 2000. Vanilla diseases in East New Britain. In *Prospects for Vanilla Development in PNT*. Thiagalingam, K. and Wiles, G.C. (eds.), Tavilo, Papua New Guinea: NARI, PNG, pp. 57–63.

Kondo, H., Maeda, T., and Tamada, T. 2009. Identification and characterization of structural proteins of orchid fleck virus. *Archives of Virology* 154(1):37–45.

Kondo, H., Maeda, T., Shirako, Y., and Tamada. 2006. Orchid fleck virus is a rhabdovirus with an unusual bipartite genome. *Journal of General Virology* 87:2413–2421.

Langeveld, S.A., Dore, J.-M., Memelink, J., Derks, A.F.L.M., Van der Vlugt, C.I.M., Asjes, C.J., and Bol, J.F. 1991. Identification of potyviruses using the polymerase chain reaction with degenerate primers. *Journal of General Virology* 72:1531–1541.

Lavina, A., Aramburu, J., and Moriones, E. 1996. Occurence of tomato spotted wilt and cucumber mosaic viruses in field grown tomato crops and associated weeds in northern spain. *Plant Pathology* 45(5):837–842.

Lawson, R.H. and Hearon, S.S. 1974. Distribution and staining properties of cymbidium mosaic virus in buds of cattleya orchids. *Acta Horticulturae* 36:195–202.

Leclercq Le Quillec, F., Riviere, C., and Lagorce, A. 2001. Spread of *Cymbidium mosaic potexvirus* and potyviruses in vanilla plants grown in shade houses in Reunion Island. *Fruits (Paris)* 56:249–260.

Leclercq-Le Quillec, F. and Nany, F. 1999. *Compte rendu de mission à Madagascar* 22 juin–2 juil., CIRAD/FED/FOFIFA, 12pp. +annexes pages.

Leclercq-Le Quillec, F. and Nany, F. 2000a. *Compte rendu de mission à Madagascar* 18 fev.–15 mars 2000, CIRAD/FED/FOFIFA, 11pp.

Leclercq-Le Quillec, F. and Nany, F. 2000b. *Compte rendu de mission à Madagascar* 5–23 mai 2000, CIRAD/FED/FOFIFA, 10pp.

Leclercq-Le Quillec, F. and Servé, G. 2001. *Diagnostic et variabilité du Cymbidium Mosaic Potexvirus*. CIRAD, Saint Pierre–Réunion, 43pp.

Lee, S.-C. and Chang, Y.-C. 2008. Performances and application of antisera produced by recombinant capsid proteins of *Cymbidium mosaic virus* and *Odontoglossum ringspot virus*. *European Journal of Plant Pathology* 122:297–306.

Liao, L.-J., Pan, I.-C., Chan, Y.-L., Hsu, Y.-H., Chen, W.-H., and Chan, M.-T. 2004. Transgene silencing in *Phalaenopsis* expressing the coat protein of *Cymbidium mosaic virus* is a manifestation of RNA-mediated resistance. *Molecular Breeding* 13:229–242.

Lim, S.-H., Moon, K.K., Sang, J.L., Yong, J.L., and KIM, B.-D. 1999. *Cymbidium mosaic virus* coat protein gene in antisense confers resistance to transgenic *Nicotiana occidentalis*. *Molecules and Cells* 9:603.

Lim, S.T., Wong, S.M., and Goh, C.J. 1993a. Elimination of *Cymbidium mosaic virus* and *Odontoglossum ringspot* virus from orchids by meristem culture and thin section culture with chemotherapy. *Annals of Applied Biology* 122:289–297.

Lim, S.T., Wong, S.M., Yeong, C.Y., Lee, S.C., and Goh, C.J. 1993b. Rapid detection of cymbidium mosaic virus by the polymerase chain reaction (PCR). *Journal of Virological Methods* 41:37–46.

Lockard, B.E.L. 2000. Virus diseases of musa in Africa: Epidemiology, detection and control. International Conference on Banana and Plantain for Africa, Kampala, Uganda, ISHS.

Madhubala, R., Bhadramurthy, V., Bhat, A.I., Hareesh, P.S., Retheesh, S.T., and Bhai, R.S. 2005. Occurence of *Cucumber mosaic virus* on vanilla (*Vanilla planifolia* Andrews) in India. *Journal of Biosc*ience 30(3):339–350.

Maeda, T. and Inouye, N. 1991. Rapid detection of cucumber mosaic virus by a simplified F(ab')2 ELISA using PEG. Berichte des Ohara Instituts für Landwirtschaftliche Biologie, Okayama Universität 20:19–26.

Magnaye, L.V. and Valmayor, R V. 1995. BBTV, CMV and other viruses affecting banana in Asia and the Pacific. Extension Bulletin. Taipei, Taiwan, Food & Fertilizer Technology Center. 2008.

Marie-Jeanne, V., Ioos, R., Peyre, J., Alliot, B., and Signoret, P., 2000. Differentiation of Poaceae Potyviruses by reverse transcription-polymerase chain reaction and restriction analysis. *Journal of Phytopathology* 148:141–151.

Martos, F. 2005. Détection spécifique du *Cymbidium mosaic virus* chez le vanillier par "Quantitative-reverse transcription polymerase chain reaction" (Q-RT-PCR): *Biologie des populations et des écosystèmes*, Université de La Réunion, Saint Denis, 26pp.

McKirdy, S.J. and Jones, R.A.C. 1994. Infection of alternative hosts associated with annual medics (Medicago spp.) by alfalfa mosaic virus and its persistence between growing seasons. *Australian Journal of Agricultural Research* 45(7):1413–1426.

Minoo, D., Pillai, G.S., Babu, K.N., and Peter, K.V. 2008. Isolation and fusion of protoplasts in *Vanilla* species. *Current Science* 94:115–120.

Moles, M., Delatte, H., Farreyrol, K., and Grisoni, M. 2007. Evidence that *Cymbidium mosaic virus* strains segregate into two groups based on coat protein and replicase genes. *Archives of Virology* 152:705–715.

Mongredien, G. 2002. Research of *Orchid fleck virus* and molecular typing *Cucumber mosaic virus* in French Polynesia. M.Sc. *Biotechnology*, ISAB/University, Wageningen, 27pp.

Moury, B. 2004. Differential selection of genes of *Cucumber mosaic virus* subgroups. *Molecular Biology and Evolution* 21:1602–1611.

Ooi, D.J.M., Dzulkurnain, A., Othman, R.Y., Lim, S.H., and Harikrishna, J.A. 2006. Use of superparamagnetic beads for the isolation of a peptide with specificity to cymbidium mosaic virus. *Journal of Virological Methods* 136:160–165.

Palukaitis, P., Roosinck, M.J., Dietzgen, R.G., and Francki, R.I.B., 1992. Cucumber mosaic virus. *Advances in Virus Research* 41:281–338.

Pappu, S.S., Brand, R., Pappu, H.R., Rybicki, E.P., Gough, K.H., Frenkel, M.J., and Niblett, C.L. 1993. A polymerase chain reaction method adapted for selective amplification and cloning of 3′ sequences of potyviral genomes: Application to dasheen mosaic virus. *Journal of Virological Methods*, 41(1):9–20.

Pearson, M.N. 1997. Assessment of vanilla virus problem in Ile de la Reunion, The University of Auckland, Auckland, 17pp.

Pearson, M.N. and Cole, J.S. 1991. Further observations on the effects of *cymbidium mosaic virus* and *odotonglossum ring spot* virus on growth of Cymbidium orchids. *Journal of Phytopathology* 131:193–198.

Pearson, M.N. and Pone, S.P. 1988. Viruses of vanilla in the kingdom of Tonga. *Australasian Plant Pathology* 17:59–60.

Pearson, M.N., Brunt, A.A., and Pone, S.P. 1990. Some hosts and properties of a potyvirus infecting *Vanilla fragrans* (Orchidaceae) in the Kingdom of Tonga. *Journal of Phytopathology* 128:46–54.

Pearson, M.N., Jackson, G.V.H., Pone, S.P., and Howitt, R.L.J. 1993. Vanilla viruses in the South Pacific. *Plant Pathology* 42:127–131.

Rassaby, L. 2003. Rapport de mission d'expertise phytosanitaire: Virus/vanille/Maurice, CIRAD, Saint Pierre, Réunion, 12pp.

Read, A., Bushe, B., and Shintaku, M. 2007. Characterization of a *Cymbidium mosaic virus* isolate that is undetectable by a commercial ELISA virus detection kit. *Phytopathology* 97:S97–S97.

Revers, F., LeGall, O., Candresse, T., and Maule, A.J. 1999. New advances in understanding the molecular biology of plant/potyvirus interactions. *Molecular Plant Microbe Interactions* 12:367–376.

Richard, A., Farreyrol, K., Rodier, B., Leoce-Mouk-San, K, Wong, M., Pearson, M., and Grisoni, M. 2009. Control of virus diseases in intensively cultivated vanilla plots of French Polynesia. *Crop Protection* 28:870–877.

Rist, D.L. and Lorbeer, J.W. 1991. Relationship of weed reservoirs of cucumber mosaic virus (CMV) and broad bean wilt virus (BBWV) to CMV and BBWV in commercial lettuce field in New York. *Phytopathology* 81(4):367–371.

Roossinck, M.J., Zhang, L., and Hellwald, K. 1999. Rearrangements in the 5′ nontranslated and phylogenetic analysis of *Cucumber mosaic virus* RNA3 indicate radial evolution of three subgroups. *Journal of Virology* 73:6752–6758.

Ryu, K.H., Yoon, K.E., and Park, W.M. 1995. Detection by RT-PCR of Cymbidium mosaic virus in orchids. *Journal of Phytopathology* 143(11–12): 643–646.

Seoh, M.L., Wong, S.M., and Zhang, L. 1998. Simultaneous TD/RT-PCR detection of *cymbidium mosaic potexvirus* and *odontoglossum ringspot tobamovirus* with a single pair of primer. *Journal of Virological Methods* 72:197–204.

Sherpa, A., Hallan, V., Pathak, P., and Zaidi, A. 2007. Complete nucleotide sequence analysis of *Cymbidium mosaic virus* Indian isolate: Further evidence for natural recombination among potexviruses. *Journal of Biosciences* 32:663–669.

Sherpa, A.R., Hallan, V., Pathak, P., and Zaidi, A.A. 2006. Characterization of the coat protein gene of *Cymbidium mosaic virus* isolates from India. *Journal of Phytopathology* 154:275–280.

Shukla, D.D., Ward, C.W., and Brunt, A.A. 1994. The Potyviridae, CABI.

Skoric, D., Krajacic, M., Perica, M.C., Halupecki, E., Topic J., and Igrc-Barcic, J. 2000. Cucumber mosaic virus (Cucumovirus) and associated satRNA in weed species under the natural epidemic conditions of tomato lethal necrosis in Croatia. *Zeitschrift fur Pflanzenkrankheiten und Pflanzenschutz* 107(3):304–309.

Stukenbrock, E.H. and McDonald, B.A. 2008. The origins of plant pathogens in agro-ecosystems. *Annual Review of Plant Pathology* 46:75–100.

Tanaka, S., Nishii, H., Ito, S., KameyaIwaki, M., and Sommartya, P. 1997. Detection of *Cymbidium mosaic potexvirus* and *Odontoglossum ringspot tobamovirus* from Thai orchids by rapid immunofilter paper assay. *Plant Disease* 81:167–170.

Thomas, J., Vijayan, A.K., and Bhai, S.R. 2002. Vanilla diseases in India and their management. *Indian Journal of Arecanut, Spices & Medicinal Plants* 4:143–149.

Urcuqui-Inchima, S., Haenni, A.-L., and Bernardi, F. 2001. Potyvirus proteins: A wealth of functions. *Virus Research* 74(1–2):157–175.

Varveri, C. and Boutsika, K. 1999. Characterization of cucumber mosaic cucumovirus isolates in Greece. *Plant Pathology* 48(1):95–100.

Vaughan, S., Grisoni, M., Kumagai, M., and Kuehnle, A. 2008. Characterization of Hawaiian isolates of *Cymbidium mosaic virus* (CymMV) co-infecting *Dendrobium* orchid. *Archives of Virology* 153:1185–1189.

Vejaratpimol, R., Channuntapipat, C., Liewsaree, P., Pewnim, T., Ito, K., Iizuka, M., and Minamiura, N. 1998. Evaluation of enzyme-linked immunosorbent assay for the detection of *Cymbidium mosaic virus* in orchids. *Journal of Bioscience and Bioengineering* 86:65–71.

Wang, Y.Y. and Pearson, M.N. 1992. Some characteristics of potyvirus isolates from *Vanilla tahitensis* in French Polynesia and the Cook Islands. *Journal of Phytopathology* 135:71–76.

Wang, Y.Y., Beck, D.L., Gardner, R.C., and Pearson, M.N. 1993. Nucleotide sequence, serology and symptomatology suggest that vanilla necrosis potyvirus is a strain of watermelon mosaic virus II. *Archives of Virology* 129:93–103.

Wang, Y.Y., Gardner, R.C., and Pearson, M.N. 1997. Resistance to vanilla necrosis potyvirus in transgenic *Nicotiana benthamiana* plants containing the virus coat protein gene. *Journal of Phytopathology Phytopathologische Zeitschrift* 145:7–15.

Wannakrairoj, S. 2008. Impact of *Cymbidium mosaic virus* on growth of *Dendrobium* Jaq-Hawaii cv. Uniwai Pearl. Poster presented at the Intern. Workshop on Ornamental Plants. *Acta Horticulturae*.

Webster, C.G., Coutts, B.A., Jones, R.A.C., Jones, M.G.K., and Wylie, S.J. 2007. Virus impact at the interface of an ancient ecosystem and a recent agroecosystem: Studies on three legume-infecting potyviruses in the southwest Australian floristic region. *Plant Pathology* 56(5):729–742.

Wei, T., Pearson, M.N., Blohm, D., Nölte, M., and Armstrong, K. 2009. Development of a short oligonucleotide microarray for the detection and identification of multiple potyviruses. *Journal of Virological Methods* 162:109–118.

Wisler, G.C., Zettler, F.W., and Mu, L. 1987. Virus infections of Vanilla and other orchids in French Polynesia. *Plant Disease* 71:1125–1129.

Wisler, G.C., Zettler, F.W., and Purcifull, D.E. 1982. A serodiagnostic technique for detection of *Cymbidium mosaic* and *Odotonglossum Ringspot* viruses. *Phytopathology* 72:835–837.

Wong, S.M., Mahtani, P.H., Lee, K.C., Yu, H.H., Tan, Y., Neo, K.K., Chan, Y., Wu, M., and Chng, C.G. 1997. *Cymbidium mosaic potexvirus* RNA: Complete nucleotide sequence and phylogenetic analysis–Brief report. *Archives of Virology* 142:383–391.

Wren, J.D., Roossinck, M.J., Nelson, R.S., Scheets, K., Palmer, M.W., and Melcher, U. 2006. Plant virus biodiversity and ecology. *PLoS Biology* 4(3):e80.

Wylie, S., Wilson, C.R., Jones, R.A.C., and Jones, M. 1993. A polymerase chain reaction assay for cucumber mosaic virus in lupin seeds. *Australian Journal of Agricultural Research* (44):41–51.

Yamane, K., Oyama, K., Iuchi, E., Ogawa, H., Suzuki, T., and Natsuaki, T. 2008. RT-PCR detection of *Odontoglossum ringspot virus*, *Cymbidium mosaic virus* and tospoviruses and association of infections with leaf-yellowing symptoms in *Phalaenopsis*. *Journal of Phytopathology* 156:268–273.

Yang, S.Y., Jian, Z.F., Chieh, J.J., Horng, H.E., Yang, H.C., Huang, I.J., and Hong, C.-Y. 2008. Wash-free, antibody-assisted magnetoreduction assays of orchid viruses. *Journal of Virological Methods* 149:334–337.

Yu, C., Wu, J., and Zhou, X. 2005. Detection and subgrouping of Cucumber mosaic virus isolates by TAS-ELISA and immunocapture RT-PCR. *Journal of Virological Methods* 123(2):155–161.

Yuen, C., Kamemoto, H., and Ishii, M. 1979. Transmission of cymbidium mosaic virus through seed propagation in *Dendrobium*. *American Orchid Society Bulletin* 48:1245–1247.

Zettler, F.W., Ko, N.J., Wisler, G.C., Elliott, M.S., and Wong, S.M. 1990. Viruses of orchids and their control. *Plant Disease* 74:621–626.

8 Fungal Diseases of Vanilla

Mesak Tombe and Edward C.Y. Liew

CONTENTS

INTRODUCTION

Diseases of vanilla crop caused by pathogenic fungi have been reported ever since vanilla was commercially cultivated. In vanilla-producing countries such as Indonesia, Madagascar, India, Papua New Guinea, Mexico, and Puerto Rico, diseases caused by fungi are significant constraints to the cultivation of this plant. Until now several fungal genera have been reported to infect vanilla, including *Fusarium, Phytophthora, Sclerotium*, and *Colletotrichum*. These attack or infect the stems, roots, leaves, shoots, and fruit of vanilla. However, the disease caused by *Fusarium* is the most serious and has resulted in crop devastation of epidemic proportion in vanilla farms or plantations in many of the producing countries.

FUSARIUM ROT OF VANILLA

Fusarium rot occurs on the roots, stems, fruit, leaves, and shoots of the plant. However, stem and root infections cause the greatest amount of damage, leading to significant yield loss, and hence the disease is often referred to as stem rot, foot rot, or stem and root rot. In Indonesia, Fusarium rot can infect all parts of the plant from nursery to productive plants but the attack on the stem is most harmful. In light of its significance, a greater emphasis on this disease is placed in this chapter.

Historical Review and Distribution

Fusarium rot, attacking the vanilla crop, has been reported since the crop was cultivated commercially in producing countries, including Indonesia, India, Puerto Rico, the Seychelles, Reunion Island, Madagascar, and Polynesia.

A disease on vanilla was first recorded by Zimmermann in Indonesia in 1903 (Tombe, 1994) when infections on stems and leaves were noted. In 1918, symptoms of a root disease were first noticed in vanilla plantations in Mayagüez, Puerto Rico (Alconero and Santiago, 1969). Tucker (1927) reported that the vanilla root rot in Puerto Rico was caused by a soilborne pathogen called *Fusarium batatatis* var. *vanillae* Tucker. In 1925, van Hall (in Tucker, 1927) mentioned the occurrence of a root fungus that caused death of vanilla cuttings in Java, Indonesia. However, the earliest reference associating a *Fusarium* species to vanilla stem rot in Indonesia was by Soetono (1962), who isolated a species of *Fusarium* from infected vanilla plants and identified it as *F. batatatis* Wollenweber.

In the Seychelles Islands, where vanilla was formerly an important crop, the same root disease was present according to a report made by Dupont (1921). The disease description coincided quite closely with that in Puerto Rico (Tucker, 1927). Meinecke (in Tucker, 1927) also observed an undescribed *Fusarium* attacking the tender tips and young pods of vanilla. Averna-Sacca (1930) reported that the same species of *Fusarium* was isolated from the diseased vanilla growing in Brazil. However, he stated that the *Fusarium* pathogen only attacked leaves and stems without any mention of its occurrence on the roots.

Lealy (1970) and Owino (2008) reported that root rot of vanilla caused by *Fusarium oxysporum* (syn. *F. batatatis* var. *vanillae*) was the most serious disease in vanilla plantations in Uganda. This and other species of *Fusarium*, including *Fusarium solani* have been reported in vanilla plantations in India (Balagopal et al., 1974a, 1974b; Philip, 1980; Anandaraj et al., 2005), Thailand (Ratanachurdchai and Soytong, 2008), Tonga (Stier, 1984), and China (Ruan et al., 1998). The pathogen is observed in Reunion Island and Madagascar today, but was probably present as far back as 1871 and 1902, respectively (Bouriquet, 1954).

Symptoms and Disease Development

Symptoms of stem and root rot may appear at any growth stage of the vanilla plant: cuttings, young vines, and mature productive vines in the field. In addition to the stems and roots, the disease attacks other plant parts such as shoots and beans at any time of the year (Tombe et al., 1992a). In Indonesia, most cases of infection occur initially on the stems, followed by roots and shoots, and occasionally on beans and young leaves. Foliar infection is more commonly found on young leaves. During the rainy season infection on shoots is more prevalent than on other parts of the plant, although the damage is not as severe as on the stems.

Under adverse disease development conditions, symptoms appear as black spots on the stems with limited progress and obvious brown margins. On the other hand, under favorable conditions brown to dark brown lesions with less clearly defined margins enlarge and extend very rapidly and spread along the whole stem internode. A chlorotic zone is often observed between the lesion and healthy tissue (Figure 8.1a) (Soetono, 1962; Tombe, 1993a; Anandaraj et al., 2005). Consequently, the infected stem internode constricts, turns brown to dark brown and finally becomes dry and necrotic (Figure 8.1b). However, disease progress appears to be inhibited by nodes along the vine (Figure 8.1c). On the rotted and constricted parts yellowish-orangey-white spore masses are formed, consisting of fungal conidiophores and conidia. When the infected stem is longitudinally cut, necrosis is apparent, as indicated by a brown discoloration, developing from the inner to the peripheral tissue.

On roots the symptoms initially appear in the form of browning, followed by eventual death (Rachmadiono et al., 1982; Tombe, 1993a; Anandaraj et al., 2005). Aerial roots die promptly after entering the soil (Figure 8.1c), resulting in flaccidity and shriveling of the stem and consequently the vine droops. It turns dark brown and decays, and the rot is either soft and watery or somewhat dry, depending on the existing moisture conditions (Alconero, 1968). Fusarium rot of aerial roots is often difficult to distinguish from anthracnose as both have very similar symptoms and the respective pathogens are often coisolated when present in the same farm. Tucker (1927) recorded in some instances that as much as the lower 3 m of a vine may rot away while the upper part remains green and continues its existence. It is, however, important to note that the pathogen can survive within green healthy internodes with no apparent internal or external symptoms.

FIGURE 8.1 **(See color insert following page 136.)** Symptoms of Fusarium rot on vanilla vines: (a) dark brown lesion on stem internode with a chlorotic zone; (b) advanced necrotic stem internode and root; (c) numerous aerial roots are produced at the nodes but rot after reaching the soil.

Causal Organism

Tucker (1927) was the first to describe the pathogen of vanilla root rot by detailing the morphology of various spore types and identifying the pathogen as *F. batatatis* var. *vanillae*. This name has undergone multiple nomenclatural changes since then. This together with the association of the disease with various plant parts, resulted in the occasional confusion as to the true etiology of the disease. The species name given by Tucker (1927) was on the basis of morphological similarity to *Fusarium batatis*, the wilt pathogen of sweet potato, but the two pathogens from the respective hosts were not found to be cross-pathogenic, hence, a variety name was used for the vanilla strain. The host specialization of morphologically identical strains of *Fusarium* species led to the concept of *forma specialis* (Snyder and Hansen, 1940), whereby many *Fusarium* species previously named on the basis of host pathogenicity, despite morphological similarity to *F. oxysporum* Schlechtendahl, were renamed *F. oxysporum* with different *forma specialis* epithets according to the hosts. *Fusarium oxysporum* Schlecht. f. sp. *batatis* (Wollemw.) Snyder et Hansen was synonymized with *F. batatis* (Snyder and Hansen, 1940), but it was sometime later that *F. batatatis* var. *vanillae* (Tucker) was renamed *F. oxysporum* Schlecht. f. sp. *vanillae* (Tucker) Gordon (Gordon, 1965).

F. oxysporum f. sp. *vanillae* produces various asexual structures: the microconidia, macroconidia, and chlamydospores (Figure 8.2). Mycelia are immersed, sometimes running over the surface of lesions, hyaline, slender; sporodochia on decaying

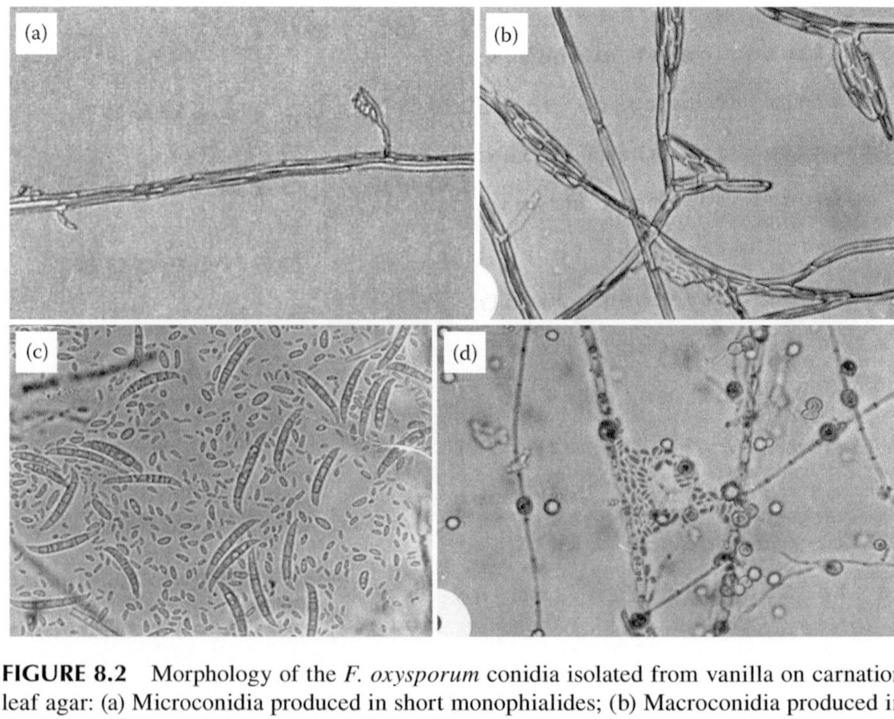

FIGURE 8.2 Morphology of the *F. oxysporum* conidia isolated from vanilla on carnation leaf agar: (a) Microconidia produced in short monophialides; (b) Macroconidia produced in aerial mycelium; (c) Chlamydospores formed singly or in short chains; (d) Macroconidia and microconidia.

infected part of vanilla; microconidia in false heads and short conidiophores, abundant, single-celled, oval, 4–9 × 2–3.5 μm; macroconidia usually 3-septated, occasionally 1- to 2-septated, rarely 4- to 5-septated, abundant, hyaline, curved, pedicellate, 20–46 × 2.4–8 μm; chlamydospores present, singly or in pairs, thick-walled when old, brown, 6.5–10 μm (mean 7 μm). On potato dextrose agar (PDA) medium the growth of colony is rapid and the white aerial mycelia may become tinged with pale purple.

The general morphological features of the *Fusarium* isolates from vanilla isolated in Indonesia, Reunion Island, the Comoros, and Central America today are all similar to those of *F. oxysporum* described by Messiaen and Cassini (1981). They also fit the key descriptions of *F. oxysporum* (Matuo, 1972). Morphologically, these isolates are also similar to those isolated from vanilla in India (Philip, 1980).

Inoculation tests on five plants with *F. oxysporum* indicated that the vanilla isolates were pathogenic to vanilla and not pathogenic to melon, cucumber, tomato, and cymbidium, whereas isolates from tomato were nonpathogenic to vanilla (Nurawan, 1990). In a recent study, Xia-Hong (2007) isolated 87 strains of *F. oxysporum* from vanilla showing stem rot in seven provinces of China. Among these strains, 81 were tested for pathogenicity and only 43 were found pathogenic.

Vegetative compatibility groups (VCG), sometimes called heterokaryon compatibility groups, are a useful tool to identify different genetic groupings in a population of fungi. Each VCG is unique in the sense that the members of one group are not compatible with the members of any other groups. The Indonesian isolates of *F. oxysporum* f. sp. *vanillae* have been grouped into two VCGs with another four VCGs represented by single isolates (Tombe et al., 1994b). In Indonesia, isolates within the same VCG were not necessarily restricted to the same region or location (Tombe et al., 1994). These results were similar to those reported in *F. oxysporum* f. sp. *tuberasi* from potato (Venter et al., 1992) and *Fusarium proliferatum* from asparagus (Elmer, 1991).

Studies are currently being conducted on the genetic diversity and population structure of *F. oxysporum* f. sp. *vanillae* throughout Indonesia, including reference isolates from various parts of the world. These include the use of PCR-based fingerprinting markers as well as phylogenetic analysis of multiple gene regions. Preliminary results indicate the presence of a relatively high number of haplotypes, some of which form large clonal groups. The 87 strains of *F. oxysporum* f. sp. *vanillae* isolated in China (Xia-Hong, 2007) belonged to 12 different VCGs with no correlation between VCG and virulence.

The isolation of nonpathogenic strains of *F. oxysporum* from vanilla roots led to investigations on the possibility of utilizing some of these strains as biocontrol agents against Fusarium rot of vanilla. Promising results have been observed under experimental conditions, but field efficacy has yet to be verified.

Control Measures

Although several vanilla relatives, such as *Vanilla phaeantha*, *Vanilla aphylla*, and *Vanilla andamanica*, have been shown to be resistant to Fusarium rot (Theis and Jiménez, 1957; Minoo et al., 2008), commercial varieties resistant or tolerant to Fusarium rot have not yet been described. Owing to the fact that Fusarium rot can infect various plant parts at all stages of growth, integrated control measures against this disease are paramount and should be implemented right from the cutting preparation stage, throughout the vegetative and productive phases in the field, until senescence (Hadisutrisno, 1996; Tombe et al., 1997; Anandaraj et al., 2005; Tombe, 2008). Several components of reported control measures among others are the application of benomyl, carbendazim, and mancozeb fungicides (Matsumoto et al., 1992; Anandaraj et al., 2005; Bhai et al., 2006), and biological agents such as *Bacillus* sp., *Pseudomonas fluorescens*, and avirulent strains of *F. oxysporum* (Tombe et al., 1992c, 1997; Hadisutrisno, 1996; Anilkumar, 2004).

Integrated control measures have been developed in Indonesia, which involve the use of (a) pathogen-free and disease-tolerant cuttings inoculated with a nonpathogenic *F. oxysporum* strain 10A–M (Figure 8.3) to induce resistance (Bio-FOB cuttings), (b) fungicides, such as benomyl or mancozeb, by dipping cuttings for 20–30 min, (c) biological agents *Trichoderma lactae* and *Bacillus pantothenticus* (Figure 8.3), *Bacillus firmus and T. lactae* or *P. fluorescens* (Tombe, 2008) premixed with organic material, and (d) botanical fungicide eugenol extracted from clove leaves or clove fruit stalks. Table 8.1 outlines the recommendations for the control of

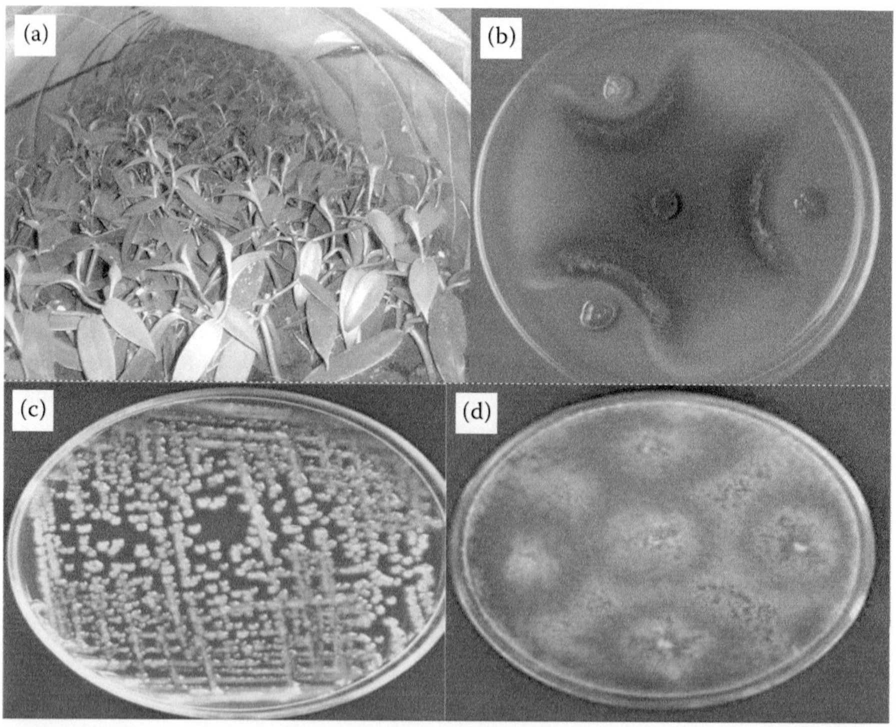

FIGURE 8.3 Biological agents used in the integrated control of Fusarium rot in Indonesia: (a) approximately 3-month-old vanilla cuttings treated with nonpathogenic *Fusarium oxysporum* strain 10A–M; (b) antagonistic test of *B. pantothenticus* to *F. oxysporum* f. sp. *vanillae*; (c) *B. pantothenticus* on sucrose peptone agar; (d) *Trichoderma lactae* on potato dextrose agar.

TABLE 8.1
IDM Recommendations for *Fusarium* Rot of Vanilla in Indonesia

IDM Recommendations	New Plantation	Plantation without Any Disease Symptoms	Plantations with Low Disease Incidence and Mild Severity	Plantations with High Disease Incidence and Severity (before Replanting)
Plant pathogen-free cuttings (Bio-FOB)	+			
Dip cuttings with fungicides (benomyl or mancozeb)	+			
Use organic fertilizer amended with biocontrol agents	+	+	+	
Apply organic mulch, for example, hay, coconut husks, clove leaves	+	+	+	
Follow local recommendations for land tilling, plant spacing, irrigation, and shade trees	+			
Regularly monitor for early symptoms and physically remove any disease tissue. Disinfect cutting tool	+	+	+	
Regularly apply botanical or synthetic fungicide, especially after fertilizing, pruning, weeding, and harvesting	+	+	+	
Regularly prune shade trees to control humidity and shade	+	+	+	
Improve drainage especially during rainy season	+	+	+	
Physically remove disease tissue and apply botanical or synthetic fungicide to cut wounds. Disinfect cutting tools			+	+
Introduce crop rotation with other plants				+
Completely eradicate diseased plants and implement rigorous sanitation measures				+
Plough land thoroughly to improve soil solarization				+
Grow annual crops such as corn and beans as well as crops believed to maintain antagonistic microorganisms (e.g., Welsh and red onion, Chinese chive, garlic)				+

Fusarium rot of vanilla in various scenarios based on an integrated disease management (IDM) approach in Indonesia.

ANTHRACNOSE OF VANILLA

Although anthracnose of vanilla is generally regarded to be present in most growing regions today, including Indonesia, India, the Comoros, Madagascar, Mexico, Uganda (Bouriquet, 1954; Tombe, 1993b Augstburger et al., 2000; Thomas et al., 2002; Magala, 2008), there are no good records of its distribution and history of its detection and etiology.

SYMPTOMS AND DISEASE DEVELOPMENT

Anthracnose symptoms on vanilla appear, especially on older leaves and stems, initially as small gray or black spots, which coalesce or increase in size to become dark brown to black patches or lesions of varying sizes (Taufik and Manohara, 1998) (Figure 8.4a). The disease is also reported to infect young shoots, inflorescence, and beans in India (Anandaraj et al., 2001; Anilkumar, 2004). In advanced disease stages, these lesions may contain the fruiting structures (*acervuli*), which appear as tiny black dots (Tombe, 1993b). Under humid conditions infective propagules (*conidia*) are produced and released from the fruiting structures, observed as masses of salmon-pink spore ooze on the surface of the disease tissue (Taufik and Manohara, 1998). These conidia are dispersed by rain splash or carried on various body parts of insects that come into contact with the spore ooze. Infection on the leaves may occur by spore penetrating through stomata or direct penetration. Latent infection occurs when spores penetrate only into a few epidermic cells and does not develop continuously. Only under favorable environmental conditions the fungus will develop and form necrotic spots. Infection on older or senescing leaves does not usually cause

FIGURE 8.4 (See color insert following page 136.) Symptom of (a) Anthracnose on leaf; (b) sclerotium rot (brown water-soaked leaf tissue and white mycelium and *sclerotia* on soil; (c) Phytophthora stem rot on naturally infected shoot (Inset: symptom after artificial inoculation). (From Andriyani, N. et al. *Jurnal Biologi Indonesia*, 5, 227–234, 2008. With permission.)

significant damage to the crop. The best way to distinguish Fusarium rot from anthracnose is to examine the internal tissue where external symptoms are observed. The presence of internal discoloration, in particular within the vascular tissue is indicative of Fusarium rot. Early symptoms of anthracnose are confined to the epidermis or superficial layers while the vascular tissue appears healthy.

Causal Organism

The causal organism of anthracnose on vanilla is *Colletotrichum gloeosporioides* Penzig et Saccardo (syn. *Colletotrichum vanillae* Scalia, *C. vanillae* Verplancke et Claess, *Gloeosporium vanillae* Cooke) (Tombe, 1993b; Anilkumar, 2004; Ratanachurdchai and Soytong, 2008). The pathogen is sometimes reported as the teleomorph *Glomerella cingulata* (Stoneman) Spauld et Schrenk [syn. *G. vanillae* (Stoneman) Sacc et Traverso], but the sexual stage has never been observed on the plant host. In brief, the morphological description of *C. gloeosporioides* includes (1) *acervuli* disc or cushion-shaped, subepidermal, epiphyllous, 30–345 μm in diameter; conidiophores simple, short, hyaline, setae brown slightly swollen at the base and tapered to the apex on which conidia are occasionally borne 22.5–75 × 2.5–5 μm; and (2) *conidia* hyaline, unicellular, oval to oblong, 10–17.5 × 3–5 μm.

C. gloeosporioides has quite a wide host range and commonly exists in various kinds of conditions and places. However, the relationship between the vanilla isolates and those from other hosts has yet to be investigated.

Control Measures

Anthracnose on vanilla until now has not been reported to cause great losses to farmers (Taufik and Manohara, 1998; Magala, 2008). However, the disease indirectly affects the overall health of vanilla plants and hence vanilla production. This disease can be controlled using several methods, examples of which are outlined below.

1. Regular pruning of shade trees to control humidity and sunlight. Pruning is preferably carried out at the beginning of the rainy season. Drainage should also be improved at the same time.
2. Eradication of diseased plant parts (by burning for example) to prevent disease spread.
3. Application of fungicides (e.g., benomyl, mancozeb, and carbendazim) especially when disease is recurrent (Taufik and Manohara, 1998; Bhai et al., 2006).
4. Improve plant vigor (e.g., nutrition with organic matter).

SCLEROTIUM ROT OF VANILLA

This disease generally attacks during the rainy season when the humidity is high and is often found together with Fusarium rot. Sclerotium rot can be a serious problem in the nursery and, whenever associated with Fusarium rot, can result in the death of productive plants in the field (Tombe and Sitepu, 1987; Taufik and Manohara, 1998). The disease is reported in India and Indonesia (Tombe and Sitepu, 1987; Anandaraj et al., 2005), but there are hardly any records on the history and distribution of this disease in other growing regions, probably due to its frequent association and perhaps confusion with the more significant Fusarium rot.

Symptoms and Damage

The disease occurs on the stem base of vanilla, commonly restricted to 5 cm above the soil surface. The diseased stem base becomes water-soaked, brown to dark brown in color, then becoming necrotic and dies. White fluffy mycelium is often found on the infected stem and the surrounding soil surface (Matsumoto, 1993) (Figure 8.4b). Small spherical structures (*sclerotia*), light brown in color, 0.5–2 mm in diameter, are often observed in association with the mycelium. If the disease occurs in the cutting propagation nursery bed, damage is usually very heavy (Taufik and Manohara, 1998). Sclerotium rot is occasionally found on the vanilla bean, characterized by rotting of bean tips with thick white mat of fungal mycelium, which eventually covers the whole bean. Excess shade, continuous heavy rains, overcrowding of vines, waterlogged conditions, and the presence of the pathogen inoculum in the field are the predisposing factors for bean rot (Anandaraj et al., 2005).

Causal Organism

The causal organism of Sclerotium rot, *Sclerotium rolfsii*, does not form *conidia* or any other reproductive structures. However, dormant structures, the *sclerotia*, are commonly formed (Taufik and Manohara, 1998), which persist in the soil for years. These survival structures are easily spread by rain water splash and run off, contaminated soil, animals and farming equipment (Tombe and Sitepu, 1987), and thus also represent the dispersal and infective propagules of the disease. Under favorable environmental conditions, the *sclerotia* germinate and form mycelial mats, which colonize host tissue. Characteristic diagnostic symptoms and signs of the disease are the white mycelium around infested plant parts and the light brown *sclerotia* produced on basal stems and surrounding soil surface. The causal fungus grows well on artificial medium in the laboratory but the teleomorph is never observed. Numerous *sclerotia* are formed on the agar surface after 10–14 days. Deriving from the mycelium, these *sclerotia* are initially white in color and turn pale to darker brown as they mature and form an outer melanized rind.

Control Measures

The following is an outline of recommended control measures for Sclerotium rot of vanilla in Indonesia. Most of these measures are generic for *Sclerotium*-induced diseases and are hence applicable in other regions.

1. Regular monitoring and early and accurate detection of the disease.
2. Physical removal of diseased tissue, including surrounding plant parts and soil which may contain the *sclerotia* of the pathogen.

3. Application of biological control agents *Trichoderma harzianum, T. lactae*, and *B. pantothenticus* (Tombe, 2007), mixed with organic mulch or compost. The application of *T. harzianum* mixed in rice grain and sterilized soil, spread on the soil surface of vanilla nursery, was shown to suppress the disease by up to 54% (Kasim and Prayitno, 1993).
4. The use of eugenol as a natural fungicide has also been reported to suppress the disease (Sukamto et al., 1996; Tombe et al., 1997). Ground leaves and flower buds of cloves were spread around the base of vanilla stems or mixed with growing media in the nursery. Eugenol (approx. 10–20% active ingredient) can also be applied by spraying or stem dipping before planting.
5. Application of synthetic fungicides, such as carbendazim, mancozeb, and benomyl, especially in conjunction with the above measures.

PHYTOPHTHORA ROT OF SHOOT AND BEAN

This disease has been reported in several vanilla-producing regions in the world, including Indonesia, Madagascar, Puerto Rico, Polynesia, but is generally not as significant as Fusarium rot and anthracnose. However, in Polynesia shoot rot is a serious threat to vanilla production, causing seedling death (Tsao and Mu, 1987). In Indonesia, Phytophthora shoot and bean rot occurs in nurseries and plantations in Bali, Sumatra, and Java (Manohara, 1993), but the disease has not become a serious problem.

Symptoms and Damage

Infected shoots turn light yellowish brown and become completely necrotic in later stages (Figure 8.4c). In Indonesia, disease lesions are usually restricted to the young parts without progressing to the older internodes. The diseased young shoots often die and fall off, after which the older parts are often observed to be further infected with Fusarium rot, which proceeds to cause serious damage (Manohara, 1993). On the shoot, Phytophthora rot is usually characterized by a yellowish brown color on diseased tissue, while Fusarium rot is more typically darker brown (Taufik and Manohara, 1998).

On the bean, Phytophthora rot develops from the tip, forming a water-soaked lesion, which slowly extends toward the pedicel, becoming darker green. The eventual necrosis extends to the whole bunch of beans, sometimes exhibiting abundant external growth of mycelium (Anandaraj et al., 2005). In India and Indonesia, the occurrence of this disease on vanilla plants is heavier during the rainy season (Rachmadiono et al., 1982; Manohara, 1993; Bhai and Thomas, 2000).

Causal Organism

There have been three species of *Phytophthora* reported to attack vanilla plants, namely *Phytophthora palmivora* in Polynesia (Tsao and Mu, 1987) and Thailand (Sangchote et al., 2004), *Phytophthora capsici* in Indonesia (Andriyani et al., 2008), and *Phytophthora meadii* in India (Bhai and Thomas, 2000).

Morphological characteristics of *P. capsici* from vanilla in Indonesia include the presence of sporangia, formed sympodially, which vary in shape from ovoid to obpyriform (Figures 8.5a and b), 35–125 μm long and 17–58 μm wide, and clearly papillate at the tip. Chlamydospores are also present and formed in the middle of the hypha (Figure 8.5c). Colony morphology varies and vegetative growth (Figure 8.5d) is observed at an optimum temperature range of 25–35°C (Andriyani et al., 2008). The isolates found in Indonesia are heterothallic, belonging to the A1 mating type. The oospores are produced in oogonia with amphygenous antheridia. The three species of *Phytophthora* reported on vanilla are most easily distinguished from each other based on the sporangial pedicel length: *P. capsici* with the longest pedicel (10–200 μm) (Figure 8.5a), *P. meadii* intermediate (10–20 μm) and *P. palmivora* the shortest (<5 μm) (Figure 8.5b) (Tsao and Mu, 1987; Erwin and Ribeiro, 1996). Further studies on the etiology of this disease are warranted due to the taxonomic confusion of this genus based solely on morphology.

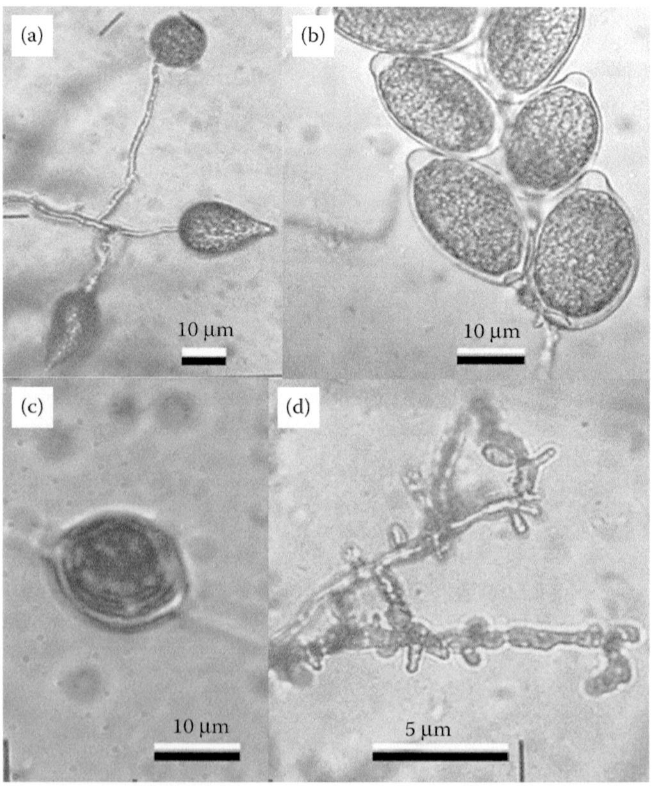

FIGURE 8.5 Morphology of *Phytophthora* isolates: (a) Sporangial long-branching pattern of *P. capsici*; (b) sporangial short-branching pattern of *P. palmivora*; (c) chlamydospore; and (d) mycelium of *P. capsici* generated on V8 juice agar. (From Andriyani, N. et al. *Jurnal Biologi Indonesia*, 5, 227–234, 2008. With permission.)

Control Measures

Various recommendations for the control of Phytophthora shoot and bean rot of vanilla include phytosanitation (removal of infected plant parts, implementation of farm hygiene, etc.), control of humidity by pruning shade trees, application of botanical fungicide, such as eugenol (Tombe et al., 1993; Andriyani et al., 2008) and the use of phosphonate compounds, systemic fungicides effective against a wide range of plant diseases caused by many *Phytophthora* species (Drenth and Guest, 2004).

CONCLUDING REMARKS

There are other minor fungal diseases reported on vanilla in various growing regions but these are not well documented, not verified, or only reported in isolated cases. Some of these diseases include dry rot (caused by *Rhizoctonia* sp.) (Anandaraj et al., 2001), brown rot (caused by *Cylindrocladium quinqueseptatum*) (Bhai and Anandaraj, 2006), vanilla rust (caused by *Puccinia* sp. and *Uredo scabies*) (Correll, 1953; Augstburger et al., 2000), and other foliar diseases caused by *Vermicularia vanillae*, *Guignardia traverse*, *Macrophoma vanillae*, *Pestalospora vanillae*, and *Physalospora vanillae* (Correll, 1953). Reports on vanilla diseases are often confined within local scientific journals and extension reports, which are not easily accessible internationally. Fungal diseases of vanilla are generally not well studied. Much work is required to understand the biology, distribution, pathogen structure and spread, and disease management of these diseases.

REFERENCES

Alconero, R. 1968. Infection and development of *Fusarium oxysporum* f. sp. *vanillae* in vanilla roots. *Phytopathology* 58:1281–1283.

Alconero, R. and A.G. Santiago. 1969. *Fusarium* pathogenic to vanilla. *Plant Disease Reporter* 53 (11):854–857.

Anandaraj, M., J. Rema, and B. Sasikumar. 2001. *Vanilla*. Indian Institute of Spices Research, Calicut, Kerala, India. 8pp.

Anandaraj, M., J. Rema, B. Sasikumar, and R.S. Bhai. 2005. *Vanilla*. Indian Institute of Spices Research, Calicut, Kerala, India. 11pp.

Andriyani, N., D. Wahyuno, D. Manohara, and A.W. Gunawan. 2008. *Phytophthora capsici* is the causal agent of shoot rot disease of vanilla in Indonesia. *Jurnal Biologi Indonesia*, 5 (2):227–234. (in Indonesian, English Abstract).

Anilkumar, A.S. 2004. Vanilla cultivation: A profitable agri-based enterprise. *Kerala Calling* 24:26–30.

Augstburger, F., J. Berger, U. Censkowsky, P. Heid, J. Milz, and C. Streit. 2000. *Organic Farming in the Tropics and Subtropics: Vanilla. Exemplary Descriptions of 20 Crops*. Naturland e.V. http://www.naturland.de/fileadmin/MDB/documents/Publication/English/vanilla.PDF (Accessed May 30, 2009).

Averna-Sacca, R. 1930. Algunas das molestias cryptogamicas mais comuns da baunilha (*Vanilla planifolia*) nos estados de Sao Paulo E.S. *Catharina Bol. Agr.* (Sao Paulo) 23: 282–305.

Balagopal, C., D.S. Bhavani, A. Muralledaran, and E.V.G. Nair. 1974a. Wilt disease, a serious threat to vanilla plantation. *Arecanut and Spices Bulletin* 5:116–117.

Balagopal, C., G. Indarasen, D.S. Bhavani, and V.A. Sukumara. 1974b. Wilt of *Vanilla planifolia. Agricultural Research Journal of Kerala* 12:70–71.

Bhai, R.S. and M. Anandaraj. 2006. Brown spot—A new disease of vanilla (*Vanilla planifolia*) from India. *Journal of Spices and Aromatic Crops* 15:139–140.

Bhai, R.S., A.I. Bhat, and M. Anandaraj. 2006. Yellowing and premature bean dropping in vanilla (*Vanilla planifolia* Andrews). *Journal of Plantation Crops* 34 (2):115–117.

Bhai, R.S. and J. Thomas. 2000. Phytophthora rot, a new disease of vanilla (*Vanilla planifolia* Andrews) in Indian. *Journal of Spices and Aromatic Crops* 9:73–75 (Abstract).

Bouriquet, G. 1954. Pathologie. In: G. Bouriquet, ed. *Le vanillier et la vanille dans le monde*. Editions Paul Lechevalier, Paris, 459–487.

Correll, D.S. 1953. Vanilla—Its botany, history, cultivation and economic importance. *Journal of Economic Botany* 7:291–358.

Drenth, A. and D.I. Guest. 2004. Principles of Phytophthora disease management. In: A. Drenth and D.I. Guest, eds. *Diversity and Management of Phytophthora in Southeast Asia*. Australian Centre for International Agricultural Research Monograph No. 114, 154–160.

Dupont, P.R. 1921. The vanilla industry. Seychelles Agr. and Crown Lants Ann. Report 1920:3.

Elmer, W.H. 1991. Vegetative compatibility groups of *Fusarium proliferatum* from asparagus and comparisons of virulence, growth rates and colonization of asparagus residues among groups. *Phytopathology* 81:852–857.

Erwin, D.C. and O.K. Ribeiro. 1996. *Phytophthora Diseases Worldwide*. APS Pr, St Paul.

Gordon, W.L. 1965. Pathogenic strains of *Fusarium oxysporum. Canadian Journal of Botany* 43:1309–1318.

Hadisutrisno, B. 1996. Integrated control of vanilla stem rot disease (*Fusarium oxysporum* f. sp. *vanillae*). *Proceedings Seminar on Integrated Control of Main Diseases of Industrial Crops*. Bogor, March 13–14, 1996. AARD-JICA. 95–102. (in Indonesian, English Abstract).

Kasim, R. and S. Prayitno. 1993. Effect of application methods and rates of *Trichoderma* spp. on infection of *Sclerotium rolfsii* on vanilla cuttings. *Journal of Spice and Medicinal Crops* II (1):1–4.

Lealy, C.L.A. 1970. The role of fungi and cultural factors in the etiology of root rot of vanilla *V. fragrans* (Salisb) ames. Uganda and *East African Agriculture Journal* 36:207–210.

Magala, W.C. 2008. High value crop production ecofarming: Vanilla. Case in Uganda. Crop production-Necofa. http://www.necofa.org/194.0.html. (Accessed May 17, 2009)

Manohara, D. 1993. *Shoot and Fruit Rot (Phytophthora rot) Diagnostic Manual for Industrial Crop Diseases in Indonesia*. Japan International Cooperation Agency and Research Institute and Medicinal Crops, Indonesia, p. 78.

Matuo, T. 1972. Taxonomic studies of Phytopathogenic Fusaria in Japan. *Reviews of Plant Protection Research* 5:34–45.

Matsumoto, K. 1993. *Southern Sclerotium Rot. Diagnostic Manual for Industrial Crop Diseases in Indonesia*. Japan International Cooperation Agency and Research Institute and Medicinal Crops, Indonesia, p. 78.

Matsumoto, K., M. Tombe, A. Nurawan, and S.B. Nazarudin. 1992. Stem rot of vanilla: Integrated control measures based on the use of fungicides. *Proceedings of the Final Seminar of the Joint-Study Programme RISMC-JICA, ATA-380*. Bogor, Indonesia, 67–72.

Messiaen, C.M. and R. Cassini. 1981. Taxonomy of *Fusarium*. In P.E. Nelson, T.A. Toussoun, and R.J. Cook, eds. *Fusarium: Disease, Biology, and Taxonomy*. Pennsylvania State University Press, University Park, PA, 428–445.

Minoo, D., V.N. Jayakumar, S.S. Veena, J. Vimala, A. Basha, K.V. Saji, K. Nirmal Babu, and K.V. Peter. 2008. Genetic variations and interrelationships in *Vanilla planifolia* and few related species as expressed by RAPD polymorphisms. *Genetic Resources and Crop Evolution* 55:459–470.

Nurawan, A. 1990. Pathogenicity test of *Fusarium oxysporum* in different crops. Research News of Industrial Crops. Bogor, Indonesia. XVI (2):50–52 (in Indonesian, English Summary).

Owino, K. 2008. Vanilla. Available from http://www.worknets.org/wiki.cgi?action=browse&id=Vanilla&dgt=111xdk786 (Accessed May 17, 2009)

Philip, S. 1980. Wilt of *Vanilla planifolia* caused by *Fusarium oxysporum* f. sp. vanillae. *Agricultural Research Journal of Kerala* 18 (1):139–140.

Rachmadiono, S., G. Kartono, Sudjindro and H. Hariadi. 1982. Observation on vanilla disease in Bali. Research News of Industrial Crops. Bogor, Indonesia. VIII (42): 21–24 (in Indonesian, English Summary).

Ratanachurdchai, K. and K. Soytong. 2008. A survey of vanilla diseases. Department of Plant Pest Management Technology. Faculty of Agricultural Technology, King Mongkut's Institute of Technology, Ladkrabang, Bangkok 10520, Thailand. http://www.thaiscience.info/Article%20for%20ThaiScience/Article/2/Ts-2%20survey%20of%20vanilla%20diseases.pdf (Accessed September 29, 2009).

Ruan, X.Y., J.B. Chen, and Y.Y. Zhu. 1998. Renew on vanilla diseases researches. *Journal of Yunnan Agricultural University* 13 (1):139–144 (in Chinese, English Abstract).

Sangchote, S., S. Poonpolgul, R. Sdoodee, M. Kanjanamaneesathian, T. Baothong, and P. Lumyong. 2004. Phytophthora diseases in Thailand. In A. Drenth and D.I. Guest, eds. *Diversity and Management of Phytophthora in Southeast Asia*, 77–82. Australian Centre for International Agricultural Research. Monograph 114.

Snyder, W.C. and H.N. Hansen. 1940. The species concept in *Fusarium. American Journal of Botany* 27 (2):64–67.

Soetono. 1962. *Stem rot disease of Vanilla.* National Congress of Science in Indonesia, Yogyakarta, 24pp. (in Indonesian).

Stier, H.Z. 1984. *Pratylenchus brachyurus* associated with vine-dying of vanilla plants in Tonga. Disease Note. *Plant Disease* 68:628.

Sukamto, M. Tombe, and S. Mugi, 1996. Clove by-products to control stem rot of vanilla. *Proceedings of the Seminar on Integrated Control on Main Diseases off Industrial Crops*. Bogor. March 13–14, 1996 AARD – JICA. 77–86 (in Indonesia, English Abstract).

Taufik, E. and D. Manohara. 1998. *Other Vanilla Diseases*. Research Institute for Spices and Medicinal Crops, Bogor, Indonesia. Monograf No. 4: 96–101 (in Indonesia, English Abstract).

Theis, T. and F.A. Jiménez. 1957. A vanilla hybrid resistant to root rot. *Phytopathology* 47:579–581.

Thomas, J., A.K. Vijayan, and R.S. Bhai. 2002. Vanilla diseases in India and their management. *Indian Journal of Arecanut, Spices and Medicinal Plants* 4:143–149.

Tombe, M. 1993a. *Stem and Foot Rot on Vanilla. Diagnostic Manual for Industrial Crop Diseases in Indonesia*. Japan International Cooperation Agency and Research Institute and Medicinal Crops, Indonesia, 76–77.

Tombe, M. 1993b. *Anthracnose of Vanilla. Diagnostic Manual for Industrial Crop Diseases in Indonesia*. Japan International Cooperation Agency and Research Institute and Medicinal Crops, Indonesia, 78pp.

Tombe, M. 1994. Studies on the stem rot disease of Vanilla (*Vanilla planifolia* Andrews) in Indonesia. Faculty of Agriculture, Hokkaido University, Japan. Ph.D. Thesis (unpublished).

Tombe, M. 2007. Introduction of Vanilla Stem Rot Disease (VSRD) and its control. *Vanilla Malaysian for The World Course. Vanilla Biomatrix* Sdn Bhd., Quality Hotel, Kuala Lumpur, Malaysia, December 14, 2007.

Tombe, M. 2008. Cultivation of Vanilla by using technology of BioFOB. *Vanilla Course*, Rural Development Corporation, Sabah, Malaysia. Beringgis Beach Resort, Kota Kinabalu Sabah, February 28–29, 2008, 13pp.

Tombe, M., K. Kobayashi, Ma'mun, Triantoro, M. Oniki, and K. Matsumoto. 1993. The role of eugenol in disease suppression of stem rot of vanilla. *Annals of Phytopathological Society of Japan* 59: 282 (Abstract).

Tombe, M., K. Kobayashi, and A. Ogoshi. 1994. Vegetative compatibility grouping of *Fusarium oxysporum* f. sp. *vanillae* in Indonesia. *Indonesian Journal of Crop Science* 9:29–39.

Tombe, M., K. Matsumoto, A. Nurawan, Sukamto, and S.B. Nazarudin. 1992a. Strains, morphology. Physiology and ecology of causal fungus and disease damages of vanilla. *Proceedings of the Final Seminar of the Joint Study Programme RISMC-JICA, ATA-380*. Bogor, Indonesia, 50–57.

Tombe, M., K. Matsumoto, A. Nurawan, Sukamto and S.B. Nazarudin. 1992b. Strains, morphology, physiology and ecology of causal fungus and disease damages of vanilla. *Proceedings of the Final Seminar of the Joint Study Programme RISMC-JICA, ATA-380*. Bogor, Indonesia, 50–57.

Tombe, M. and D. Sitepu. 1987. Diseases of vanilla in Indonesia. *Edsus. Littri, Balittro* 3(2):103–108 (in Indonesian).

Tombe, M., D. Sitepu, and S. Mogi. 1997. Present status of biological control research of vanilla stem rot disease in Indonesia. *Proceedings of the Fourth International Workshop on Plant Growth Promoting Rhizobacteria*. Japan-OECD, Workshop, 13–17.

Tombe, M., K. Tsuchiya, A. Nurawan, S.B. Nazarudin, M. Oniki, and K. Matsumoto. 1992c. Experiment on the introduction of biological and culture control of stem rot disease of vanilla. *Industrial Crops Research Journal, Bogor, Indonesia* 4 (2):20–26.

Tsao, P.H. and L. Mu. 1987. Involvement of *Phytopthora* in vanilla root rot (abstract). *Phytopathology* 77:1704.

Tucker, C.M. 1927. Vanilla root rot. *Journal of Agricultural Research* 35 (12):1121–1136.

Venter, S.L., D.J. Theron, P.J. Steyn, D.I. Ferreira, and A. Eicker. 1992. Relationship between vegetative compatibility and pathogenicity of isolates of *Fusarium oxysporum* f. sp. *tuberosi* from potato. *Phytopathology* 82:858–862.

Xia-Hong, H. 2007. Biocontrol of root rot disease in *Vanilla*. Ph.D. thesis, the University of Wolverhampton, UK, 224pp.

9 Bio-Ecology and Control of an Emerging Vanilla Pest, the Scale *Conchaspis angraeci*

Serge Quilici, Agathe Richard, and Kenny Le Roux

CONTENTS

A few years after it was recorded on Reunion Island, a French island in the southwestern part of the Indian Ocean, the scale *Conchaspis angraeci* Cockerell, 1893 was of increasing concern for the vanilla producers on this island because of the damage it caused to this crop. By absorbing the sap of its host plants, this armored scale first leads to the formation of chlorotic spots on the leaves, then to a drying out of the leaf, and eventually, in extreme cases, to the death of the vine itself. As this pest has a wide geographical distribution over the recent years, we review its taxonomic position, its biology, ecology, natural enemies, and control.

TAXONOMY AND MORPHOLOGY

The Angraecum scale or vanilla scale, *C. angraeci*, belongs to the family Conchaspididae (Hemiptera: Coccoidea) whose members closely resemble Diaspididae, with which they have long been confused. The family Conchaspididae is now regarded as a separate group within the Coccoidea. Female species in this family have a feature in common with members of the family Diaspididae—they build an armor that is independent from the rest of the insect body and their last abdominal segments are fused to form a pygidium, which is involved in the secretion of the armor. The genus *Conchaspis* contains 22 species (BenDov, 1981). One of the main morphological characteristics of *C. angraeci* female species is the aspect of the armor, circular in outline and conical in shape. The cone presents 6–8 ridges, radiating from the apex but ending without reaching the margin (Hamon, 1979). The armor is usually white, but may look grayish or yellowish in some cases.

GEOGRAPHICAL DISTRIBUTION

This species, common in tropical and subtropical areas, was first described from Jamaica (Cockerell, 1893). For a long time, it was mainly restricted to the Nearctic and Neotropical regions, whereas it is known in Mexico, the United States, Bermuda, Barbados, Cuba, Brazil, Ecuador, Nicaragua, Panama, Peru, Puerto-Rico, Surinam, Trinidad and Tobago, and Venezuela (BenDov, 1981). It was only in 1973 that it was also recorded in the Afrotropical region, in Angola and Cameroon, on *Acalypha* spp. (BenDov, 1974). It has since been found in Hawaii (Beardsley, 1993), in Australia (BenDov et al., 1985), in the Fiji Islands (Williams and Watson, 1990), in Malaysia (Takagi, 1992, 1997), and in French Polynesia (Quilici, 2002). In the southwestern Indian Ocean, it was first recorded in Reunion Island in 1997 (Richard et al., 2003), then in Mayotte, a French island belonging to the Comoros archipelago (Richard, 2003), and more recently in Mauritius in 2007 (S. Quilici and J.C. Streito, unpubl. data).

HOST PLANTS

The species was described from material collected as *Angraecum eburneum* Bory var. *virens* (Orchidaceae) (Cockerell, 1893) but appears polyphagous and is recorded on numerous host plants belonging to various families among which the Euphorbiaceae and Orchidaceae are the most common (BenDov, 1981; Hamon, 1979; Mamet, 1954). In Florida, its host plant has a wide range and includes about 30 species of woody plants and orchids such as *Plumeria* sp. (Apocynaceae), *Stephanotis* sp. (Asclepiadaceae), *Rhododendron* sp. (Ericaceae), *Acalypha wilkesiana* Müll., *Codiaeum variegatum* (L.), *Euphorbia lactea* Haw. cv. "cristata" (Euphorbiaceae), *Flacourtia indica* (Burm. F.) Merrill (Flacourtiaceae), *Hibiscus rosa-sinensis* L. (Malvaceae), *Ficus sagitata* Vahl (Moraceae), *Bougainvillea* sp. (Nyctaginaceae), *Angraecum superbum* Thouars, *Epidendrum secundum* Jacq. (Orchidaceae), *Pittosporum tobira* Ait. (Pittosporaceae), *Coccoloba uvifera* (L.) (Polygonaceae) … (Hamon, 1979). In Reunion Island, apart from vanilla, it was also found on an exotic vine belonging to the Asclepiadaceae: *Hoya bella* Hooker (Richard et al., 2003), on *Angraecum eburneum* Bory and *Bulbophyllum prismaticum* Thouars (Orchidaceae). In Mauritius, very high populations were observed on *Euphorbia lactea* Haw. (S. Quilici, unpubl. data).

On vanilla, it is recorded on *Vanilla planifolia* (G. Jackson) in Fiji (Williams and Watson, 1990), Reunion Island (Richard et al., 2003), Mauritius (S. Quilici, unpubl. data), and Mayotte (Richard, 2003) and also attacks *Vanilla tahitensis* J.W. Moore in French Polynesia and Mauritius (M. Grisoni, pers. comm.). In the island of Mayotte, it has also been observed on *Vanilla humblotii* Rchb. F. (R. Gigant, pers. comm.).

SYMPTOMS AND DAMAGE

On vanilla crops, on which the scale causes damage, it is eurymerous and may develop on leaves, stems, aerial roots, or pods (Richard et al., 2003). With its sucking mouthparts, the scale probably injects a toxin through leaves or stems, which produce chlorotic spots evolving into necrosis, and leading to the weakening of the vine and eventually its death (Figure 9.1). In Florida, damage leading to plant weakening, has been observed to be severe on Hibiscus, Pittosporum, orchids, and seagrapes (Hamon, 1979).

In Reunion Island, though no precise economic study has been conducted to evaluate the damage caused by this scale, many producers claim that they observed a decrease in their production over the years following the arrival of the scale (Richard et al., 2003). A survey conducted in 2002 showed heavy infestations in the vanilla plots of the eastern area of the island and in 2003, the scale extended its distribution area and damage up to the southeastern part of the island.

In the Comoros archipelago, the scale was detected in 1999 on the French island of Mayotte, where it rapidly spread; a survey conducted in 2007, showed high infestations, with up to 50% of the vines attacked on particular plots, while the scale was not detected in Grande Comore and Moheli, belonging to the Islamic Republic of Comoros (Grisoni and Abdoul-Karime, 2007). In Mauritius, limited damage was observed in 2007 in commercial shade houses in the north of the island (Grisoni, 2007). In French Polynesia, the damage observed in 2002 was apparently limited to localized outbreaks in a few plantations in Moorea and Tahaa (M. Grisoni,

FIGURE 9.1 (See color insert following page 136.) Damage caused by the Angraecum scale: (a) on a vanilla plant; (b) detail of damage on a vanilla leaf.

pers. comm.). Although extensive surveys were conducted since 1998 in the major vanilla-growing areas of Madagascar, *C. angraeci* has not been reported in this country (Alabouvette and Grisoni, 2009).

LIFE-HISTORY TRAITS AND ECOLOGY

Few data are available on the biology of this pest species. Takagi (1992) showed that the females in the genus *Conchaspis* have only three instars (L1, L2, and adult), while males have five instars (L1 mobile then fixed, L2, prepupa, pupa, and adult). In *C. angraeci*, males are unknown and this species probably reproduces mostly by parthenogenesis. The female lays a maximum of 80 eggs under its armor (Figure 9.2). After hatching, the mobile first instar larva (L1), which is the only mobile instar (crawler), quickly gets out of the armor. After its mobile phase, the L1 then settles and starts secreting its own armor. After some growth, it transforms into an L2, which changes into an adult female, beginning to lay eggs at the end of its pre-oviposition period. During rearing trials on young vanilla vines, the duration of the preimaginal cycle appeared to be long (90 days at $T = 25°C$; RH = 70%; photoperiod: L12:D12).

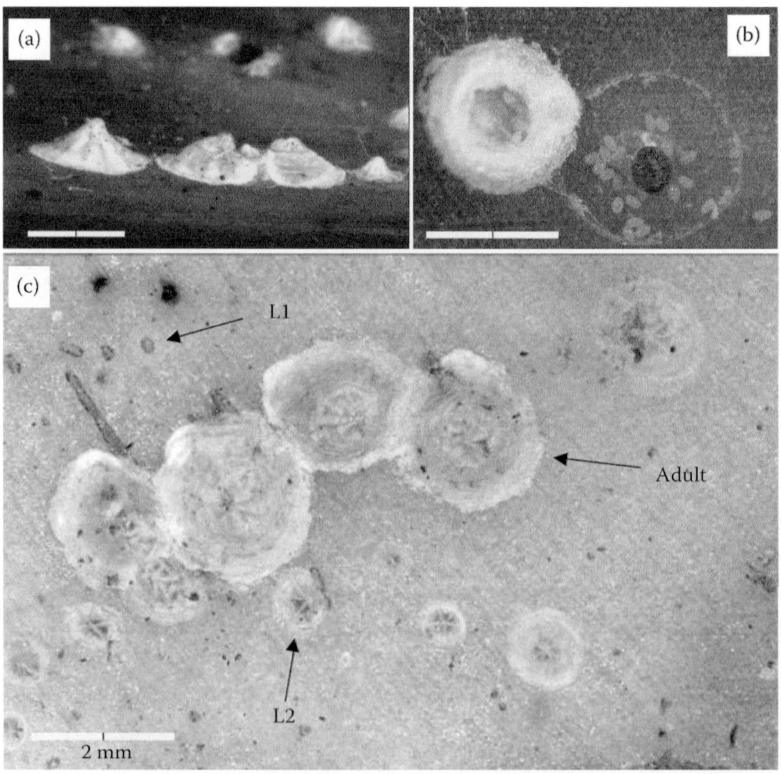

FIGURE 9.2 Different instars of *Conchaspis angraeci*: (a) adult female; (b) female with armor removed, showing eggs; (c) first instar (L1), second instar (L2), and adult female.

In Reunion Island, while vanilla crops are localized in the eastern, the south-eastern and the southern part of the island, surveys carried out in 2001–2003 showed that the species was more prevalent in the eastern and south-eastern areas, which are also the areas with a higher rainfall. A study conducted in four vanilla plots in 2002–2003 showed that larval stages (L1 and L2) were more abundant during the warm season (December to April). The adult stage dominated compared to the other stages, whatever the cultural mode (intensive crop, intercropped with sugarcane, under trees, shade house) or the season.

NATURAL ENEMIES

Records of natural enemies of this scale are recent. The first parasitoid was recorded in Hawaii by Beardsley and Tsuda (1990) who mention that, some ten years after its arrival in 1980, the scale was heavily parasitized by the Aphelinidae *Marietta pulchella* Howard, whose larvae develop in the same way as *Aphytis* spp. do on Diaspididae. Though the species in the genus *Marietta* are most frequently hyperparasitoids, the authors showed that *M. pulchella* does develop as a primary parasitoid in this case, which is the first record for a species in this genus. In Reunion Island, three parasitoid species have been found associated with the populations of the scale: one endoparasitoid, *Encarsia lounsbury* Berlese and Paoli (Aphelinidae), and two ectoparasitoid: *Aphytis africanus* Quednau (Aphelinidae) and *Cephaleta* sp. (Pteromalidae). While the first two species attack L1 and L2, the third one parasitizes the female adults (Richard et al., 2003). Globally, the parasitism rate remained very low, among the studied biotopes and the recorded parasitoids are probably not specific to the vanilla scale.

In Reunion Island, a few predatory species, mostly thrips and mites, are also associated with high levels of population of the scale. Among the five thrips species found, three are known as scale predators: *Aleurodothrips fasciapennis* (Franklin), *Karnyothrips flavipes* (Jones), and *Karnyothrips melaleucus* (Bagnall), belonging to the Phlaeothripinae. Among the mites, two species have been collected, which are known to feed on eggs or mobile young instars of scale insects: *Saniosulus nudus* Summers (Eupalopsellidae) and *Bdella distincta* Baker and Balock (Bdellidae). Other species have also been collected, which are known to have predaceous habits: *Cheletogenes* sp. and *Hemicheyletia wellsi* (Baker) (Cheyletidae), *Bdellodes* sp. and *Spinibedella depressa* (Ewing) (Bdellidae) and *Asca* sp. (Ascidae). In Hawaii, the predatory mite *Cheletogenes ornatus* (Canestrini and Fanzago) has also been found preying on *C. angraeci* on the host Hibiscus (Goff and Conant, 1985). In Reunion Island, an entomopathogenous fungus, *Fusarium coccidicola* P. Henn, has been recorded, which is one of the four known species of *Fusarium* that is pathogenic on scale insects (Richard et al., 2003). Globally, the predation rate appeared low (2.5–6.5%), with a higher rate observed in plots situated under the cover of a forest.

CONTROL METHODS

In Hawaii, biological control solved the problems linked with the heavy infestations of *C. angraeci*: soon after its discovery in 1989, the Aphelinidae *Marietta pulchella* exerted a complete control on the scale, which is now considered to be rare (Beardsley, 1993).

Very few data are available on the results of chemical control trials against this scale. Experiments conducted in 2004 by the Plant Protection Service of Reunion Island showed a high effectiveness of imidacloprid (C.P.: Confidor; Bayer CropScience) and a good effectiveness, though delayed and inferior to that of imidacloprid, is of mineral oil (C.P.: Ovipron; Cerexagri) (SPV Réunion, 2004). Further experiments using mineral oils are required, with applications at different periods of the year, to determine its effectiveness in controlling the scale. As mobile instars are more sensitive to chemical control, such treatments, if not phytotoxic, should preferentially be applied during summer when the first instar is more abundant and active.

The vanilla scale has demonstrated, at least in certain areas, its ability to cause significant damage to vanilla production. Despite a few studies devoted to this scale, our knowledge is limited about the bio-ecology of this pest. Additional studies would be helpful to improve our knowledge on its natural enemies and their impact, as well as on possible control methods.

ACKNOWLEDGMENTS

The authors express their sincere thanks to the taxonomists who identified the parasitoids (G. Delvare, CIRAD, Montpellier, France), the predatory mites (E.A. Ueckermann, PPRI, Pretoria, South Africa), and the thrips (B. Michel, CIRAD, Montpellier, France).

REFERENCES

Alabouvette, C. and M. Grisoni. 2009. Mission d'expertise en phytopathologie du vanillier dans la SAVA. Rapport provisoire CTHT/FED-Stabex, Toamasina (Madagascar), 35pp. (+ Annexes).

Beardsley, J.W. 1993. Exotic terrestrial arthropods in the Hawaiian islands: Origins and impacts. *Micronesica Suppl.* 4:11–15.

Beardsley, J.W. and D.M. Tsuda. 1990. *Marietta pulchella* (Howard) (Hymenoptera: Aphelinidae), a primary parasite of *Conchaspis angraeci* Cockerell (Homoptera: Conchaspididae). *Proc. Hawaiian Entomol. Soc.* 30:151–153.

Ben Dov, Y. 1974. On the species of Conchaspididae (Homoptera: Coccoidea) from Africa and Madagascar with description of a new species. *Rev. Zool. Afr.* 88 (2):363–373.

Ben Dov, Y. 1981. A catalogue of the Conchaspididae (Insecta: Homoptera: Coccoïdea) of the world. *Annales Soc. Entomol. France* 17 (2):143–156.

Ben Dov, Y., Y. Shanuni, M. Mizrahi, M. Weiss, Y. Feldman, and M. Feldmann. 1985. Interceptions of scale insects (Coccoïdea) at port of entry in Israël. *Hassadeh* 85:2120–2121, 2128.

Cockerell, T.D.A. 1893. Coccoïdea or scale insects. II. *Bull. Bot. Dept. Jamaica* 40: 9.

Goff, M.L. and P. Conant. 1985. Mites feeding on scale insects. *Proc. Hawaiian Entomol. Soc.* 25:4.

Grisoni, M. 2007. Rapport de mission à l'île Maurice pour la Compagnie d'Exploitation de Vanille Naturelle Limitée (CEVNL), du 5 au 6 mai 2008. *Doc. CIRAD*, 16pp.

Grisoni, M. and A.L. Abdoul-Karime. 2007. Epidémiosurveillance de Mayotte et des Comores. Rapport technique. Décembre 2007. *Doc. CIRAD - Délégation de Mayotte*, 34pp.

Hamon, A.B. 1979. Angraecum scale, *Conchaspis angraeci* Cockerell (Homoptera: Coccoidea: Conchaspididae). Entomology Circular, Division of Plant Industry, Florida Department of Agriculture and Consumer Service, 2pp.

Mamet, R. 1954. A monograph of the Conchaspididae Green (Hemiptera: Coccoidea). *Transactions of the Royal Entomological Society of London* 105:189–239.

Quilici, S. 2002. Rapport de mission en Polynésie du 29/10 au 12/11/2002. *Doc. CIRAD Réunion/3P*, 24pp.

Richard, A., C. Rivière, P. Ryckewaert, B. Come, and S. Quilici. 2003. Un nouveau ravageur de la vanille, la cochenille *Conchaspis angraeci.* Etude préliminaire à la mise en place d'une stratégie de lutte raisonnée à La Réunion. *Phytoma – La Défense des Végétaux* 562:36–39.

Richard, A. 2003. Rapport de mission à Mayotte du 14 au 18/03/2003. *Doc. CIRAD Réunion/3P*, 9pp.

SPV Réunion. 2004. Expérimentation sur la cochenille de la vanille. *Conchaspis angraeci* Cockerell. Ministère de l'Agriculture et de la Pêche, Direction de l'Agriculture et de la Forêt de La Réunion, 11pp.

Takagi, S. 1992. A contribution to Conchaspidid systematics (Homoptera: Coccoidea). *Insecta Matsumurana* 46:1–71.

Takagi, S. 1997. Further material of *Conchaspis* from Southeast Asia (Homoptera: Coccoidea: Conchaspididae). *Insecta Matsumurana* 53:27–79.

Williams, D.J. and G.W. Watson. 1990. The scale insects of the tropical south Pacific region. Part 3. The soft scales (Coccidae) and other families. *CAB International Institute of Entomology Ed*, 205–207.

10 Anatomy and Biochemistry of Vanilla Bean Development (*Vanilla planifolia* G. Jackson)

Fabienne Lapeyre-Montes, Geneviève Conéjéro, Jean-Luc Verdeil, and Eric Odoux

CONTENTS

INTRODUCTION

The fruit of the vanilla plant is commonly known as a "bean" being similar in shape with those of legumes, which stems from the evolution of a single carpel. In fact, from a botanical viewpoint, it is not actually a bean as it results from the evolution of three fused carpels, but a capsule (dry, dehiscent fruit resulting from the evolution of several carpels). In the orchid family, which includes the genus *Vanilla*, the capsule is composed of three valves that are delimited by six dehiscence splits (two per carpel) situated at either end of the placentas (paraplacental dehiscence, Figure 10.1a) (Dupont and Guignard, 2007).

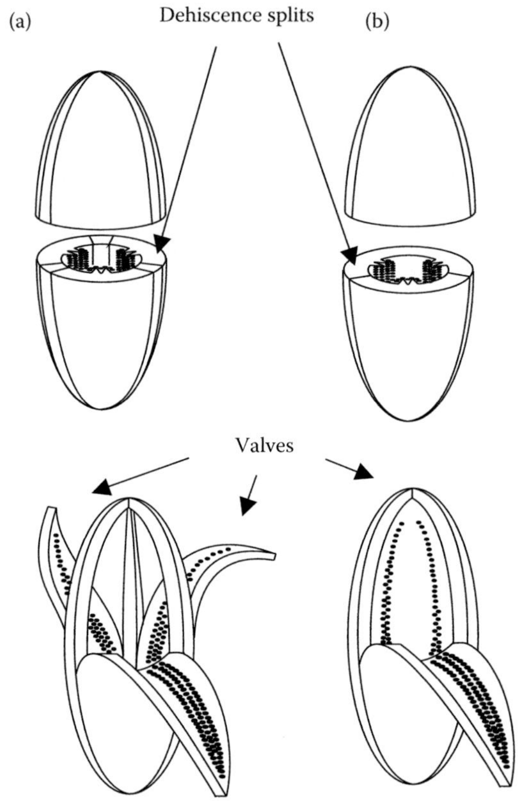

FIGURE 10.1 Paraplacental dehiscence: (a) general case of an orchid family capsule (six dehiscence splits resulting in three valves); (b) specific case of a *V. planifolia* capsule (two dehiscence splits resulting in two valves).

The members of the subfamily *Vanilloideae*, however, have certain characteristics that are unusual in the orchid family (Cameron and Chase, 2000), including those relating to dehiscence. In the case of the genus *Vanilla*, the capsule has only two dehiscence splits, corresponding to the central axis of two of the three fused carpels. At maturity, the capsule opens along the two dehiscence splits (for *Vanilla planifolia* and *Vanilla pompona*), thereby showing two valves (Figure 10.1b); this is dorsal dehiscence. Some species may even be indehiscent (for *Vanilla tahitensis*).

The vanilla capsule is unilocular (a single central cavity in the ovary that contains all the seeds). Later in the chapter we no longer use the botanical word "capsule," but the more commonly used "bean" or "pod."

For *V. planifolia* G. Jackson (which, unless otherwise indicated, is the only species referred to in the rest of the chapter), the flowers of the vanilla plant are grouped in inflorescences (Figure 10.2) that resemble a cluster, as the inferior ovary simulates a floral pedicel that is absent.

After pollination, the ovary develops very rapidly, doubling in length in a few days, and curves downward (Figure 10.2). Fertilization occurs 1.5–2 months later

FIGURE 10.2 *V. planifolia* inflorescence. Before pollination, ovaries face upward with their flower, and then curve downward.

(Childers and Cibes, 1948; Roux, 1954). Each fertilized ovary produces a bean. The bean reaches its full size and weight between 10 and 15 weeks after pollination (Gregory et al., 1967; Brodélius, 1994; Havkin-Frenkel et al., 1999) (Figure 10.3). As the bean matures, the moisture content decreases from around 90–92% to 82–85% toward the harvesting time (Shankaracharya and Natarajan, 1973; Brodélius, 1994; Ranadive, 1994), in other words, around 8–9 months after pollination (Sreekrishna Bhat and Sudharshan, 2002).

Mature beans are long, green, and curved at the end of the peduncle. They are around 15 cm long and may reach a width up to 15 mm, and weigh around 10–15 g. These figures are only orders of magnitude and may vary considerably depending on genetic factors, physiology of the plant, agronomic or environmental conditions, and so on.

After this period of maturation, whether harvested or not, the fruit becomes pale green in color and begins to turn yellow from its floral tip; the dehiscent fruits (the proportion varies according to the species, and also within the same species) split into two from the floral tip. The beans then darken, some very markedly, again from the floral tip. They then lose their initial turgidity and become completely flexible, marking the senescence of the fruit.

It appears that the fruit releases ethylene (Ducamp et al., 2000); however, some authors have not been able to measure this during the maturation of the bean (Havkin-Frenkel et al., 2005). However, the fruit's sensitivity to ethylene has been clearly observed during different scientific works; in particular, ethylene increases the rate of dehiscence in beans (Balls and Arana, 1941; Arana, 1944; Havkin-Frenkel et al., 2005).

It would be worthwhile conducting some in-depth research to determine whether or not the fruit is climacteric and, if so, at what stage of its development the respiratory climacteric takes place. These data would be helpful in controlling the harvesting stage for beans and their conservation before curing in a better way.

FIGURE 10.3 Immature vanilla beans.

There are very little data on the composition of mature green beans. Table 10.1 provides values obtained by Garros-Patin and Hahn (1954) for beans from Madagascar that were analyzed in 1950.

The results of the analyses conducted by CIRAD on mature green beans (usual harvesting time) from Madagascar are presented in Table 10.2. These results are not intended to be representative of the average composition of mature green beans from the vanilla plant, but should be taken only as indications. They are, moreover, relatively consistent with the results obtained by Garros-Patin and Hahn.

Fibers (hemicelluloses, cellulose, and lignin) constitute nearly 8% of the fresh weight, which is high when compared with fruits and vegetables (between 1 and 4%).

Sugars are mainly composed of sucrose and their total content is comparable with that found in many vegetables (<4%) and far less than that found in most fruits (12%).

Proteins are present in relatively small quantities, which are commonly observed in most fruits and vegetables (<1%). Among these proteins, we note the presence of exceptionally high glucosidase activity (on an average, around 1000 nkatal/g of fresh fruit) closely linked to the aromatic development of vanilla; this will be discussed in detail at the end of the chapter.

Lipids represent 2% of the fresh weight, which is high compared to many fruits and vegetables (excluding oleaginous fruits), where the content commonly observed is

TABLE 10.1
Typical Composition of a Mature Green Bean

	Fresh Weight %		Dry Weight %	
	A	B	A	B
Water	79.6	75		
Ash	0.75	0.97	3.68	3.88
Cellulose	0.79	2.65	3.87	10.60
Reducing sugars	1.42	1.01	6.96	4.04
Nonreducing sugars	3.03	2.45	14.85	9.72
Nonnitrogenous substances	10.85	14.41	53.19	57.7
Ether extract	1.58	2.14	7.74	8.56
Proteins	1.75	1.37	8.58	5.50
Acidity	0.23		1.13	

Source: Data from Garros-Patin, J. and Hahn, J. 1954. In: G. Bouriquet, ed. *Le vanillier et la vanille dans le monde*. Paul Lechevalier, Paris, 559–615.
A and B are two different samples of mature green beans from Madagascar.

around 0.2–0.4%. Unsurprisingly, it is known that during the "killing" of beans, a considerable oily phase appears on the surface of water. However, a great deal of research study has been carried out on this lipid fraction (Ramaroson-Raonizafinimanana et al., 1997, 1998a, 1998b, 1999, 2000; Maestro et al., 2007).

TABLE 10.2
Typical Composition of a Mature Green Bean (Normal Harvest Stage)

	(g/100 g FW)	(g/100 g DW)	Percentage of Each Compound in Relation to Total		
Water	83.0	0			
Fibers	7.6	45	Lignin	Cellulose	Hemicelluloses
			62	27	11
Sugars	1.7	10	Sucrose	Glucose	Fructose
			80	15	5
Lipids[a]	2.0	12	C18:2	C18:1ω9	C16
			54	10	10
Proteins	0.5	3			
Organic acids	0.9	5	Citric acid	Malic acid	
			50	30	
Mineral elements	1.7	10	K	Ca	Mg
			28	10	2
Glucovanillin	1.7	10			

[a] Percentages of fatty acids are in relation to saponifiable fraction (not to total lipid fraction).

Organic acids are mainly represented by malate and citrate (80% of all organic acids), and their total content is similar to that found in low-acid fruits and vegetables.

The main mineral elements are potassium and calcium; their respective contents (470 mg and 170 mg) are high when compared with most fruits and vegetables.

Finally, the glucovanillin content represents about 1.7% of the fresh weight of the green fruit (and sometimes more), which is really exceptional and characteristic of *V. planifolia*, and the reason for its commercial value. Glucovanillin is dealt with in greater detail in the final part of the chapter.

These different data on the basic composition of the green fruit were confirmed on the whole in a recent publication (Odoux and Brillouet, 2009); they nevertheless need to be supplemented by more systematic studies, which are currently absent.

ANATOMY AND MORPHOGENESIS OF VANILLA BEAN

Morphology, Anatomy, and Histology of Mature Vanilla Bean

Part of the information presented below is largely taken from the following references: De Lanessan, 1886; Villiers et al., 1909; Swamy, 1947; Roux, 1954; Odoux et al., 2003; and French, 2005.

The mature green vanilla bean has a roughly triangular transverse section with a central cavity containing numerous black seeds (Figure 10.4). From an anatomical and histological viewpoint, from the outer to the inner part of the fruit we find the epicarp, the mesocarp, and the endocarp (Figure 10.5d).

The epicarp is made up of a layer of contiguous cells, which vary in length, are polygonal in shape, and run parallel to the long axis of the bean. These thick-walled cells, which are stained in intense pink by Schiff reagent (Figure 10.5d) due to their biochemical composition (cellulose, hemicelluloses, and pectic substances),

FIGURE 10.4 Cross section of mature vanilla bean.

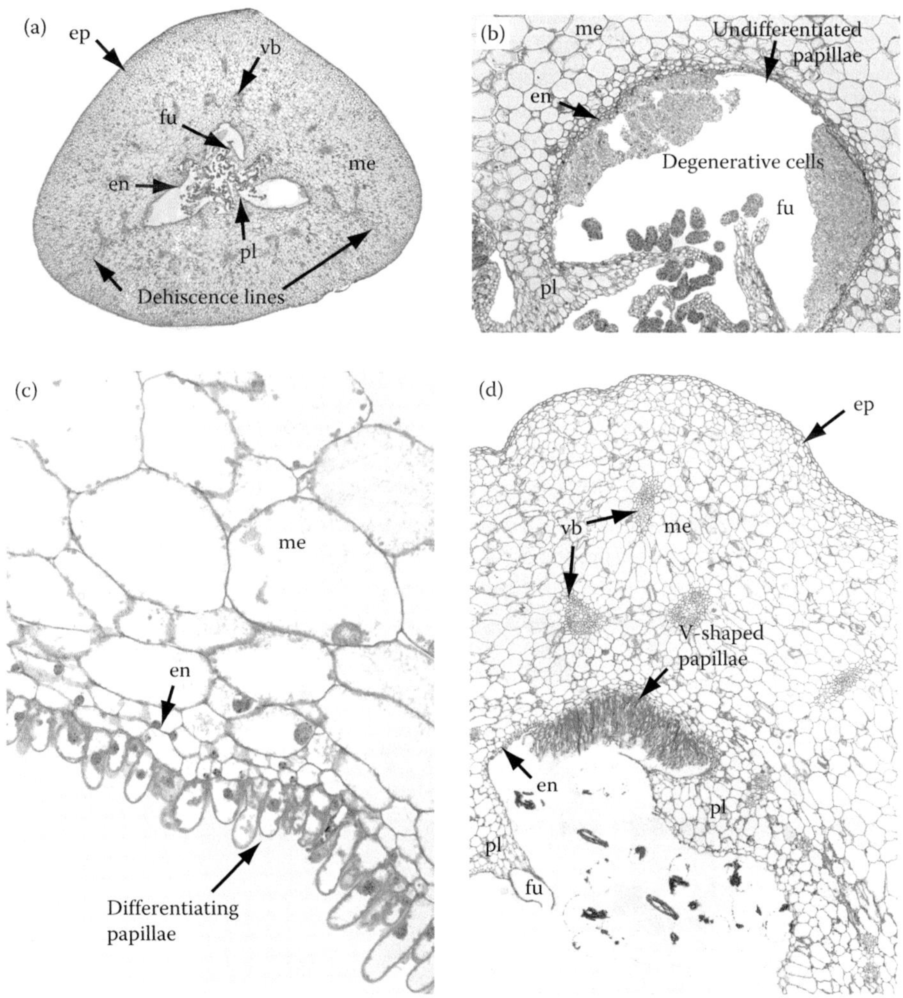

FIGURE 10.5 **(See color insert following page 136.)** From the flower to the mature bean. Cross sections (3 μm) of a vanilla bean embedded in Technovit 7100 resin, at different stages after staining with periodic acid-Schiff (PAS)–Naphthol Blue Black: (a) 9 days after pollination (dap); (b) 14 dap; (c) 60 dap; (d) 8 months after pollination. (Data from Odoux et al., 2003. *Annals of Botany* 92: 437–444.) The walls and the storage sugars are stained in pink; the proteins in blue. en, endocarp; ep, epicarp; fu, funicle; me, mesocarp; pl, placenta; s, seed; vb, vascular bundle.

differentiate into a thick cuticle on the outer part of the fruit. The layer of the cells that forms the epicarp provides a protective layer for the bean.

The mesocarp makes up the majority of the fruit's volume (Figure 10.5d). It consists of parenchyma cells. It is composed of vacuolated cells that increase in size from the epicarp or endocarp toward the central part of the mesocarp, where their size may reach 300 μm. This considerable increase in size appears to be accompanied by an increase in ploidy through endoreplication (S. Brown, pers. comm.).

The mesocarp is vascularized. There are three groups of three triangularly arranged vascular bundles (Figure 10.5d). They mark the center of each carpel and, from an evolutionary viewpoint, represent the main vascular bundle of the macrophylla. Among these triangularly arranged groups of bundles, we note the presence of three additional vascular bundles located at the center of the mesocarp, midway between the epicarp and the endocarp. The vascular bundles of the bean are of the closed collateral type, as with most monocotyledons.

Inside the mesocarp, across two of the three groups of three vascular bundles, a radial layer of specialized cells can be observed on cross histological bean sections (Figure 10.5a). The cells that make up this layer are aligned along the ray of the bean and contain carbohydrate reserves in the form of starch granules. These two layers that radiate out from the inner to the outer part of the mesocarp mark the location of the two future dehiscence lines (Figure 10.5a).

The endocarp is made up of one or two layers of small cells that cover the inside of the fruit's cavity (Figure 10.5c).

At the center of the carpellary leaf are found specialized cells, the papillae. The papillae are intensely stained by Schiff reagent (Figure 10.5d). In agreement with the previous observations (De Lanessan, 1886), the papillae are also intensely stained by Nile Red (Figure 10.6), indicating the presence of high concentration of storage lipids. Some proteinaceous material, stained greenish-blue by Naphthol Blue Black, can be seen in the papillae and in the central cavity in the immediate vicinity of the apical ends of the papillae (Figures 10.5c and d).

Each side of the pod bears a placenta composed of four to five layers of parenchyma cells and covered by an epidermis. The placenta is divided into two longitudinal

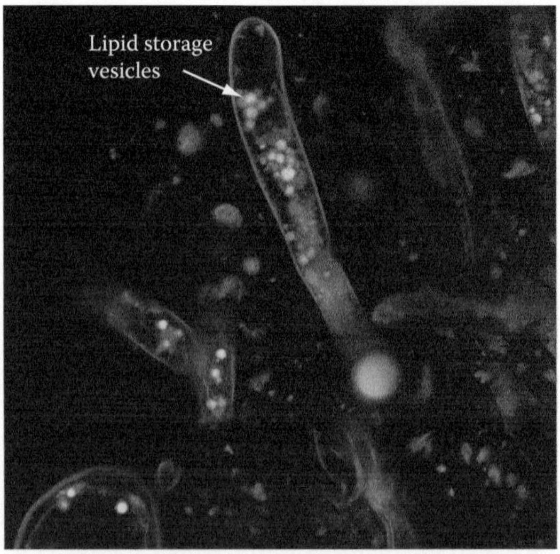

FIGURE 10.6 (See color insert following page 136.) Visualization of lipid storage in the papillae of a mature vanilla bean after Nile Red staining (imaging with confocal microscope Zeiss 510 Meta, laser 488 nm and 405 nm, yellow: Nile Red staining, blue: autofluorescence of walls and papillae).

placental laminae-bearing funicles, made up of three or four layers of cells (Figure 10.5b), to which seeds are attached. On cross-sectioning, each pair of placental laminae appears as finger-shaped lobes bent inside the central cavity (Figure 10.5d). The cells of the epidermis of the placental laminae contain numerous lipid storage vesicles, as in the papillae.

At maturity, the cavity of the fruit contains numerous small seeds (0.2 μm on average) that are oblong in shape with a dark-colored integument. Each seed is attached to a long, narrow funicle. The seeds are held in mucilage that was found to be essentially polysaccharidic in nature (Odoux and Brillouet, 2009); remnants of this mucilage stained in light pink are shown in Figure 10.5d around the holes previously occupied by the seeds.

Vanilla Bean Ontogenesis: From the Flower to the Mature Pod

The vanilla flowers, in groups of 10 or 15, form small bunches at the leaf axil. White, greenish, or pale yellow in color, they have the typical structure of orchid flowers, the most evolved of all the flowers in the plant kingdom. The perianth of these flowers is made up of three sepals and three petals. The lowest petal of the flower, the lip, is usually large, and is spurred. Under the perianth is a very long ovary, which ends with a short pedicel attaching the flower to the inflorescence axis (Figure 10.2). Before pollination, the vanilla ovary is far from being fully developed. After pollination—natural pollinators are not very well known (Lubinsky et al., 2006)—the perianth withers and falls off (Figure 10.2), while the wall of the inferior ovary progressively evolves to form the fruit's pericarp (capsule), and the ovules inside the cavity of the ovary develop into seeds. The first sign of the ovary developing into a fruit is a considerable and rapid increase in its size. From an anatomical and histocytological viewpoint, the most spectacular change concerns the inner part of the ovary. After fertilization, a polarized elongation of the endocarp cells toward the cavity of the ovary can be observed. These cells will develop into secretory trichomes, the papillae. At 9 or 14 days after pollination, the papillae remain undifferentiated (Figures 10.5a and b). At this stage, the upper part of the central cavity of the pollinated ovary contains a tissue composed of degenerative cells deeply stained in pink by Schiff reagent (Figure 10.5b). This tissue could correspond to the tissues termed "transmitting tissues" by Arber (1937), a parenchyma that provides a nutrient substrate, which aids the pollen tube to grow through the style and inside the ovary cavity (Figures 10.5a and b).

The differentiation zone of the papillae is not continuous; it is located at the center of the carpel leaf, in the pericarp zone situated under the three triangularly arranged vascular bundles (Figure 10.5b). Two months after pollination, the papilla cells begin their elongation and differentiation. At this stage, their length can reach 20 μm (Figure 10.5c). At maturity, the papillae are around 200 μm in length (Figure 10.5d).

Under ultraviolet light, the papillae autofluoresce in white (Figure 10.7a). The cell wall of the mature papillae, which thickens in an uneven manner, undergoes lysis in its distal extremity, facilitating the secretion of its content into the cavity of the capsule (Figure 10.5d).

A white fluorescent substance (Figures 10.7b and 10.8) can also be observed, which surrounds the seeds and partially fills the bean cavity. This substance—called

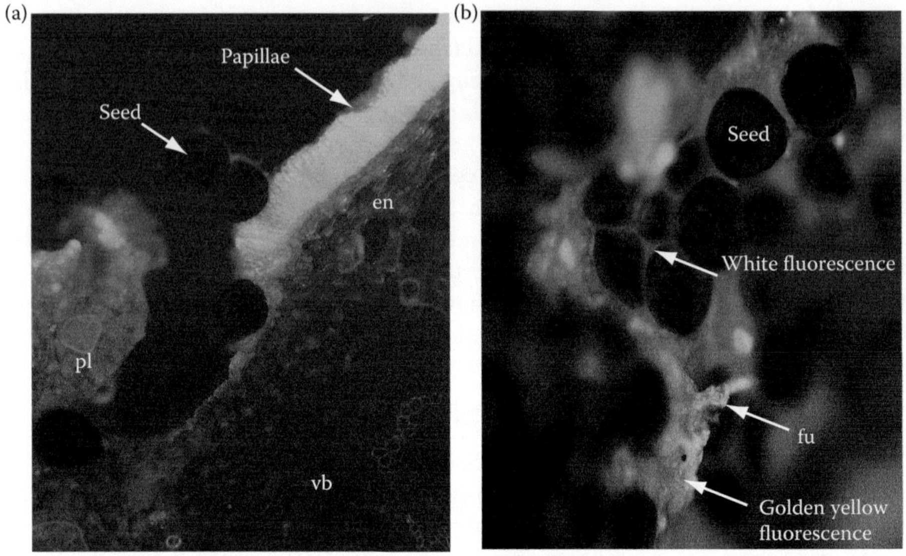

FIGURE 10.7 (See color insert following page 136.) Fresh cross sections (100 µm) of mature vanilla bean (8 months after pollination) observed with epifluorescence microscope (Leica DM6000, filter A: 340–380 nm excitation, 425–800 emission). (a) a general view of placenta and papillae; (b) magnification of funicle, seeds, and "matrix." en, Endocarp; fu, funicle; pl, placenta; vb, vascular bundle.

the "matrix" by French (2005)—does not appear to be of cellular origin (i.e., the extremity of funicles), but rather appears to be amorphous and different from the polysaccharidic mucilage.

The placentas (especially the placental lamina extremity and funicles) also autofluoresce (golden yellow fluorescence, Figures 10.7a and b).

In the mesocarp, white fluorescent globules can be observed in the cells, which may correspond to polyphenols. The lignified tissues of the vascular bundles (xylem and sclerenchyma fibers) emit a bluish autofluorescence linked to the lignin contained in their walls (Figure 10.7a).

Ovule Ontogenesis: From the Ovule to the Seed

In vanilla plants, there is a considerable interval between microsporogenesis, which produces the pollen, and macrosporogenesis, which results in the embryo sac. Surprisingly, the differentiation of the ovules at the top of the placentas mainly occurs after pollination. At two days after pollination, the ovules remain undifferentiated (Figure 10.9a). The placental laminae branch out into a large number of funicles that constitute the placentas (Figure 10.9a). They are made up of four layers of cells that gradually vacuolate from the base (the placental lamina side) to the top. The end of each placenta contains a pool of meristematic cells (actively dividing cells that have dense cytoplasm with a centrally positioned nucleus) (Figure 10.9a). It is this meristematic end that assures the growth of the placenta, and later the shift of the ovules

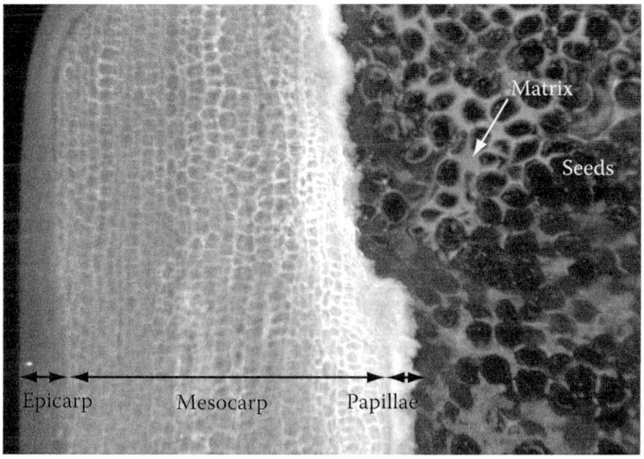

FIGURE 10.8 Longitudinal view of a mature bean opened in one of the three corners (in the papillae area) with a razor blade and observed with stereomicroscope Zeiss Lumar V12 [white fluorescence of walls (in the mesocarp), papillae and "matrix" = autofluorescence through UV excitation].

into the terminal position. This is clear at 15 days after pollination (Figure 10.9b). The outer integument and the inner integument of the ovule differentiate almost simultaneously with the individualization of the spore mother cell, which will undergo meiosis (Figure 10.9b). The outer integument is made up of four layers of cells and up to eight at its base. This characteristic differs from the other orchids, whose inner integument is made up of a single layer of cells. The inner integument and the outer integument of the ovule are not fused and there is a gap between the two.

The nucellus shows a considerable development in relation to other orchids; it persists in the form of several cells at the base of the seeds when they are disseminated.

Most commonly in orchids, endosperm development is limited to 10 nuclei on an average (up to a maximum of 12). This endosperm is rapidly digested, from the first divisions of the zygote. Unlike the endosperm, the nucellus persists until the embryo development ends, particularly at the chalaza, when the seed is mature. The zygote is surrounded by a thickened outer cell wall and has a central star-shaped nucleus with a single nucleolus stained in black by Naphthol Blue Black (Figure 10.9c). The embryo, which ceases to develop very early on (just after the globular stage but before the torpedo stage), accumulates lipid and protein reserves in the form of aleurone grains that are intensely stained black by Naphthol Blue Black (Figure 10.9d).

The vanilla seed coat is the result of the evolution of the outer integument and the inner integument of the ovule. The outermost layer of cells in the outer integument of the ovule becomes sclerous (Figure 10.9d). After fertilization, these cells undergo a polarized elongation following a perpendicular axis at the surface of the ovule. They accumulate brown compounds along their wall, which gradually make them opaque and give the seed coat a reticulated ornamentation. This change differs from those generally observed in other orchids, whose seed coats are finer and translucent.

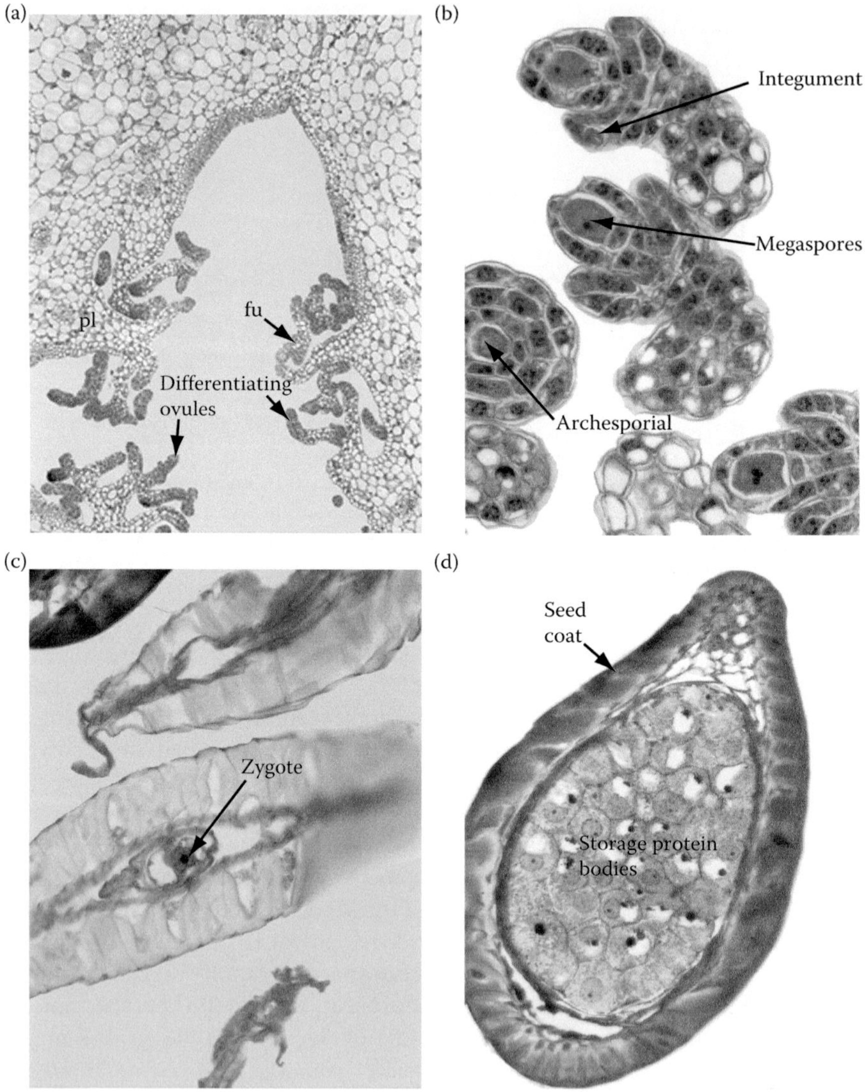

FIGURE 10.9 **(See color insert following page 136.)** From the ovule to the seed: Cross sections (3 μm) of vanilla beans at different stages of development, embedded in Technovit 7100 resin after staining with PAS–Naphthol Blue Black. (a) 2 dap; (b) 15 dap; (c) 20 dap; (d) 200 dap. The walls and the storage sugars are stained in pink, the proteins in blue. fu: Funicle; pl: placenta.

β-GLUCOSIDASE AND GLUCOVANILLIN METABOLISM IN THE VANILLA BEAN

As noted in the introduction, one of the most important characteristics of *V. planifolia* is that its glucosidase activity and its glucosylated precursor content (especially its glucovanillin content) are all exceptionally high. Inasmuch as the aromatic quality of the vanilla is closely linked to the hydrolysis of these glucosylated precursors by the β-glucosidase(s) present in the bean (see also Chapters 11 and 12), various researches have been carried out to determine their accumulation and biosynthesis sites in the fruit.

Tissue and Cellular Localization of Glucosidase Activity and Glucovanillin in the Mature Vanilla Bean

Finally, very little research has been conducted to try to identify the parts of the fruit that contain the glucoside precursors of the aroma components and the glucosidase activity.

However, for a long time, De Lanessan (1886) had implicitly suggested the hypothesis that the aroma precursors and the glucosidase activity occur in the central placental region, because he observed that only this part of the fruit had a characteristic smell after the bean had been cut in fine longitudinal slices from the external part toward the internal part.

However, Arana (1943) and Jones and Vicente (1949) found that most of the glucovanillin (60–80%) was present in the fleshy part of the bean (external mesocarp), and the rest was present in the internal placenta, whereas the enzymatic activity occurs exclusively in the external part of the bean (Arana, 1943) (Figure 10.10a). Arana concludes that the glucovanillin in the internal part of the fruit has to spread to the external part where the enzyme is found or vice versa, in order to be hydrolyzed during vanilla

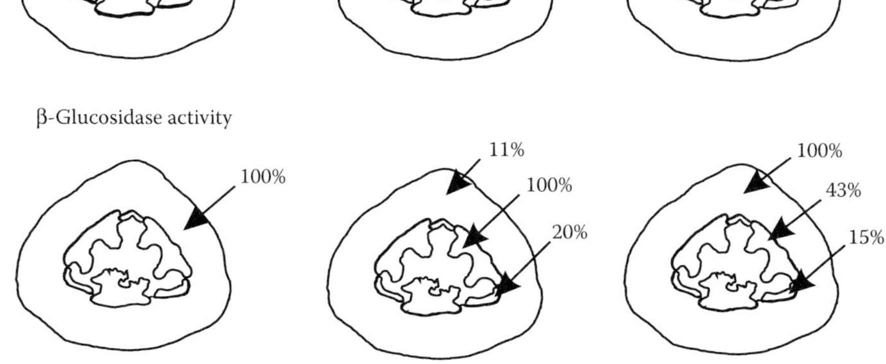

FIGURE 10.10 Tissue localization of glucovanillin and β-glucosidase activity in vanilla bean according to (a) Arana (1943) and Jones and Vicente (1949), (b) Odoux et al. (2003), (c) Joel et al. (2003) and Havkin-Frenkel et al. (2005). (Data from Odoux, E., *Fruits*. 61, 171–184, 2006.)

curing or when the fruit matures on the vine. This hypothesis has been maintained by all the authors who have published work on vanilla during the past 60 years.

Odoux et al. (2003), on the other hand, showed that glucovanillin was found only in the internal part of the fruit and that it is mainly present in the placentas and, to a lesser extent, in the papillae (Figure 10.10b). In a more detailed study, Odoux and Brillouet (2009) found that, given the mass ratios of the different tissues and of their respective glucovanillin contents, 92.2% of glucovanillin was found in the placentas, compared to 7% in the papillae and 0.8% in the mesocarp. They found no glucovanillin in the intralocular space around the seeds, except as traces (which could be artifacts).

These authors (Odoux et al., 2003) found that glucosidase activity was much higher in the placentas than in the mesocarps or the papillae (expressed as total activity per mass unit of fresh tissue); the distribution of β-glucosidase activity expressed as a percentage of the maximum value was found as follows (Odoux and Havkin-Frenkel, 2005): 11% in the mesocarp, 100% in the placentas, and 20% in the papillae (Figure 10.10b); in other words, there is a near-perfect superposition between the distribution of glucovanillin and the enzymatic activity. As a result, the enzyme and the glucovanillin do not need to spread in the fruit tissues for hydrolysis to occur (see also Chapter 11).

Other research studies (Joel et al., 2003; Havkin-Frenkel et al., 2005) confirmed that glucovanillin was present in the white, inner part of the fruit (placentas and papillae). They also suggested the presence of glucovanillin in the intralocular space (Figure 10.10c), a result obtained by staining with catechin-HCl, after its biosynthesis (see below) and excretion by the papillae.

However, Havkin-Frenkel et al. (2005) found a decreasing gradient of enzymatic activity (expressed as specific activity) from the external part toward the internal placental region. The distribution of β-glucosidase activity expressed as a percentage of the maximum value was found as follows (Odoux and Havkin-Frenkel, 2005): 100% in the green outer fruit tissue, 43% in the placental tissue, and 15% in the hair cells (Figure 10.10c). Havkin-Frenkel et al. (2005), whose results are diametrically opposed to those of Arana (1943), also conclude that glucovanillin or the enzyme must spread through the bean tissues for its complete hydrolysis.

It is important to note that the results obtained by Havkin-Frenkel et al. (2005) and Odoux et al. (2003), concerning tissue localization of enzymatic activity, are not necessarily contradictory. Indeed, the specific activity is the ratio between the total activity and the protein content, and this protein content is much higher in the placentas than in the mesocarps (Odoux and Brillouet, 2009). In such a study, expressing enzymatic activity as specific activity—as Havkin-Frenkel et al. (2005) did—is not useful and may lead to incorrect interpretations.

At cellular level, the glucosidase activity is located in the cytoplasm or the apoplasm (Figure 10.11). However, it is neither vacuolar nor parietal (Odoux et al., 2003). Glucovanillin was not positively present, but different considerations (concentration commonly around 300 mM and volume ratios of cellular compartments) suggest that it may be present in the vacuole (Figure 10.11), which is the preferred compartment for storing secondary metabolites (Boudet et al., 1984; Wink, 1997; Beckman, 2000; Bartholomew et al., 2002). It can also be present in the extracellular region around the seeds, as suggested by Joel et al. (2003), but the results obtained by Odoux and Brillouet (2009) contradict this hypothesis.

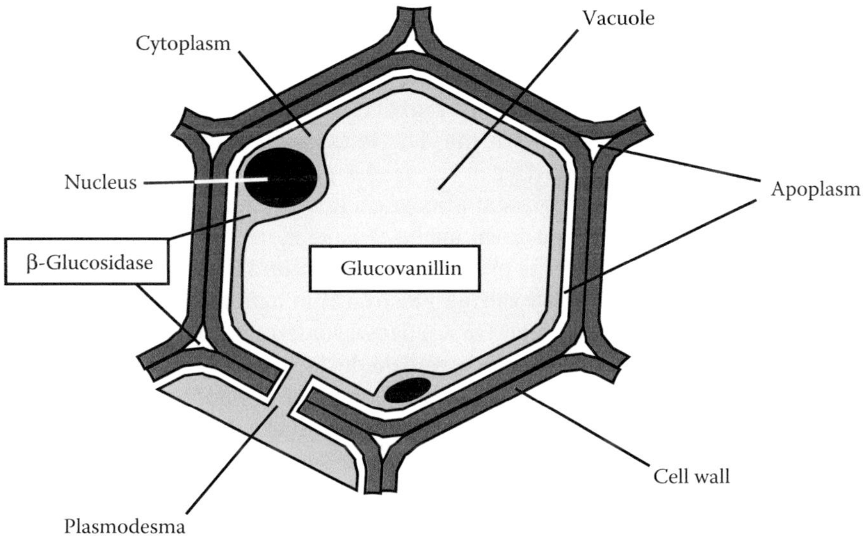

FIGURE 10.11 Cellular localization of glucovanillin and β-glucosidase activity according to Odoux et al. (2003).

Despite considerable controversy and confusion, the localization of glucovanillin and glucosidase activity at the tissue level has now been clarified. It remains to be determined whether glucovanillin is present in the intralocular space around the seeds, which may be important in confirming a possible tissue specialization in the biosynthesis of glucovanillin (see below).

At cellular and subcellular levels, the localization of β-glucosidase can be clarified using techniques such as immunolocalization; similarly, the cellular localization of glucovanillin remains to be determined.

Accumulation of Glucovanillin and β-Glucosidase during Vanilla Bean Development

Another area where very little research has been carried out is the evolution of glucovanillin and of β-glucosidase activity during bean development, with even less research been done on their evolution by tissue type.

The only point of agreement that has emerged from the various research studies on the evolution of glucovanillin (Ranadive et al., 1983; Sagrero-Nieves and Schwartz, 1988; Kanisawa et al., 1994; Brodélius, 1994; Havkin-Frenkel et al., 1999) is that the accumulation of vanillin or glucovanillin in the fruit starts from the 15th week after pollination and continues until around the 30th week.

However, the form in which vanillin is accumulated (free or glucosyl form) may lead to confusion. Ranadive et al. (1983) and Sagrero-Nieves and Schwartz (1988) show the evolution of free vanillin without prior hydrolysis; Ranadive's results even show that free vanillin represents between 50% and 90% of the potential total vanillin. Brodélius (1994) considers that most of the vanillin is in glucosyl form with the free form not exceeding 15% of the potential total amount. Kanisawa et al. (1994) do not report the presence of vanillin in its free form during the development of the green fruit, and Havkin-Frenkel et al. (1999) indicate that vanillin is only accumulated in its glucosyl form. The latter point is confirmed by Leong (1991), who does not find the free form in the green beans. Arana (1943) had already found that vanillin is present almost exclusively in glucosyl form. Except in certain exceptional cases, our own analyses have always shown that in mature green fruits, the glucosyl form is predominant (at around 95% of the total), if we prevent any accidental hydrolysis during glucovanillin extraction (e.g., by conducting extraction in pure methanol at −18°C). However, a more timely study (unpublished results) on the evolution of glucovanillin during the development of the fruit showed that for beans around 3, 5, 7, and 9 months of development after pollination, the percentage of free vanillin in relation to the total (glucosyl plus free forms) was 33, 6, 1.5, and 0.2%, respectively. It would be interesting to get a confirmation of this evolution, which raises questions on the role of glucosylation of vanillin in the vanilla bean.

According to different researchers who monitored the evolution of glucosidase activity during bean development on the vine (Wild-Altamirano, 1969, Ranadive et al., 1983; Kanisawa et al., 1994), it would appear that this activity is measurable at all stages of fruit growth. However, the enzyme activity increases considerably between the third and the fourth month after pollination, reaching a maximum at around the fifth month. Therefore, the evolution of β-glucosidase activity during bean growth is on par with that of glucovanillin. The assays (unpublished results) conducted by the authors for glucosidase activity in fruits harvested during February 2005 in Madagascar at a developmental stage estimated at less than two months after pollination (highly asymmetrical fruit shape with a floral part that is far more rounded than the peduncular part) showed a glucosidase activity of around 650 nkatal/g of fresh weight for the floral part, compared to 230 nkatal/g of fresh weight for the peduncular part. These activities were already high and suggested an activity gradient in phase with the fruit development. For fruits from the same batch that had reached full size, but had not developed for more than 5 months after pollination, the glucosidase activities for the floral parts and peduncles were almost identical, at around 1100 nkatal/g of fresh weight, or the mean value obtained for fruits at the usual harvesting time (see below).

Glucovanillin contents obtained for the green fruit after eight months of development differ greatly from one research study to the other. If we convert the different values given in the literature into grams of glucovanillin per 100 g of dry weight, they range between 2% (Sagrero-Nieves and Schwartz, 1988) and 12% (Havkin-Frenkel et al., 1999). Further research studies have confirmed that mature green beans could comprise of glucovanillin around 10–15% of the dry weight (Ansaldi et al., 1988; Leong, 1991; Brunerie, 1993; Odoux, 2000; Havkin-Frenkel et al., 2005). Determination of glucovanillin from 70 green beans of seven different batches during the year 2000 in Madagascar (unpublished results) showed that glucovanillin contents could vary from 1.5% to 12% of dry weight depending on the fruit, with the majority of individual beans presenting a glucovanillin content of around 10%. Other determinations conducted in 2006 on batches from Papua New Guinea even showed maximum glucovanillin contents of more than 20% of dry weight for certain fruits, confirming the extreme variability that may exist in the glucovanillin content of *V. planifolia* fruits.

For β-glucosidase activities, it is impossible to compare the values given in the bibliography because of the means of expression (units) used, the protocols for obtaining enzyme extracts, the nature of the buffers used (pH, ionic strength, etc.), the molarity of the substrate (usually pNPG), and so on. Our own experience in this field has shown that in mature green beans with a physiologically healthy appearance and with a standardized, accurate protocol (Odoux, 2004), this activity could also vary considerably. Based on around 100 fruits from the Madagascan harvest in 2000 (unpublished results), the β-glucosidase activities ranged from around 100 to 2000 nkatal/g of fresh weight, with the majority of individual beans presenting an activity of around 1000 nkatal/g of fresh weight.

More systematic studies on the evolution of glucovanillin (and other glucosides) and of glucosidase activity during bean development are essential in order to confirm or refute the research already published. In the case of the aroma components, very strict analytical protocols must be established in order to remove any ambiguity regarding the form in which they are present at the different stages of development; it would also be useful to standardize the assays for glucosidase activity, for which the results can almost never be compared from one study to another.

These evolutions should also be measured by tissue type; in the case of glucosides, this could make it possible to obtain additional information on the biosynthetic pathways and sites for these components (see the following section).

Biosynthetic Site and Pathway for Glucovanillin

In his work on biosynthesis of vanillin, Lecomte (1901, 1913) concluded that it involved "coniferoside" (coniferin) that produced coniferyl alcohol through enzymatic hydrolysis, which was then turned into vanillin through the action of an "oxidase." Goris, who isolated "vanilloside" (glucovanillin) in 1924, suggested that a second possible pathway consisted of imagining the action of the "oxidase" before that of the "hydrolase" (Figure 10.12). Unable to isolate either the "coniferoside" or the coniferyl alcohol, he finally concluded that these hypotheses were unconfirmed (Goris, 1947). However, they were later resumed by Anwar (1963).

It is now accepted that vanillin is a product of the biosynthetic pathway of shikimic acid, via phenylalanine, which leads to the phenylpropane compounds through enzymatic deamination, and primarily to cinnamic acid (Figure 10.13). Successive enzymatic hydroxylations and methylations then lead to the formation of *p*-hydroxycinnamic acids and, notably, coumaric, caffeic, ferulic, and sinapic acids.

Most of the research published in an attempt to clarify the subsequent stages of the biosynthetic pathway of vanillin was conducted using cell cultures and showed that different pathways could be activated, depending on the experimental conditions (reviewed by Dignum et al., 2001; Walton et al., 2003). It is therefore difficult to draw any definitive conclusions and even more difficult to attempt to extrapolate the results obtained in these conditions to the plant.

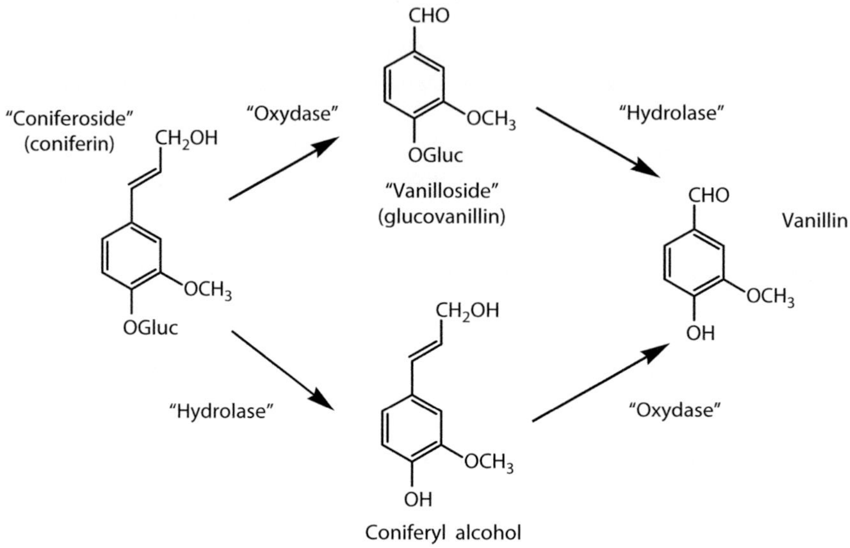

FIGURE 10.12 Biosynthetic pathway of vanillin proposed by Lecomte (1901, 1913) and Goris (1947).

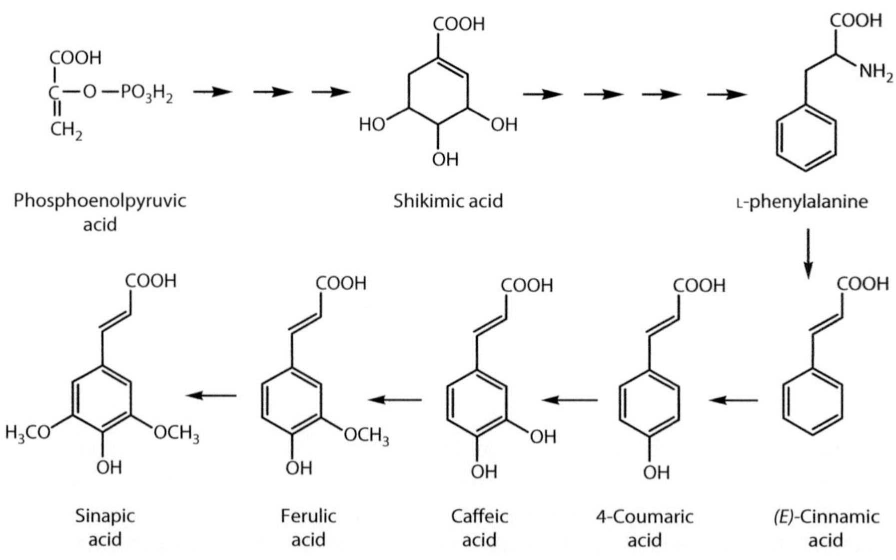

FIGURE 10.13 Presumed biosynthetic pathway of the phenylpropane compounds via shikimic acid and L-phenylalanine. (Data from Odoux, E., *Fruits*, 61, 171–184, 2006.)

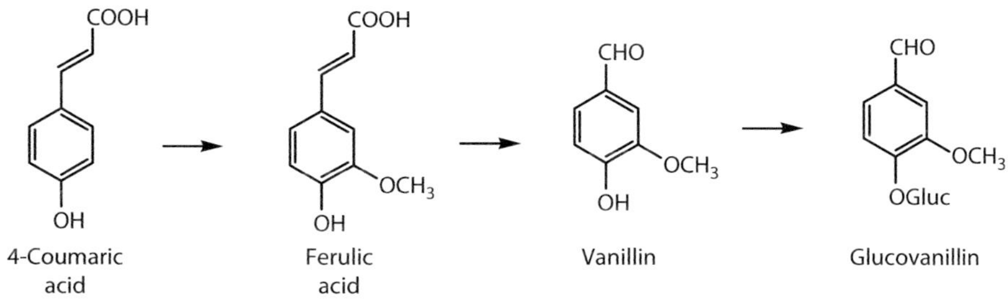

FIGURE 10.14 Biosynthetic pathway of vanillin proposed by Zenk (1965) and Negishi et al. (2009).

What emerges from the research conducted on the fruits, of which there is very little information (Zenk, 1965; Kanisawa, 1993; Kanisawa et al., 1994; Negishi et al., 2009), is that two biosynthetic pathways are suggested:

- The first one suggests that the direct precursor of vanillin is ferulic acid (C6–C3 compound), which means that the C3 side chain of the molecule is later shortened to give vanillin (C6–C1 compound) (Figure 10.14). This is the argument put forward by Zenk (1965) and confirmed by the recent work of Negishi et al. (2009). In both cases, the results were obtained after incorporating ^{14}C-labeled molecules on green vanilla discs and monitoring their conversion.

The results obtained by Negishi et al. (2009) also suggest that biosynthesis does not cause glucosylated intermediate compounds to intervene; vanillin is therefore synthesized in the aglycon form, and then glucosylated once produced.

- The second one suggests that shortening of part C3 occurs higher up at the level of 4-coumaric acid (C6–C3 compound) to give 4-hydroxybenzaldehyde (C6–C1 compound), which is then hydroxylated and methylated to give vanillin (Figure 10.15). This argument is favored by Kanisawa et al. (1994)—although he also suggests a pathway via diglucosides and does not

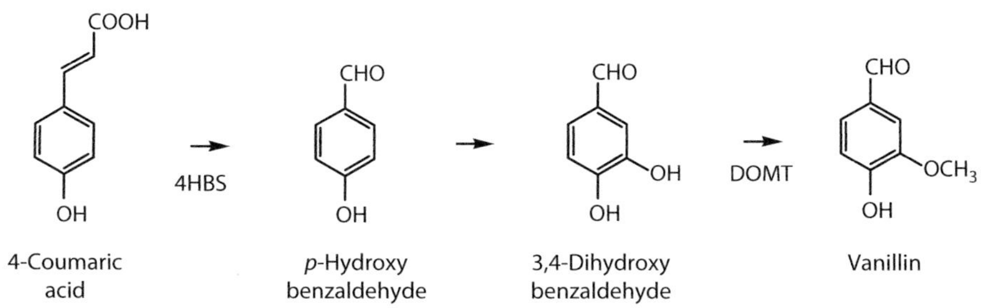

FIGURE 10.15 Biosynthetic pathway of vanillin proposed by Kanisawa et al. (1994) and Podstolski et al. (2002). In the pathway suggested by Kanisawa et al. (1994), the compounds involved are glucosylated from 4-coumaric acid (not shown in the figure).

> rule out the possibility of ferulic acid as a precursor—based on the different compounds identified in the fruit at different stages of development. This is also the argument defended by the team at Rutgers University, Princeton, USA, who purified a 4-hydroxybenzaldehyde synthase (4HBS) (from cell cultures) and a methyltransferase (DOMT) (from the bean) that can catalyze, respectively, the conversion of 4-coumaric acid into 4-hydroxybenzaldehyde and 3,4-dihydroxybenzaldehyde (protocatechuic aldehyde) into vanillin (Podstolski et al., 2002; Pak et al., 2004). The existence of an enzyme able to hydroxylate 4-hydroxybenzaldehyde into 3,4-dihydroxybenzaldehyde remains to be proven, preferably in the fruit. The results obtained by Negishi et al. (2009) do not show any conversion of 4-hydroxybenzaldehyde into vanillin.

In the pathway suggested by Kanisawa et al. (1994), glucosylation takes place as soon as 4-coumaric acid appears; the subsequent reactions would therefore involve glucosylated intermediates up to glucovanillin.

As discussed in the previous sections, glucovanillin is present in fully mature fruits mainly in the placentas and, to a lesser extent, in the papillae. According to Joel et al. (2003), papillae are the site of biosynthesis of glucovanillin in the bean, based on the presence of 4-hydroxybenzaldehyde synthase (4HBS) in the cytoplasm of the cells. The presence of 4HBS was revealed by immunolocalization, but the research note that supported the presentation of the results gives no details of the methodology used. Glucovanillin (or precursors of this molecule) is then secreted by these papillae in the extracellular space around the seeds. French (2005) believes that the presence of other enzymes involved in the biosynthesis of glucovanillin must also be shown in order to confirm this hypothesis.

Furthermore, if papillae were the only site of biosynthesis of glucovanillin, this would raise the question of how it is transported to the placentas (Odoux and Brillouet, 2009), which are the preferred sites of accumulation.

A great deal of research remains to be carried out to clarify the biosynthetic pathway(s) of glucovanillin in the vanilla bean, and also to determine the tissue(s) involved.

REFERENCES

Ansaldi, G., Gil, G., and Le Petit, J. 1988. Procédé d'obtention d'arôme naturel de vanille par traitement des gousses de vanille verte et arôme obtenu. Brevet, INPI. No. 2, 634 979, p. 7.

Anwar, M.H. 1963. Paper chromatography of monohydroxyphenols in vanilla extract. *Analytical Chemistry* 35:1974–1976.

Arana, F.E. 1943. Action of a β-glucosidase in the curing of vanilla. *Food Research*. 8:343–351.

Arana, F.E. 1944. Vanilla curing and its chemistry. Federal Experiment Station of the US Department of Agriculture, Mayaguez, Puerto Rico, Bulletin No. 42, 17pp.

Arber, A. 1937. The interpretation of the flower: A study of some aspects of morphological thought. *Biology Review* 12:157–184.

Balls, A.K. and Arana, F.E. 1941. The curing of vanilla. *Industrial and Engineering Chemistry* 33(8):1073–1075.

Bartholomew, D.M., Van Dyk, D.E., Cindy Lau, S.-M., O'keefe, D.P., Rea, P.A., and Viitanen, P.V. 2002. Alternate energy-dependent pathways for the vacuolar uptake of glucose and glutathione conjugates. *Plant Physiology* 130:1562–1572.

Beckman, C.H. 2000. Phenolic-storing cells: Keys to programmed cell death and periderm formation in wilt disease resistance and in general defence responses in plants? *Physiological and Molecular Plant Pathology* 57:101–110.

Boudet, A.M., Alibert, A., and Marigo, G. 1984. Vacuoles and tonoplast in the regulation of cellular metabolism. In: A.M. Boudet, A. Alibert, G. Marigo, and P.J. Lea, eds. *Membranes and Compartmentation in the Regulation of Plant Functions*. Clarendon Press, Oxford, 29–47.

Brodélius, P.E. 1994. Phenylpropanoid metabolism in suspension cultures of *Vanilla planifolia* Andr. V. High performance liquid chromatographic analysis of phenolic glycosides and aglycones in developing fruits. *Phytochemical Analysis* 5:27–31.

Brunerie P.M. 1993. Procédé d'obtention d'arôme naturel de vanille par traitement enzymatique des gousses de vanille verte et arôme obtenu. International Patent Application, 1993, No PCT/FR92/00837, 11pp.

Cameron, K. and Chase, M. 2000. Nuclear 18S rDNA sequences of Orchidacea confirm the subfamilial status and circumscription of Vanilloideae. In: K.L. Wilson and D.A. Morrison, eds. *Monocots—Systematics and Evolution, Vol. 1 of Proceedings of the Second International Conference on the Comparative Biology of the Monocots*. CSIRO, Melbourne, 457–464.

Childers, N.F. and Cibes, H.R. 1948. Vanilla culture in Puerto Rico. Federal Experiment Station of the US Department of Agriculture, Mayaguez, Puerto Rico, Circular No. 28, 94pp.

De Lanessan, J.L. 1886. Vanille. In: J.L. De Lanessan, ed. *Les plantes utiles des colonies françaises*. Imprimerie Nationale, Paris, 62–74.

Dignum, M.J.W., Kerler, J., and Verpoorte, R. 2001. Vanilla production: Technological, chemical and biosynthetic aspects. *Food Reviews International* 17:199–219.

Ducamp, M.N., Lebrun, M., and Thi, H.T. 2000. Intensité respiratoire de certaines productions tropicales et facteurs pouvant interférer. *Fruits* 55(6):401–408.

Dupont, F. and Guignard, J.L. 2007. Botanique: systématique moléculaire – 14e ed. (Collection Abrégés). Masson, Paris.

French, J.C. 2005. Development of vanilla-bearing placental trichomes in *Vanilla planifolia*. In: D. Havkin-Frenkel, ed. *Vanilla, Proceedings of the First International Congress*. Allured Publishing, Carol Stream, 71–77.

Garros-Patin, J. and Hahn, J. 1954. La chimie de la vanille. In: G. Bouriquet, ed. *Le vanillier et la vanille dans le monde*. Paul Lechevalier, Paris, 559–615.

Goris, A. 1924. Sur la composition chimique des fruits verts de vanille et le mode de formation du parfum de la vanille. *Comptes Rendus de l'Académie des Sciences* 179:70–72.

Goris, A. 1947. Formation du parfum de la vanille. *Industrie de la Parfumerie* 2:4–12.

Gregory, L.E., Gaskins, M.H., and Colberg, C. 1967. Parthenocarpic pod development by *Vanilla planifolia* Andrews induced with growth-regulating chemicals. *Economic Botany* 21(4):351–357.

Havkin-Frenkel, D., Podstolski, A., Witkowska, E., Molecki, P., and Mikolajczyk, M. 1999. Vanillin biosynthetic pathways, an overview. In: T. J. Fu, G. Singh, and W. R. Curtis, eds. *Plant Cell and Tissue Culture for the Production of Food Ingredients*. Kluwer Academic/Plenum Publishers, New York, pp. 35–43.

Havkin-Frenkel, D., French, J.C., and Frenkel, C. 2005. Inside vanilla. *Perfumer and Flavorist* 30:36–55.

Joel, D.M., French, J.C., Graft, N., Kourteva, G., Dixon, R.A., and Havkin-Frenkel, D. 2003. A hairy tissue produces vanillin. *Israel Journal of Plant Sciences* 51:157–159.

Jones, M.A. and Vicente, G.C. 1949. Quality of cured vanilla in relation to some natural factors. *Journal of Agricultural Research* 78(11):445–450.

Kanisawa, T. 1993. Flavor development in vanilla beans. *Kouryou* 180:113–123.

Kanisawa, T., Tokoro, K., and Kawahara, S. 1994. Flavour development in the beans of *Vanilla planifolia*. In: K. Kurihara, N. Suzuki, and H. Ogawa, eds. *Proceedings of the International Symposium*. Springer, Tokyo, 268–270.

Lecomte, H. 1901. Sur la formation du parfum de la vanille. *Compte Rendu des Séances de l'Académie des Sciences de Paris* T133:745–748.

Lecomte, H. 1913. Formation de la vanilline dans la vanille. *L'Agriculture Pratique des Pays Chauds* 13:3–14.

Leong, G. 1991. Contribution à l'étude des hétérosides des gousses de vanille vertes. Thèse, Université de Marseille III, Marseille, France, 153pp.

Lubinsky, P., Van Dam, M., and Van Dam, A. 2006. Pollination of Vanilla and evolution in Orchidaceae. *Lindleyana (*in *Orchids)* 75:926–929.

Maestro, Y., Bergia, D., and Lasserre, C. 2007. *Vanilla planifolia* extract, method for obtaining same, and cosmetic or dermatological composition containing same. *World Patent* WO 2007/034042 A2.

Negishi, O., Sugiura, K., and Negishi, Y. 2009. Biosynthesis of vanillin via ferulic acid in *Vanilla planifolia. Journal of Agricultural and Food Chemistry* 57:9956–9961.

Odoux, E. 2000. Changes in vanillin and glucovanillin concentrations during the various stages of the process traditionally used for curing Vanilla fragrans beans in Réunion. *Fruits* 55:119–125.

Odoux, E. 2004. Contribution à l'étude de l'hydrolyse de la glucovanilline en vanilline dans la gousse du vanillier (*Vanilla planifolia* G. Jackson). Thèse, Université de Montpellier II, Montpellier, France, 149pp.

Odoux, E. 2006. Glucosylated aroma precursors and glucosidase(s) in vanilla bean (*Vanilla planifolia* G. Jackson). *Fruits* 61(3):171–184.

Odoux, E. and Brillouet, J.M. 2009. Anatomy, histochemistry, and biochemistry of glucovanillin, oleoresin, and mucilage accumulation sites in green mature vanilla pod (*Vanilla planifolia*; Orchidaceae): A comprehensive and critical re-examination. *Fruits* 64:221–241.

Odoux, E., Escoute, J., Verdeil, J.L., and Brillouet, J.M. 2003. Localization of β-D-glucosidase activity and glucovanillin in vanilla bean (*Vanilla planifolia* Andrews). *Annals of Botany* 92:437–444.

Odoux, E. and Havkin-Frenkel, D. 2005. Hydrolysis of glucovanillin by β-glucosidase during curing of vanilla bean (*Vanilla planifolia* Andrews). In: D. Havkin-Frenkel, ed. *Vanilla, Proceedings of the First International Congress*. Allured Publishing, Carol Stream, 95–100.

Pak, F.E., Gropper, S., Dai, W.D., Havkin-Frenkel, D., and Belanger, F.C. 2004. Characterization of a multifunctional methyltransferase from the orchid *Vanilla planifolia. Plant Cell Report* 22:969–966.

Podstolski, A., Havkin-Frenkel, D., Malinowski, J., Blount, J., Kourteva, G., and Dixon, R.A. 2002. Unusual 4-hydroxybenzaldehyde synthase activity from tissue cultures of the vanilla orchid *Vanilla planifolia. Phytochemistry* 61:611–620.

Ramaroson-Raonizafinimanana, B., Gaydou, E.M., and Bombarda, I. 1997. Hydrocarbons from three *Vanilla* bean species: *V. fragrans*, *V. madagascariensis*, and *V. tahitensis*. *Journal of Agricultural and Food Chemistry* 45:2542–2545.

Ramaroson-Raonizafinimanana, B., Gaydou, E.M., and Bombarda, I. 1998a. 4-Demethylsterols and triterpene alcohols from two *Vanilla* bean species: *V. fragrans* and *V. tahitensis*. *JAOCS* 75(1):51–55.

Ramaroson-Raonizafinimanana, B., Gaydou, E.M., and Bombarda, I. 1998b. Occurrence of 5 alpha-cholesta-7, 24-dien-3 beta-ol and 23-dehydrolophenol in the bean lipids of *Vanilla madagascariensis* (collected from Madagascar). *Journal of the American Oil Chemists' Society* 75(10):1325–1328.

Ramaroson-Raonizafinimanana, B., Gaydou, E.M., and Bombarda, I. 1999. Long-chain γ-pyrones in epicuticular wax of two vanilla bean species: *V. fragrans* and *V. tahitensis*. *Journal of Agricultural and Food Chemistry* 47:3202–3205.

Ramaroson-Raonizafinimanana, B., Gaydou, E.M., and Bombarda, I. 2000. Long-chain aliphatic beta-diketones from epicuticular wax of Vanilla bean species. Synthesis of nervonoylacetone. *Journal of Agricultural and Food Chemistry* 48(10):4739–4744.

Ranadive, A.S. 1994. Vanilla—cultivation, curing, chemistry, technology and commercial products. In: G. Charalambous, ed. *Spices, Herbs and Edible Fungi*. Elsevier, Amsterdam, 517–577.

Ranadive, A.S., Szkutnica, K., Guerrera, J.G., and Frenkel, C. 1983. Vanillin biosynthesis in vanilla beans. In *Proceedings of the 9th International Congress on Essential Oils, Singapore*. Allured Publishing, Wheaton, 147–154.

Roux, P. 1954. Etudes morphologiques et anatomiques dans le genre *Vanilla*. In: G. Bouriquet, ed. *Le vanillier et la vanille dans le monde*. Paul Lechevalier, Paris, 44–92.

Shankaracharya, N.B. and Natarajan, C.P. 1973. Vanilla—chemistry, technology and uses. *Indian Food Packer* 27(3):29–36.

Sagrero-Nieves, L. and Schwartz, S.J. 1988. Phenolic content of *Vanilla planifolia* as affected by harvest period. *Journal of Food Composition and Analysis* 1:362–365.

Sreekrishna Bhat, S. and Sudharshan, M.R. 2002. Phenology of flowering in vanilla (*Vanilla planifolia* Andr.). In: K. Sreedharan, P. K. Vinod Kumar, Jayarama, and B. M. Chulaki, eds. *Proceedings of the 15th Plantation Crops Symposium: PLACROSYM XV*, Mysore, 10–13 December 2002, Director of Central Coffee Research Institute, Karnataka, India, 128–133.

Swamy, B.G.L. 1947. On the life-history of *Vanilla planifolia. Botanical Gazette* 108:449–456.

Villiers, A., Collin, E., and Fayolle, M. 1909. Vanille. In: O. Doin et fils, eds. *Traité des falsifications et alterations des substances alimentaires. II. Aliments principaux et condiments*. Paris, 339–349.

Walton, N.J., Mayer, M.J., and Narbad, A. 2003. Vanillin. *Phytochemistry* 63:505–515.

Wild-Altamirano, C. 1969. Enzymatic activity during growth of vanilla fruit. 1. Proteinase, glucosidase, peroxidase and polyphenoloxidase. *Journal of Food Science* 34:235–238.

Wink, M. 1997. Compartmentation of secondary metabolites and xenobiotics in plant vacuoles. *Advances in Botanical Research* 25:141–169.

Zenk, M.H. 1965. Biosynthese von vanillin in *Vanilla planifolia* Andr. *Zeitschrift für Pflanzenphysiologie* 53:404–414.

11 Vanilla Curing

Eric Odoux

CONTENTS

INTRODUCTION

The vanilla curing process is designed to produce an aromatically attractive and microbiologically stable product from green pods harvested before complete ripeness and which have no special aroma apart from a vague "plant" odor.

The curing techniques are of broad range, but generally involve four separate steps. Specific terms are used in reference to these steps in the different vanilla-producing countries—which are primarily Spanish-speaking in Central America (Mexico), French-speaking in the Indian Ocean region (Réunion, Comoros, Madagascar), and English-speaking (India, Uganda). Descriptions of these steps in different languages are often ambiguous and confused, further complicated by the fact that the separation between the four steps is often subjective and controversial. It is thus essential to clearly define this terminology before launching into an in-depth discussion on curing processes and their biochemical implications.

The first curing step is designed to stall pod dehiscence (especially in *Vanilla planifolia* G. Jackson, which is the most widely marketed species) by blocking the normal metabolic processes that take place during maturation. The most common techniques involve in suddenly inducing pod senescence, that is, "killing" the fruit ("mortification" in French). The main technique used for this purpose, which originates from Réunion, involves soaking vanilla pods in hot water—this is termed "scalding" and "échaudage" in French. In Mexico, the pods are traditionally placed in an oven or exposed to the sun, and the respective Spanish terms are "horneado" and "secado al sol." A common English translation of "horneado" is "oven-killing," but there is no specific term in French, while "sun-killing" is the English term for "secado al sol," but again there is no equivalent in French.

The second step involves in maintaining the heat stored during the initial step as long as possible by placing the pods in closed crates that are as heat insulated as possible (generally with gunny sacks) so as to create conditions that are similar to those of a sweating chamber. This step is called "sweating," or "étuvage" (from an etymological standpoint, this term introduces the notion of humid heat) in French, and "sudor de horno" or "sudado" in Spanish. These terms are more ambiguous and the interpretations may therefore differ markedly. Some authors (Bourriquet, 1954) report that during this step the pods can release a blackish fluid, which some authors explain as being due to drying associated with pod sweating. This drying conflicts with the conditions established during this step, and the phenomenon has not been experimentally assessed (Odoux, 2000; Perez-Silva, 2006). The blackish fluid is probably due to the result of excess condensation (mixed with bits of crushed pods), which is in line with the "sweating" and "étuvage" notions. Other authors (Ranadive, 1994; Dignum et al., 2001) consider that the sweating step also includes the first days of sun drying when moisture loss definitely occurs.

In the third step, the vanilla pods are dried to stabilize the product—this is called the "drying" step, or "séchage" in French, and "asolear" or "secado del sol" in Spanish. These are clear-cut terms describing the pod drying phenomenon, despite the fact that this step obviously involves more than just moisture loss—which is why it is hard to clearly separate the end of the sweating step from the beginning of the drying step. Moreover, "asolear" and "secado del sol" introduce the notion of sun exposure as a pod-drying technique.

The fourth step is termed "conditioning," and "conditionnement" or "affinage" in French, and "afinado" in Spanish. These terms pool two complementary notions, that is, packaging, storage, and preservation of the product in a heat-tight atmosphere, with the aim of promoting "aromatic maturation" overtime.

This chapter highlights how these techniques and their derivatives ultimately involve in maintaining vanilla pods in high temperature and humidity conditions as long as possible in order to set the stage for a number of enzymatic and chemical reactions that promote the development of the aroma typical of vanilla, while curbing microorganism proliferation (especially molds). However, as discussed here, this curing is a rather continuous process, thus hard to separate into clearly defined steps.

CONVENTIONAL CURING TECHNIQUES

Steps 1 and 2: Initial Heat Treatment

In *V. planifolia* G. Jackson, heat treatment is mainly carried out to hamper pod dehiscence (n.b. *Vanilla tahitensis* is not subjected to heat treatment because it is not a dehiscent species, see also Chapter 13), which would reduce the market quality of vanilla pods. The pods are voluntarily harvested before the ripeness stage at which dehiscence occurs, but this early harvesting does not stop the process. Different techniques have thus been developed to stop it completely. These techniques are compatible with the way vanilla is cured under local conditions in producing countries, that is, where technical resources are often very limited.

In the first step, the most common technique implemented in the Indian Ocean region (Madagascar, Comoros, Réunion, etc.) involves soaking green pods in a water-filled tank heated over an open fire (scalding process) (Figure 11.1). Then the second step involves placing the pods immediately in covered crates to avoid heat loss (sweating). In theory, scalding should be carried out for 3 min in a water bath at 65°C, while the sweating process takes 24 h. In some countries, the scalding time and temperature are adjusted according to the pod grade, so the pods are sometimes presorted.

In practice, these two steps are carried out in a variety of ways. For instance, in Madagascar, which is the top vanilla-producing country, many operators are involved, ranging from farmers who produce just a few hundred grams of vanilla beans to exporters who deal with hundreds of tons of this product, and the technical resources available to these operators also differ markedly (Odoux, 1998). Most farmers have no way to control the scalding temperature and time, which means there may be substantial variations in the treatment process. The quantity of pods cured, which

FIGURE 11.1 Killing step: green beans are soaked in hot water.

may vary considerably between operators, can have a great impact on the heat treatment quality (e.g., edge effects in crates). The sweating process can also be extended for up to a few days if the climatic conditions are unsuitable to start the drying step. In Réunion, where almost all of the vanilla crop is processed in two centralized units (a cooperative and a private company), the scalding and sweating processes are carried out twice to achieve a uniform heat treatment (Odoux, 2000). However, little is known about the actual impact of these alternatives on the end quality of the vanilla.

A second technique (conventionally used in Mexico) involves processing the pods in wood-fuelled ovens ("calorifico") for 24–48 h at around 60°C. The pods are wrapped in burlap and then in mats (the so-called "maletas"). These "maletas" are abundantly sprinkled with water so that the humidity level will remain high during the process, then they are placed in wooden crates or on shelves in the oven (Bourriquet, 1954; Théodose, 1973; Perez-Silva, 2006), and finally they are sprinkled again. The oven design is generally purely practical, so this has a substantial impact on the heat treatment uniformity and the temperature actually reached in the oven—it is therefore up to the operator to adjust the treatment time accordingly. At the end of this step, the "maletas" are taken out of the oven, the mats are removed and the pods, which are still wrapped in burlap, are placed in large crates and covered for 24–48 h (Perez-Silva, 2006). This is the so-called "sudor de horno" step, which is equivalent to the "étuvage" step in Réunion.

Another method called "secado al sol," which is also conventionally carried out in Mexico, involves spreading out the pods in the sun as an alternative heat treatment. This technique is seldom used at the present time because very large drying areas are required, and the results in terms of stopping dehiscence are much more uncertain than with other techniques (Arana, 1944).

From a more anecdotal standpoint, it should also be noted that dehiscence can be effectively stopped (Arana, 1944) by scratching the vanilla pods with a sharp object. Historically, this technique was developed in the West Indies and is only used to treat pods produced by *Vanilla pompona*, that is, very marginal production.

These treatments lead to vanilla bean senescence, resulting in pod browning (Figure 11.2) and relatively marked loss of the initial turgescence. As will be discussed in the last section of this chapter, these structural modifications are the starting point for the formation of vanilla aroma compounds. Considering the range of heat treatment techniques used, the pod color and texture may differ considerably between pods at the end of the treatment, but the long drying step that follows helps to homogenize the final pod appearance.

Step 3: Drying

The next step of this process involves intentional slow drying of vanilla pods.

Drying is obviously carried out to stabilize the pods in order to avoid infestation by microorganisms, especially molds. The pod moisture content gradually decreases from around 85% to maximum final levels of 25–38%, depending on the category. In practice, vanilla that is targeted for the agrifood industry (80% of the market) is dried to 18–20% moisture. It should be noted that, even at 18% moisture, the water

FIGURE 11.2 Vanilla beans before (in the basket) and after the sweating step.

activity (A_w) is around 0.85, which is much higher than the 0.65 A_w, which means that the level should not be surpassed to avoid microorganism development. However, vanilla can normally be stored without any problems of this sort.

Vanilla pods are usually roughly spread out (Figure 11.3) on blankets and set on racks in the sun for part of the day. After a few hours of sun exposure, the vanilla is enveloped in blankets (Figure 11.4), which in turn are piled up in the shade until the next day (Figure 11.5). This procedure is repeated daily for several weeks. Pods that are considered dry enough are removed and the shade drying phase begins; this is continued until the entire batch is sufficiently dry. During this shade drying phase, the pods are carefully laid out side by side, in a single layer (Figure 11.6), on wooden racks sheltered from the sun in a building. They are regularly monitored and turned over until they are considered dry enough to begin the next conditioning step. The level of pod drying (during either the sun drying or shade drying phase) is assessed empirically by touch, so the operator has to be quite experienced in vanilla curing to be able to pass a proper judgment on the extent of pod stability. The drying rates are highly variable between pods, even between unsplit pods of the same grade (split pods obviously dry much faster and have to be specially monitored). The whole drying process lasts 2–3 months. This curing step is managed in almost the same way in all vanilla-producing countries, irrespective of the techniques used.

The main problem with this drying procedure is that it is closely dependent on the climatic conditions, especially during the first outdoor phase when the pods are most susceptible to microorganism contamination. This step coincides with heavy rainfall periods in the main vanilla-growing areas of Madagascar (Théodose, 1973), which can upset the drying process and sometimes give rise to serious mold development

FIGURE 11.3 Drying step: vanilla beans are roughly spread out on blankets.

problems. To avoid such infestations, artificial batch drying techniques have been proposed (Théodose, 1973), which have even been successfully implemented, from a pod quality standpoint, under large-scale industrial conditions. In Madagascar and Réunion, for instance, plum drying tunnels have been used to dry vanilla.

FIGURE 11.4 After exposure to the sun, vanilla beans are enveloped in the blankets.

FIGURE 11.5 Blankets with vanilla beans are piled up in the shade.

However, such dryers are generally not used in vanilla-producing countries due to socioeconomic constraints rather than for qualitative reasons. More recently, studies have been conducted on drying vanilla in solar dryers (Kamaruddin, 1997; Ratobison et al., 1998).

FIGURE 11.6 Shade drying step: vanilla beans are laid out side by side on trays.

The slow drying technique, described in this section, is more than just a dehydration step because the reactions that give rise to the development of the typical vanilla aroma, which are initiated during the initial heat treatment, continue because of high A_w and the heat build up during the successive sun-exposure phases.

Step 4: Conditioning

The final step involves conditioning of the vanilla pods in boxes for several months. The aim is to promote full development of the vanilla aroma.

Small bunches of dried pods (generally after sorting by size and color) are tied with raffia, wrapped in waxed paper, and placed in metal or wooden boxes (Figure 11.7) to limit moisture loss as much as possible. These containers are regularly checked to make sure that there is no mold development. The vanilla pods are considered ready for marketing after a few months of storage under these conditions.

This step could be compared to wine aging, or cheese ripening, because the aromatic quality of the vanilla pods is clearly enhanced during this period.

This curing process is implemented in roughly the same way in all vanilla-producing countries worldwide. It is highly empirical, relatively uncontrolled, and closely dependent on the prevailing climatic conditions—which can lead to major pod quality defects in some circumstances—but it is tailored to the conditions in which farmers live in most producing countries, even in Madagascar—the top vanilla-producing country. Substantial handling is required throughout this vanilla curing process, thus providing a major source of income for many people.

FIGURE 11.7 Conditioning step: vanilla beans are placed in wooden boxes with waxed paper.

UNCONVENTIONAL TECHNIQUES

The different faults in the conventional curing process, already mentioned, as well as its duration (over six months before a marketable product is obtained), have encouraged many authors to try to develop alternative curing methods. Moreover, it is known that the final vanillin contents are never higher than 3% relative to the dry matter (Bayle et al., 1982; Derbesy et al., 1982; Arnaud et al., 1983; Derbesy, 1989; Falque et al., 1992; Fayet et al., 1999; Derbesy and Charvet, 2000; Saltron et al., 2002; Gassenmeier et al., 2008), whereas the initial potential is usually over 5% (Arana, 1943; Ansaldi et al., 1988; Leong, 1991; Brunerie, 1993; Odoux, 2000; Havkin-Frenkel et al., 2005). These considerations have also prompted research aimed at boosting vanillin levels in the end product.

Some authors have focused on enhancing control of the basic parameters (temperature, relative humidity, processing time, etc.) that supposedly have an impact on the final vanilla quality, while also trying to create conditions that closely match conventional pod curing conditions. Finally, the goal is to imitate the conventional process while "industrializing" the different steps.

Consequently, the techniques proposed by these authors—three of which are patented—are relatively similar, while mainly differing with respect to the initial state of the raw material, that is, depending on whether the pods are whole, in pieces or ground.

Towt (1952) proposed to grind vanilla pods to obtain a uniform purée, based on the idea that the different chemical and biochemical reactions that take place during the vanilla quality development process would be facilitated and accelerated as compared to using whole pods. This purée is subjected for 48 h to a temperature ranging from 50°C to 55°C, with air injection into the mass to promote oxidation reactions, and it is then dried at 60°C until the moisture level has dropped below 20%.

Kaul (1967) used pods that were whole or cut into pieces so as to preserve the commercial "identity" of the end product, while effectively controlling the process to curb mold development, which is a common issue under conventional curing conditions. The process proposed by this author involves incubating pods at 38°C for one week, or at 65°C for 24 h, or any other intermediary time/temperature combination, in a container that is sealed to hamper moisture loss and promote chemical and biochemical reactions. This curing step is followed by a drying process under conditions that are not outlined in the patent description. The author has noted that this drying step can, in all cases, be conducted rapidly—contrary to conventional procedures—at low or high temperature without being detrimental to the product quality. The author considers that the described curing conditions must be closely followed to obtain a top quality end product.

Karas et al. (1972) proposed to work on pods cut into pieces and placed on racks. The racks are put in an oven at 60°C for 72 h. The pods are dried at 60°C until a moisture content of 35–40% is reached, and then at ambient temperature until the pod moisture content levels off at 20–25%.

These three techniques were patented by McCormick & Company, Inc., and to our knowledge one of them (Karas et al., 1972) is used in Uganda and also by an industrial group in Madagascar.

As early as 1949, Jones and Vicente (1949a) published a study in which relatively similar procedures were compared. Whole, cut up, or ground vanilla pods were oven dried at 60°C for 24 h, and then at 45°C (until the whole pods were flexible), and finally dried and conditioned at ambient temperature. The results of aromatic tests on ice creams flavored with the corresponding vanilla extracts did not reveal any significant differences between treatments.

Bourriquet (1954) also pointed out that in 1950, a Mexican company was already producing a very flavorful product called "vanilla fruit," which was produced from a green vanilla purée that was heated for 48–60 h and then dehydrated in a dryer.

Théodose (1973) reported the findings of studies carried out at the Antalaha research station in Madagascar in the 1960s, which aimed at simplifying the vanilla curing procedure. One technique was developed that involved scalding and sweating the whole pods in the conventional way, cutting them into 2–3 cm pieces, and then drying them in batches for 12 days to obtain an end product with 20–25% moisture content. Vanilla importers considered that the results were interesting, as the vanillin contents were higher than normal and the aroma was stronger.

Other authors focused more on optimizing the reactions to hydrolyze glucovanillin into vanillin by endogenous glucosidase (see Chapter 10 and section "Relationship between Vanilla Curing and Aroma Development," of this chapter), while using a specific treatment or technology to bring these components into contact. This treatment could be the initial step before a more conventional drying step or when manufacturing a flavor extract.

Growth hormones such as ethylene were studied long back by Arana (1944) and Bourriquet (1954). It was not clearly determined whether vanilla beans are climacteric or not, but their sensitivity to ethylene as a maturation-inducing agent was clearly demonstrated by these authors. This ethylene induction process leads to pod browning and glucovanillin hydrolysis. However, ethylene induces high pod dehiscence which, according to Arana (1944), could affect up to 47.5% of all pods. More recently, Havkin-Frenkel et al. (2005) generally confirmed these high dehiscence rates in pods treated with ethylene in the presence of air or oxygen. Treatments with ethylene, naphthalene acetic acid (hormone of the auxin family), or biotic elicitors have also been tested on pods that had undergone heat treatment (63°C for 3 min), combined or not with scarification (Sreedhar et al., 2007, 2009). According to the authors, these treatments had a positive impact on the formation of vanillin and other aromatic phenolic compounds, while also reducing the vanilla curing time. The results of these experiments were somewhat inconsistent, however, so it is hard to draw clear conclusions. In addition, the physiological impact expected from these molecules on pods that had undergone a senescence-inducing heat treatment could be questioned.

Ultrasound has also been studied as a technique for bringing enzymes and substrates into contact (Obolenski, 1957, 1958, 1959) and, according to the author, this technique can boost the vanillin contents (but the results were quite confusing overall).

Cernudat and Loustalot (1948), cited in Bourriquet (1954), tested the use of infrared radiation treatments as an initial vanilla curing step, and the quality of the end product was considered fairly acceptable.

Crushing vanilla pods—without squashing or breaking them—by around 10 runs through rollers at 20 kPa pressure was also found to enhance the hydrolysis of glucovanillin into vanillin (Perera and Owen, 2010).

A technique involving freezing/thawing of green pods has been patented (Balls et al., 1942; Ansaldi et al., 1988) and has been the topic of different publications (Jones and Vicente, 1949a; Odoux et al., 2006). According to Ansaldi et al. (1988), this technique enables hydrolysis of around 80% of the glucovanillin present in green pods. In similar conditions, our findings showed that it is even possible to obtain hydrolysis rates of over 90% (Odoux et al., 2006).

Finally, other authors have proposed techniques that differ more radically from conventional processes. Since cured vanilla is mainly used in the agrifood industry to manufacture extracts, the pods have to be ground after curing. On the basis of this observation, these authors proposed methods to grind green pods and optimize enzymatic reactions by adding exogenous enzymes to the obtained purée. These mainly concern glucosidase activities, but also polysaccharide hydrolase activities (pectinase, cellulase, etc.), in order to achieve total glucovanillin hydrolysis (Graves et al., 1958; Mane and Zucca, 1992; Brunerie, 1993; Ruiz-Terán et al., 2001).

Some of these techniques have been patented. Although some of these processes will, unlikely, never be implemented, others are highly interesting as they bring definite improvements. The main shortcomings concern the fact that they are not at all tailored to the socioeconomic settings in most vanilla-producing countries, especially in Madagascar, the top vanilla producer, or to the current structure of the world vanilla market. They have therefore practically never been adopted, or only to a marginal extent. However, the vanilla market has been quite volatile in recent years, including the implementation of relatively drastic standards in importing countries (especially in terms of the microbiological quality) which, in the medium term, could represent a challenge to conventional practices.

RELATIONSHIP BETWEEN VANILLA CURING AND AROMA DEVELOPMENT

The initial heat treatments are primarily aimed at stalling pod dehiscence, as already mentioned above, but it is generally acknowledged that these treatments also have a role in initiating the aromatic development phase, which is ongoing throughout the vanilla curing process.

The hydrolysis of different glucosylated precursors is the best-known reaction in the development of the aromatic quality of vanilla. Here, we will not get into a detailed discussion on the aromatic composition of cured vanilla, since this feature will be dealt with in Chapter 12. However, it should still be noted that many aromatic compounds, such as vanillin, vanillic acid, *p*-hydroxybenzaldehyde, *p*-hydroxybenzoic acid, and so forth, are present in green vanilla beans in glucoside form, thus without any olfactory properties, and that their hydrolysis is required to enable them to release their aromatic moiety.

Concerning vanillin, the main (quantitatively and qualitatively) aromatic constituent of vanilla, it was shown (Odoux, 2000; Gatfield et al., 2007) that hydrolysis of its precursor was initiated during the scalding and sweating processes, which then continued during the slow drying and even conditioning phases. A generally similar behavior was also observed with other glucoside precursors of vanilla aroma (Dignum et al., 2002; Perez-Silva, 2006).

Since glucovanillin and glucosidase are located in the same tissues but in different cell compartments (cf. previous chapter), studies were carried out to determine in what way the curing process was associated with bringing the enzyme and substrate into contact (Odoux et al., 2006). The findings of this study clearly showed that the initial heat treatments have a definite impact on cell integrity, but the proportion of cells that maintain their compartmentation or not in the tissues could not be clearly determined. The light microscopy findings seemed to indicate that this treatment effect was only partial, which could explain the low level of glucovanillin hydrolysis during the first curing steps, that is, after scalding at 60°C for 3 min and sweating for 24 h in crates. However, the observations during postharvest pod senescence and the freezing/thawing process clearly showed that these treatments led to complete cell structure degradation, accompanied by total hydrolysis (or almost total) of glucovanillin into vanillin, despite the heavy loss of glucosidase activity. In contrast, results obtained in a similar study (Mariezcurrena et al., 2008) seemed to show that the treatment had an imperceptible effect on the tissue cell organization until the fifth or eighth day after heat treatment. In this study, it was noted that the sweating process was conducted with a small number of pods (20) so the thermal inertia was much lower than at the center of a crate, which could explain the observed differences with respect to the cell integrity. Moreover, the authors provided no indications on the glucoside hydrolysis patterns or on the level of glucosidase activity during treatment in this study. All of these results, nevertheless, seemed to indicate that the main consequence of the initial heat treatment with respect to aroma development was that it brought the glucosidase and glucosylated aroma precursors into contact via cell decompartmentation.

However, these heat treatments also have a very negative impact on glucosidase activity (Odoux et al., 2003; Marquez and Walizewski, 2008), and the different authors who had measured changes in this activity during vanilla curing noted almost total losses as of the first scalding phase (Ranadive et al., 1983; Dignum et al., 2002; Havkin-Frenkel et al., 2005; Odoux et al., 2006; Perez-Silva, 2006), which led them to question the enzymatic origin of aroma precursor hydrolysis. The possibility of chemical or exogenous enzymatic hydrolysis (e.g., due to microbiological contamination during the drying step) was considered and tested by a few authors (Dignum et al., 2002; Havkin-Frenkel et al., 2005; Odoux et al., 2006; Perez-Silva, 2006). All of the findings seemed to confirm that this hydrolysis was actually the result of endogenous glucosidase activity. Although this activity was considerably stalled by heat treatments and not measurable by enzymatic tests developed by the different authors mentioned, aroma precursors continued to be hydrolyzed, even during very advanced phases of the process. Spectacular glucosidase activity losses (Heckel, 1910; Dignum et al., 2002; Odoux et al., 2003, 2006) were also noted during pod freezing, despite the very high rate of glucovanillin hydrolysis (cf. previous paragraph). These different observations underline that efficient cell decompartmentation of glucosylated precursors and glucosidase(s) is more important than the level of enzymatic activity itself. Heat treatments are thus essential during the conventional vanilla curing process.

Another important feature of the conventional process, which has been the focus of a few research studies over the last decade, is the marked loss of vanillin noted after glucovanillin hydrolysis. Vanillin losses of around 50% are systematically observed during the curing process (Odoux, 2000; Gatfield et al., 2007). Different studies (Gatfield et al., 2006, 2007; Frenkel and Havkin-Frenkel, 2006; Perez-Silva, 2006) have revealed that vanillin released by glucosidase can serve as a basepoint for different chemical or even enzymatic reaction mechanisms. Some of these reactions, which are far from being negative, could give rise to the formation of aroma compounds that are essential in determining the end quality of cured vanilla (see Chapter 12).

It is clear that many chemical and enzymatic reactions take place, as indicated by the color, texture, and odor modifications observed during the different vanilla curing phases. Three nonglucosidase enzymatic systems have been regularly studied during vanilla processing: peroxidases, polyphenoloxidases (PPO), and proteases (Rabak, 1916; Balls and Arana, 1941; Jones and Vicente, 1949a, 1949b; Broderick, 1956; Ranadive et al., 1983; Hanum, 1997; Jiang et al., 2000; Dignum et al., 2001, 2002; Dignum, 2002; Havkin-Frenkel et al., 2005; Marquez et al., 2008; Waliszewski et al., 2009). The findings of these studies indicated that pod browning is due to enzymatic reactions and that peroxidases and PPOs, and to a lesser extent proteases, are highly resistant to heat treatment. Peroxidases and PPO are also involved in oxidation reactions leading to the formation of aromatic molecules via lipid oxidation, and so on. Proteases are involved in the inactivation of enzymes that are essential for the formation of aroma compounds.

Finally, it is noteworthy to mention that microorganism could be involved in the formation of aroma compounds (Röling et al., 2001).

The findings of studies that have been carried out to date have not revealed the exact nature of the reaction processes involved in the highly complex phenomena that take place during vanilla curing.

REFERENCES

Ansaldi, G., Gil, G., and Le Petit, J. 1988. Procédé d'obtention d'arôme naturel de vanille par traitement des gousses de vanille verte et arôme obtenu. Brevet, INPI. No. 2, 634 979, 7pp.

Arana, F.E. 1943. Action of a β-glucosidase in the curing of vanilla. *Food Research* 8:343–351.

Arana, F.E. 1944. Vanilla curing and its chemistry. Federal Experiment Station of the US Department of Agriculture, Mayaguez, Puerto Rico, Bulletin No. 42, October 1944, 17pp.

Arnaud, N., Bayle, J.C., and Troney, C. 1983. Etude des gousses de vanille de Madagascar: Critères analytiques de la récolte 1981. *Parfums, Cosmétiques, Arômes* 53:99–101.

Balls, A.K. and Arana, F.E. 1941. The curing of vanilla. *Industrial and Engineering Chemistry* 33(8):1073–1075.

Balls, A.K., Kevorkian, A.G., and Arana, F.E. 1942. Process for curing vanilla beans. United States Patent Office, No. 2 274 120, 1p.

Bayle, J.C., Nadal, G., and Derbesy, M. 1982. Contribution à l'étude des gousses de vanille: critères analytiques de différents lots de gousses de vanille Comores (récolte 1980 et récolte 1981). *Parfums, Cosmétiques, Arômes* 47:71–73.

Bourriquet, G. 1954. Altérations et défauts de la vanille préparée. In: G. Bourriquet, ed. *Le vanillier et la vanille dans le monde*. Paul Lechevalier, Paris, pp. 623–644.

Broderick, J.J. 1956. A preliminary investigation of the quick curing of vanilla beans. *Food Technology* 10:188–189.

Brunerie, P.M. Procédé d'obtention d'arôme naturel de vanille par traitement enzymatique des gousses de vanille verte et arôme obtenu. International Patent Application, 1993, No. PCT/FR92/00837, 11pp.

Derbesy, M. 1989. Gousses de vanille Comores : banque de données pour la récolte 1987. *Parfums, Cosmétiques, Arômes* 87:77–90.

Derbesy, M., Bayle, J.C., and Troney, C. 1982. Contribution à l'étude des gousses de vanille: Critères analytiques de différents lots de gousses de Madagascar. *Parfums, Cosmétiques, Arômes* 82(43):73–77.

Derbesy, M. and Charvet, S. 2000. Evolution des critères analytiques des vanilles de l'océan indien. *Arômes Ingrédients Additifs* 26:21–25.

Dignum, M.J.W., Kerler, J., and Verpoorte, R. 2001. β–Glucosidase and peroxydase stability in crude enzyme extracts from green beans of *Vanilla planifolia* Andrews. *Phytochemical Analysis* 12:174–179.

Dignum, M.J.W., Kerler, J., and Verpoorte, R. 2002. Vanilla curing under laboratory conditions. *Food Chemistry* 79:165–171.

Dignum, 2002. *Mark Biochemistry of the Processing of Vanilla Beans*. Leiden, Th: Sciences, 103pp.

Falque, V., Uzio, R., and Derbesy, M. 1992. Vanille d'Indonésie. Evolution de six années de production. *Parfums, Cosmétiques, Arômes* 105:49–53.

Fayet, B., Saltron, F., Tisse, C., and Guerere, M. 1999. Contribution à la caractérisation des gousses de vanille. *Annales des Falsifications, de l'Expertise Chimique et Toxicologique* 92(946):11–16.

Frenkel, C. and Havkin-Frenkel, D. 2006. The physics and chemistry of vanillin. *Perfumer and Flavorist* 31:28–36.

Gassenmeier, K., Riesen, B., and Magyar, B. 2008. Commercial quality and analytical parameters of cured vanilla beans (*Vanilla planifolia*) from different origins from the 2006–2007 crop. *Flavour and Fragrance Journal* 23(3):194–201.

Gatfield, I., Hilmer, J.M., Weber, B., Hammerschmidt, F., Reiβ, I., Poutot, G., and Bertram, H.-J. 2007. Chemical and biochemical changes occurring during the traditional Madagascan vanilla curing process: Effects on the sensory qualities of cured beans. *Perfumer and Flavorist* 32(7):20–28.

Gatfield, I., Reiss, I., Krammer, G., Schmidt, C.O., Kindel, G., and Bertram, H.J. 2006. Divanillin: Novel taste-active component of fermented vanilla beans. *Perfumer and Flavorist* 31(8):18–20.

Graves, R.E., Hall, R.L., and Karas, A.J. Method of producing cured vanilla extract from green vanilla beans. United States Patent Office, 1958, No. 2 835 591, 5pp.

Hanum, T. 1997. Changes in vanillin and activity of β–glucosidase and oxidases during post harvest processing of vanilla bean (*Vanilla planifolia*). *Bulletin Teknologik dan Industri Pengan* VIII(1):46–52.

Havkin-Frenkel, D., French, J.C., and Frenkel, C. 2005. Inside vanilla. *Perfumer and Flavorist* 30:36–55.

Heckel, E. 1910. De l'action du froid et des anesthésiques sur les feuilles de l'*Angaecum fragrans* Thou. (Faham) et sur les gousses vertes de vanille. *Comptes Rendus de l'Académie des Sciences (Paris)* 151:128–131.

Jiang, M., Pu, F., Xie, W.-S., Hu, Y.-Q., and Li, Y. 2000. Activity of three enzymes in Vanilla capsule. *Acta Botanica Yunnanica* 22(2):187–190.

Jones, M.A. and Vicente, G.C. 1949a. Criteria for testing vanilla in relation to killing and curing methods. *Journal of Agricultural Research* 78(11):425–434.

Jones, M.A. and Vicente, G.C. 1949b. Inactivation and vacuum infiltration of vanilla enzyme systems. *Journal of Agricultural Research* 78(11):435–444.

Kamaruddin, A. dan Mursalim. 1997. Drying of vanilla pods using a greenhouse effect solar dryer. *Drying Technology* 15(2):685–698.

Karas, A.J., Hall, R.L., and Stahl, W.H. Vanilla bean drying and curing. United States Patent, 1972, No. 3 663 238, 5pp.

Kaul, J. Curing of vanilla beans. United States Patent Office, 1967, No. 3 352 690, 3pp.

Leong, G. 1991. *Contribution à l'étude des hétérosides des gousses de vanille vertes*. Marseille III, Th: Sciences, 153pp.

Mane, J. and Zucca, J. Procédé d'obtention d'arôme naturel de vanille par traitement des gousses de vanille et arôme obtenu. Brevet, INPI, 1992, No. 2 691 880, 8pp.

Mariezcurrena, M.D., Zavaleta, H.A., Waliszewski, K.N., and Sanchez, V. 2008. The effect of killing conditions on the structural changes in vanilla (*Vanilla planifolia*, Andrews) pods during the curing process. *International Journal of Food Science and Technology* 43:1452–1457.

Marquez, O. and Waliszewski, K.N. 2008. The effect of thermal treatment on β-glucosidase inactivation in vanilla bean (*Vanilla planifolia* Andrews). *International Journal of Food Science and Technology* 43:1993–1999.

Marquez, O., Waliszewski, K.N., Oliart, R.M., and Pardio, V.T. 2008. Purification and characterization of cell wall-bound peroxidase from vanilla bean. *LWR – Food Science and Technology* 41:1372–1379.

Obolenski, G. 1957. Etude de l'action des ultra-sons sur le développement de la vanilline dans la préparation de la vanille. *Materiae Vegetabilis* II(4):336–341.

Obolenski, G. 1958. Influence des ultra-sons sur l'augmentation des taux de vanilline et d'oléorésines dans les gousses de vanille. *Qualitas Plantarum et Materiae Vegetabilis* 5(1/2):45–63.

Obolenski, G. 1959. Etude de l'extraction de la vanille par les ultra-sons. *Qualitas Plantarum et Materiae Vegetabilis* 5(4):344–366.

Odoux, E. Projet Fed-Stabex de Relance des Cultures d'Exportation dans la région du SAVA: Volet Vanille. Rapport de missions à Madagascar (du 01/07 au 27/07/98 et du 15/09 au 22/10/98), Novembre 1998, 16pp.

Odoux, E. 2000. Changes in vanillin and glucovanillin concentrations during the various stages of the process traditionally used for curing *Vanilla fragrans* beans in Réunion. *Fruits* 55:119–125.

Odoux, E., Chauwin, A., and Brillouet, J.M. 2003. Purification and characterization of vanilla bean (*Vanilla planifolia* Andrews) β–D-glucosidase. *Journal of Agricultural and Food Chemistry* 51:3168–3173.

Odoux, E., Escoute, J., and Verdeil, J.L. 2006. The relation between glucovanillin, β-D-glucosidase activity and cellular compartmentation during the senescence, freezing and traditional curing of vanilla beans. *Annals of Applied Biology* 149(1):43–52.

Perera, C.O. and Owen, E. 2010. Effect of tissue disruption by different methods followed by incubation with hydrolyzing enzymes on the production of vanillin from Tongan vanilla beans. *Food and Bioprocess Technology* 3:49–54.

Perez-Silva, A. 2006. Contribution à l'étude de la genèse des composes d'arôme au cours du procédé mexicain de transformation de la vanille (*Vanilla planifolia* G. Jackson). Montpellier II, Th: Sciences, 136pp.

Rabak, F. 1916. The effect of curing on the aromatic constituents of vanilla beans. *The Journal of Industrial and Engineering Chemistry* 8(9):815–821.

Ranadive, A.S., Szkutnica, K., Guerrera, J.G., and Frenkel, C. 1983. Vanillin biosynthesis in vanilla beans. In: *Proceedings of the 9th International Congress on Essential Oils*, Singapore, Malaysia, pp. 147–154.

Ranadive, A.S. 1994. Vanilla—Cultivation, curing, chemistry, technology and commercial products. In: G. Charalambous, ed. *Spices, Herbs and Edible Fungi*. Elsevier, Amsterdam, pp. 517–577.

Ratobison, R., Zeghmati, B., Reddy, T.A., and Draguenet, M. 1998. Sizing of solar supplemented liquid and air heating systems for the treatment of vanilla. *Solar Energy* 62(2):131–138.

Röling, W.F.M., Kerler, J., Braster, M., Apriyantono, A., Stam, H., and Van Verseveld, H.M. 2001. Microorganisms with a taste for vanilla: microbial ecology of traditional Indonesian vanilla curing. *Applied and Environmental Microbiology* 67(5):1995–2003.

Ruiz-Teran, F., Perez-Amador, I., and Lopez-Munguia, A. 2001. Enzymatic extraction and transformation of glucovanillin to vanillin from vanilla green pods. *Journal of Agricultural and Food Chemistry* 49:5207–5209.

Saltron, F., Langella, C., and Guerere, M. 2002. Evaluation de la qualité de la vanille malgache. Récolte 2000. *Annales des Falsifications, de l'Expertise Chimique et Toxicologique* 95 (958):79–105.

Sreedhar, R., Roohie, K., Maya, P., Venkatachalam, L., and Bhagyalakshmi, N. 2009. Biotic elicitor enhance flavour compounds during accelerated curing of vanilla beans. *Food Chemistry* 112:461–468.

Sreedhar, R., Roohie, K., Venkatachalam, L., Narayan, M.S., and Bhagyalakshmi, N. 2007. Specific pretreatments reduce curing period of vanilla (*Vanilla planifolia*) beans. *Journal of Agricultural and Food Chemistry* 55(8):2947–2955.

Théodose, R. 1973. Traditional methods of vanilla preparation and their improvement. *Tropical Science* 15(1):47–57.

Towt, L.V. Methods of dehydrating and curing vanilla fruit. United States Patent Office, 1952, No. 2 621 127, 2pp.

Waliszewski, K.N., Marquez, O., and Pardio, V.T. 2009. Quantification and characterisation of polyphenol oxidase from vanilla bean. *Food Chemistry* 117:196–203.

12 Developing the Aromatic Quality of Cured Vanilla Beans (*Vanilla planifolia* G. Jackson)

Eric Odoux

CONTENTS

COMPOSITION OF VOLATILE COMPOUNDS IN CURED VANILLA BEANS

The analysis of the volatile fraction of cured vanilla, which, paradoxically, has been the subject of little research, has made it possible to detect and identify over 150 molecules belonging to numerous chemical classes (Klimes and Lamparsky, 1976; Galetto and Hoffman, 1978; Schulte-Elte et al., 1978; Nakasawa et al., 1981; Vidal et al., 1989; Adedeji et al., 1993; Kiridena et al., 1994; Ramaroson-Raonizafinimanana et al., 1997; Werkhoff and Günter, 1997; Sostaric et al., 2000; Perez-Silva et al., 2006).

From a chemical viewpoint, these different compounds belong to the following classes: hydrocarbons, alcohols, aldehydes, ketones, esters, lactones, acids, terpenoids, heterocyclics, and phenols.

Among the hydrocarbons, a large number of alkanes have been identified (such as *n*-pentacosane), methylated and ethylated derivatives of these alkanes, alkenes (such as 1-hentriacontene), terpenoids (such as α-pinene or limonene), and aromatic rings (such as benzene and some of its derivatives) (Figure 12.1).

Aliphatic acids (such as acetic acid and linoleic acid) are also represented by a number of molecules, as well as aromatic acids (such as benzoic acid or cinnamic acid) (Figure 12.1).

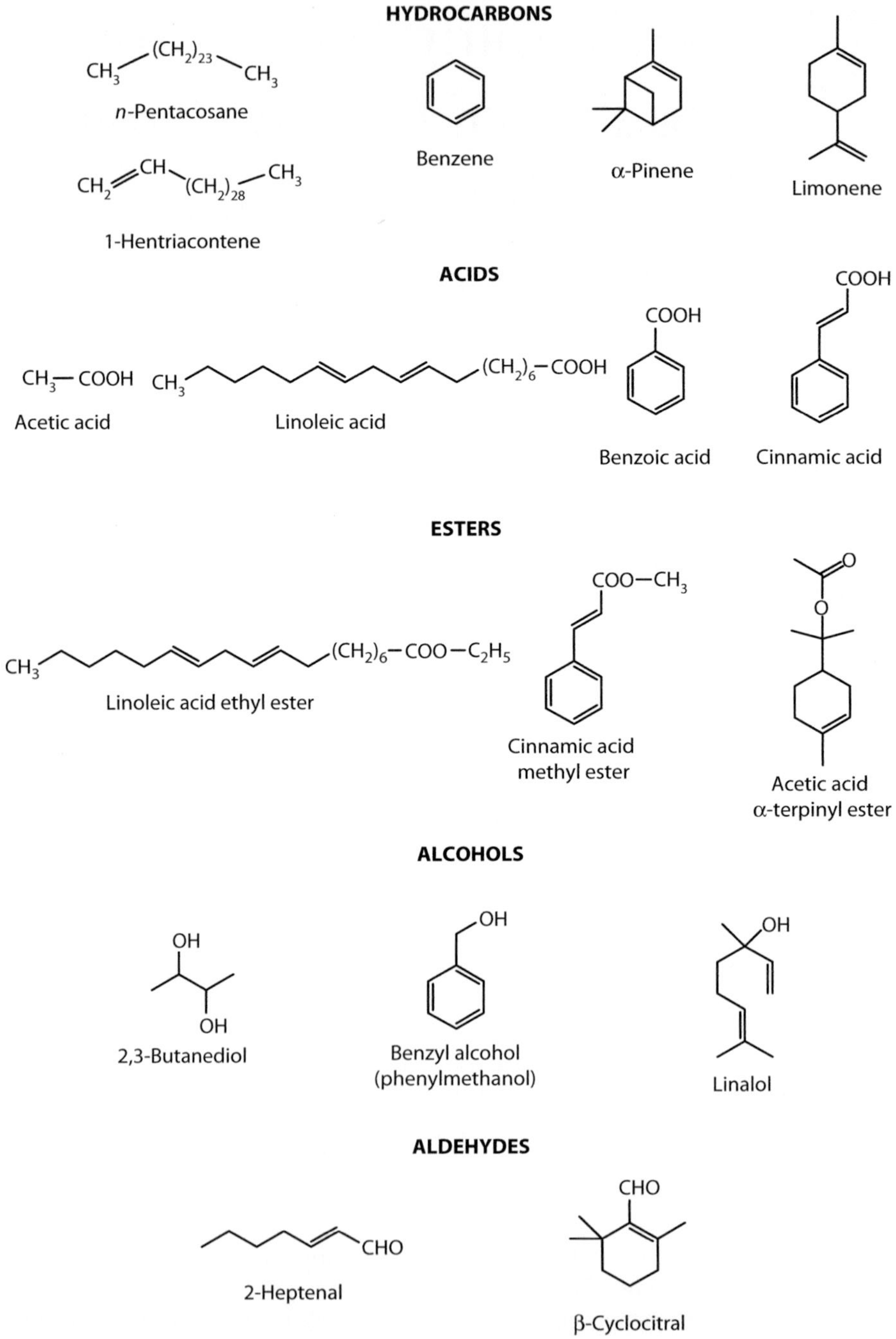

FIGURE 12.1 Chemical structures of some volatile compounds from different chemical classes identified in cured vanilla beans.

KETONES

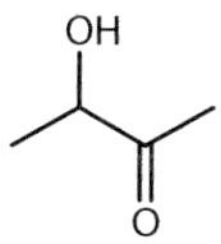

3-Hydroxy-2-butanone

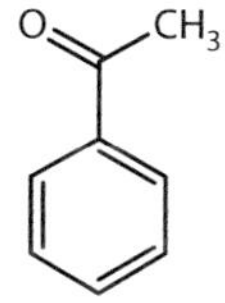

Acetophenone

LACTONE

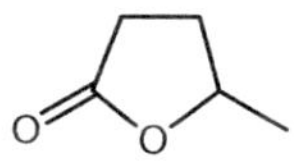

γ-Butyrolactone

HETEROCYCLICS

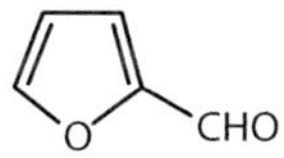

Furfural

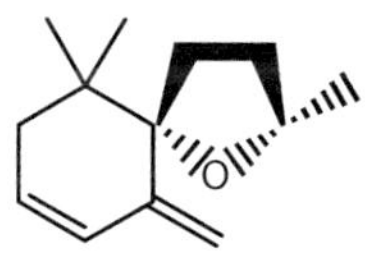

cis-Vitispirane

PHENOLS

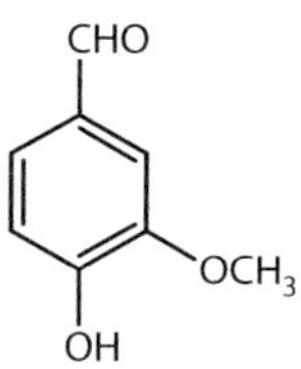

Vanillin

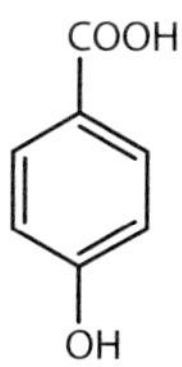

4-Hydroxybenzoic acid

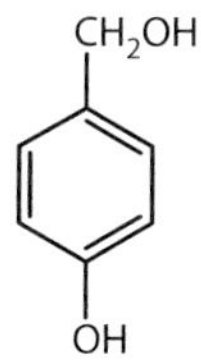

4-Hydroxybenzyl alcohol

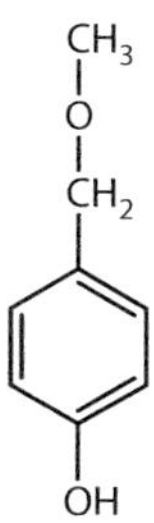

4-Hydroxy benzyl methyl ether (4-methoxy methyl phenol)

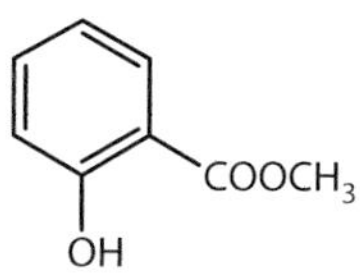

Salicylic acid methyl ester

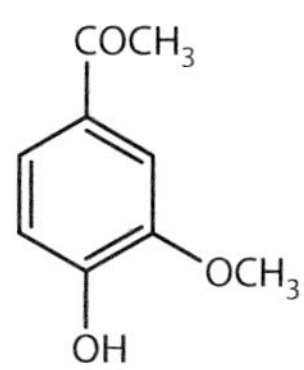

Acetovanillone

FIGURE 12.1 Continued.

Aliphatic esters and aromatic esters (such as linoleic acid ethyl ester or cinnamic acid methyl ester) are also found, and even terpene esters (such as acetic acid α-terpinyl ester) (Figure 12.1).

Aliphatic alcohols (such as 2,3-butanediol), aromatic alcohols (such as benzyl alcohol), and even terpenoids (such as linalol) have been identified (Figure 12.1).

Aliphatic aldehydes (such as 2-heptenal), terpene aldehydes (such as β-cyclocitral), aliphatic ketones (such as 3-hydroxy-2-butanone), and aromatic ketones (such as acetophenone) are also present (Figure 12.1).

Several lactones have also been reported as being a component of the volatile fraction of vanilla (such as γ-butyrolactone), along with heterocyclics (such as furfural or *cis*-vitispirane) (Figure 12.1).

Finally, phenols, which are the major representatives (both qualitatively and quantitatively) of the volatile fraction of cured vanilla, and which may also bear the aldehyde functions (such as 3-methoxy-4-hydroxy-benzaldehyde or vanillin), acid functions (such as 4-hydroxybenzoic acid), alcohol functions (such as 4-hydroxybenzyl alcohol), and even ketone functions (such as acetovanillone) (Figure 12.1); numerous esters exist (such as salicylic acid methyl ester) as well as ethers (such as 4-hydroxy benzyl methyl ether) (Figure 12.1).

A fairly exhaustive list of the compounds identified in cured vanilla can be consulted in two recent review articles by Dignum et al. (2001) and Ranadive (2006).

This composition may vary according to the geographical origin (the concept of "*terroir*" in the broad sense) and botanical origin of the samples (Adedeji et al., 1993).

Concerning the botanical aspect, only the species *V. planifolia*, *Vanilla tahitensis* (see Chapter 13), and to a lesser extent *V. pompona* (Ehlers and Pfister, 1997), have been the subject of research on their aromatic composition. But even in the case of *V. planifolia*, much research remains to be done using well-defined genetic material and a standardized curing process.

Finally, it should be remembered that extraction and analysis techniques may also result in variability in the aromatic composition (and also in artifacts).

Quantitatively, the major component in cured vanilla is vanillin (3-methoxy-4-hydroxybenzaldehyde, Figure 12.2), which may reach concentrations of several tens of grams per kilogram of dry weight (in other words, several tens of thousands of ppm) (ISO 5565-1). Three other compounds are also commonly quantified, as they are considered as indicators of quality and authenticity (see Chapter 15); these are *p*-hydroxybenzaldehyde and vanillic acid (Figure 12.2), whose concentrations are generally measured at around 1 g/kg of dry weight (or around 1000 ppm), in other words, around 10 times lower than that of vanillin, and finally *p*-hydroxybenzoic acid (Figure 12.2), whose average concentration is around 100 mg/kg of dry weight (or around 100 ppm), in other words, 100 times lower than that of vanillin. These figures may of course vary considerably, depending on the quality of the product (Gassenmeier et al., 2008; Saltron et al., 2002).

Other compounds have also been measured at concentrations of more than 100 ppm, such as acetic acid or 4-hydroxy benzyl methyl ether (Klimes and Lamparsky, 1976) and also hexadecanoic acid and linoleic acid (Perez-Silva et al., 2006) (Figure 12.2). Adedeji et al. (1993) even report a large number of compounds

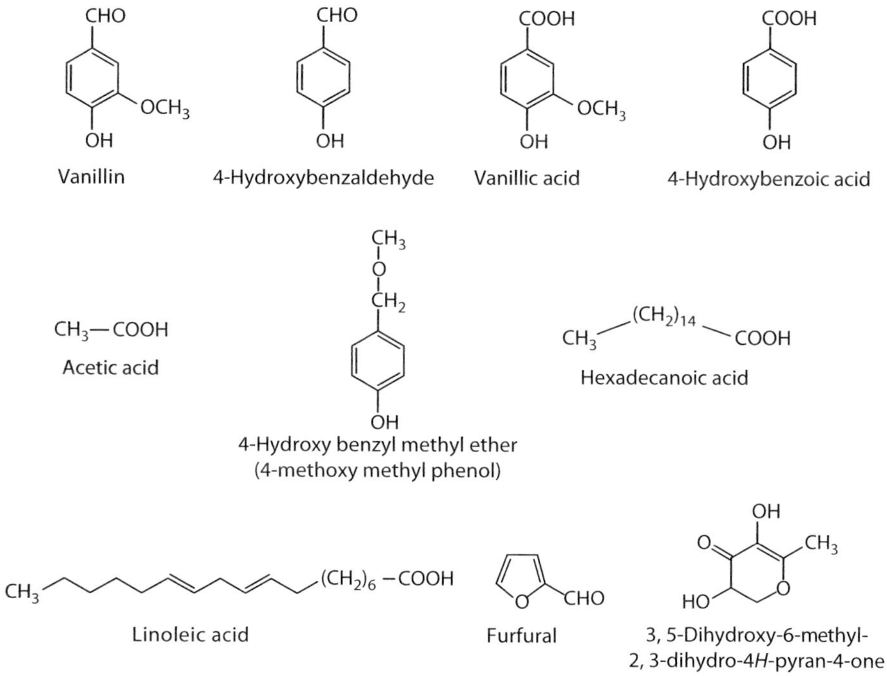

FIGURE 12.2 Chemical structures of the major compounds (quantitatively) found in the volatile fraction of cured vanilla beans.

with concentrations of more than 1000 ppm, such as 2-furfural (which is in fact the second most abundant compound after vanillin in a sample of Mexican vanilla) or 3,5-dihydroxy-6-methyl-2,3-dihydro-4*H*-pyran-4-one (Figure 12.2), which exceeds 3000 ppm in most of the samples analyzed, in other words, concentrations that are higher than those of *p*-hydroxybenzaldehyde and vanillic acid.

It should, nevertheless, be noted that techniques for analyzing and measuring these compounds vary greatly from one author to another, and are also very different from the standardized techniques (ISO 5565-2), which explains the differences observed.

In fact, it is generally acknowledged that 95% of the volatile compounds in cured vanilla are found at concentrations of less than 10 ppm (Hoffman et al., 2005).

Although vanillin is the major component quantitatively, this molecule alone does not explain the quality of the global aroma of vanilla; it is even difficult to show a correlation between the vanillin content and the sensory profile of vanilla extract (Gassenmeier et al., 2008). In fact, rather surprisingly, very few studies have been published on the contribution of different aroma compounds to the quality of the aroma and flavor of cured vanilla.

Using GC-olfactometry (but without providing details of the methodology employed), Dignum et al. (2004) found that *p*-cresol, 2-phenylethanol, guaiacol, and 4-creosol (Figure 12.3) had a considerable impact on vanilla aroma. Also using GC-olfactometry, Perez-Silva et al. (2006) detected 26 aroma-active compounds in a Mexican vanilla extract; of these compounds, 13 are derivatives of the phenylpropanoid pathway (see Chapter 10), of which some, such as guaiacol, 4-creosol, acetovanillone, and salicylic acid methyl ester (Figure 12.3), are similar in intensity to vanillin, while their concentrations are 1000 times lower (sometimes even less) in the extract. It should be noted that *p*-cresol, cinnamic acid methyl ester, and anisyl alcohol (Figure 12.3) have intermediate intensities, even though they have concentrations of just a few ppm. In general, these compounds are responsible for sweet, woody, balsamic, spicy, vanilla-like, and toasted notes. Certain aliphatic aldehydes, alcohols, and acids also have intermediate intensities despite concentrations of less than 10 ppm.

In a study aimed at correlating sensory analysis and chemical analysis in order to classify the quality of vanilla extracts, Hoffman et al. (2005) identified 14 compounds that were detected in almost all the extracts studied (55 samples in total) and contributed positively or negatively to the aroma and flavor.

Acetic acid ethyl ester, hexanoic ethyl ester, octanoic ethyl ester, nonanoic ethyl ester, and hexanal ethyl acetal (Figure 12.3) are highly correlated with the "age-related compounds" criterion; 4-hydroxy benzyl ethyl ether, vanillyl ethyl ether, and acetaldehyde ethyl acetal (Figure 12.3) are highly correlated with the "rummy/resinous" criterion; and *p*-hydroxybenzaldehyde and vanillin (Figure 12.3) are highly correlated with the "vanillin" criterion. All of these compounds are therefore associated with positive criteria, whereas *p*-hydroxybenzoic acid, vanillic acid, guaiacol, and nonanoic acid (Figure 12.3) are highly correlated with the "smoky/phenolic" criterion, in other words, a negative descriptor.

Finally, less volatile molecules should be mentioned, which are consequently involved in the quality of the flavor rather than the aroma, and are the subject of recent publications (Gatfield et al., 2006; Schwarz and Hofmann, 2009). Seven molecules (Figure 12.4), including divanillin, were identified in cured vanilla and in extracts, and are involved in the velvety mouth-coating sensation. Divanillin, for example, was found at a concentration of around 170 ppm in Madagascan vanilla (Gatfield et al., 2006). These molecules are mostly the products of condensation between two phenolic compounds.

COMPOSITION OF GLUCOSYLATED COMPOUNDS IN GREEN VANILLA BEANS

As mentioned in the previous section, the most important compounds in the volatile fraction, both quantitatively and qualitatively, have an aromatic ring and are derived from the phenylpropanoid pathway. Many of them are also found in the green fruit in the form of glycosides—in other words, bound by a glycosidic bond (such as *O*-glycoside) to one or several sugar groups—and have no olfactory properties.

Historically, the presence of glucosylated aroma precursors in the green vanilla bean was long suspected. After much debate (Lecomte, 1901, 1913; Goris, 1924), it was established that "vanilloside" (glucovanillin) was the direct precursor of vanillin, and that three other precursors also existed, but in smaller quantities. Different publications describe the attempts made to isolate and identify them (Goris, 1947; Janot, 1954). These are glucosides of vanillyl alcohol ("vanilloloside"),

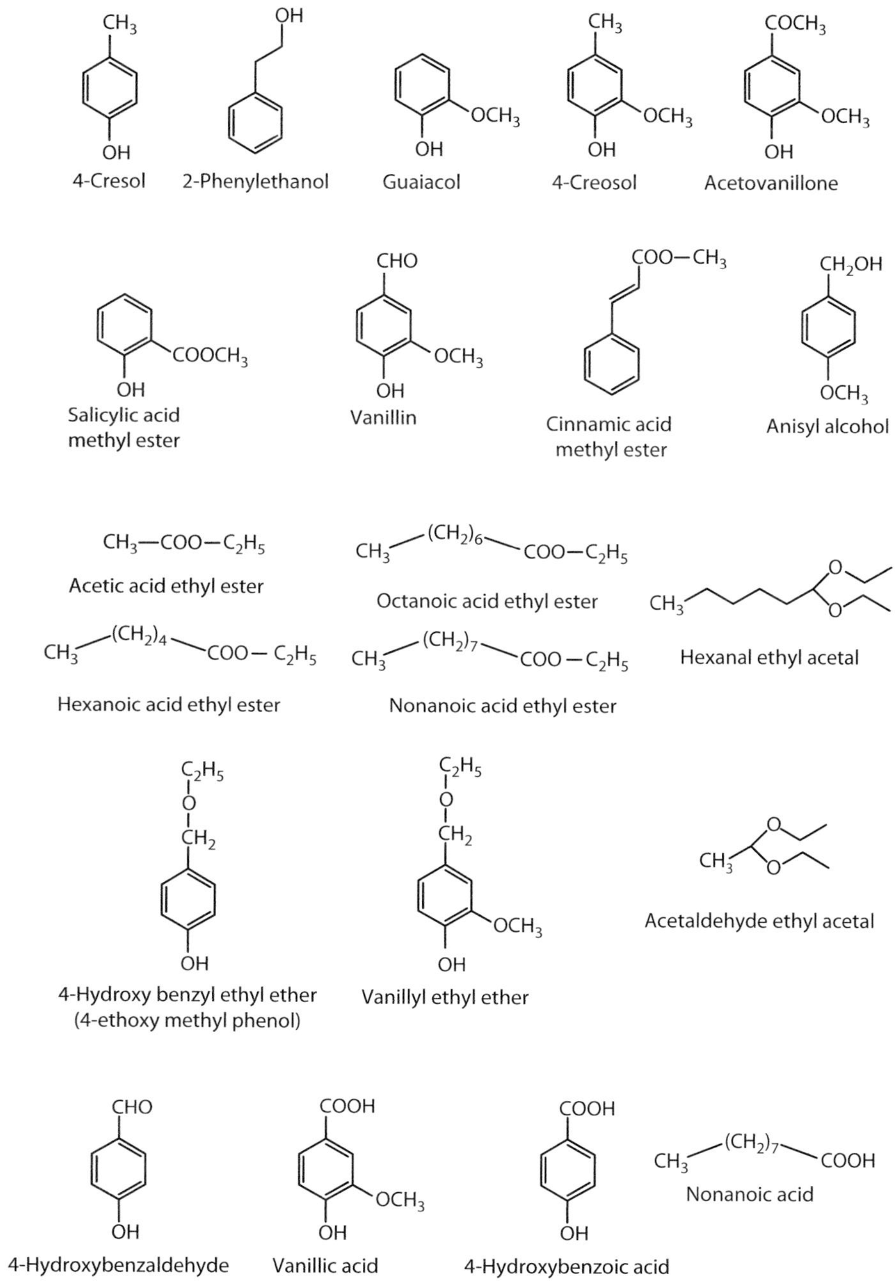

FIGURE 12.3 Chemical structures of some aroma-active compounds detected in cured vanilla beans.

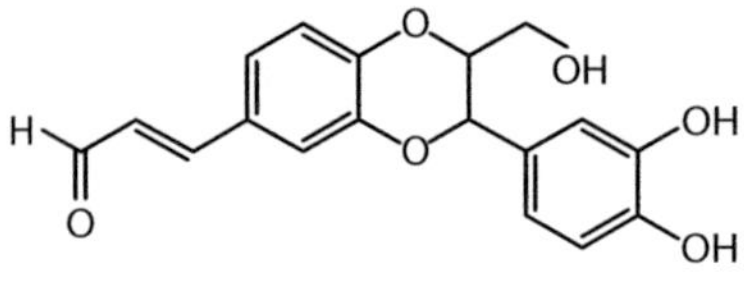

Americanin

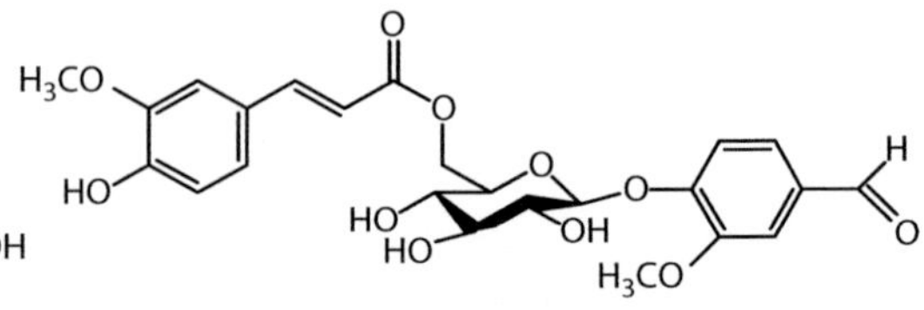

(1-*O*-Vanilloyl)-(6-*O*-feruloyl)-β-D-glucopyranoside

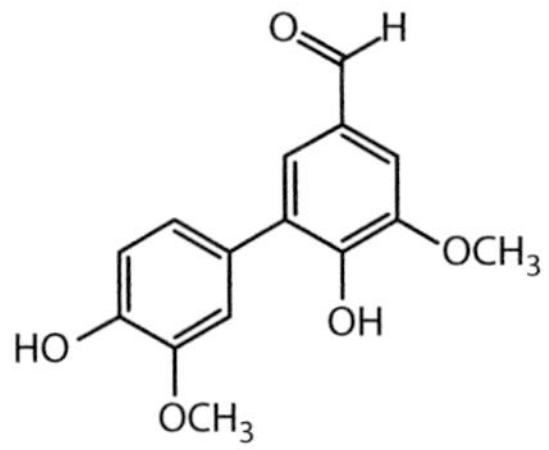

4′,6′-Dihydroxy-3′,5-dimethoxy
-[1, 1′-biphenyl]-3-carboxaldehyde

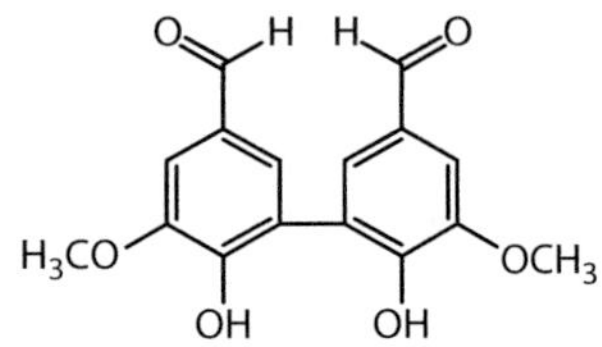

Divanillin

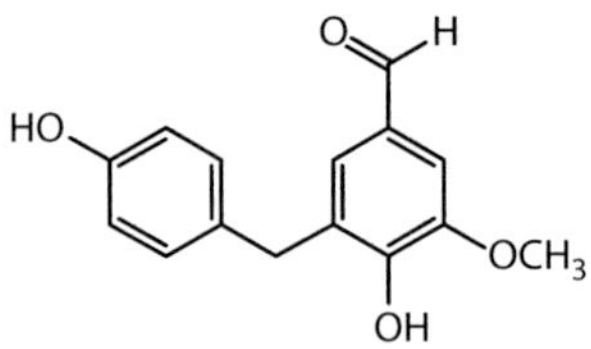

5-(4-Hydroxybenzyl)vanillin

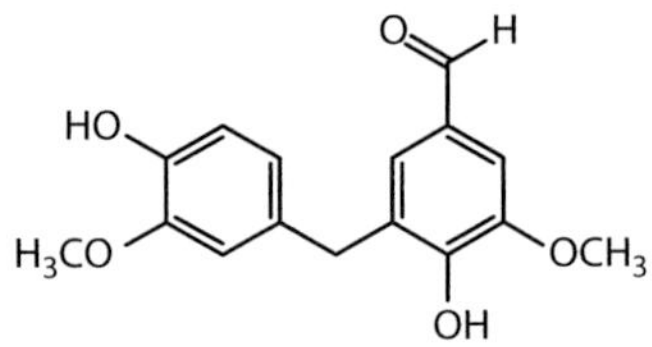

4-Hydroxy-3-(4-hydroxy-3-methoxybenzyl)
-5-methoxybenzaldehyde

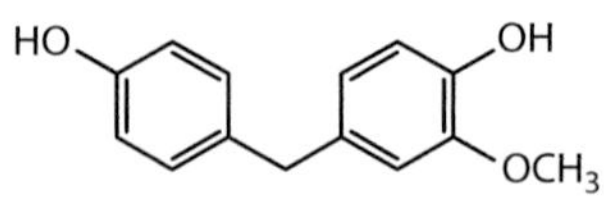

4-(4-Hydroxybenzyl)-2-methoxyphenol

FIGURE 12.4 Chemical structures of compounds involved in the velvety mouth-coating sensation identified in cured vanilla beans.

of protocatechuic aldehyde (3,4-dihydroxy benzaldehyde), and of an unidentified ester (Figure 12.5).

More recently, glucosides of vanillin, *p*-hydroxybenzaldehyde, *p*-hydroxybenzoic acid, and vanillic acid (Figure 12.5) were identified in the green bean (*V. planifolia* origin Comoros, Réunion, Madagascar, Indonesia), using modern analytical techniques (Leong et al., 1989a,b; Leong, 1991). Moreover, glucose seems to be the major sugar, since the acid hydrolysis of a prepurified extract of vanilla glycosides followed by an acetylation assay makes it possible to obtain around 20% glucose, 1% mannose, and traces of rhamnose (percentage expressed in relation to initial dry matter).

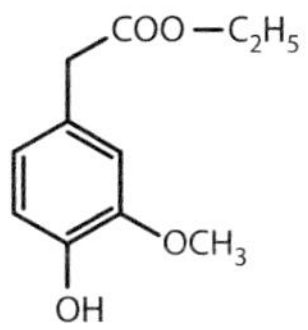

Homovanillic acid ethyl ester
(Kanisawa, 1993, 1994)

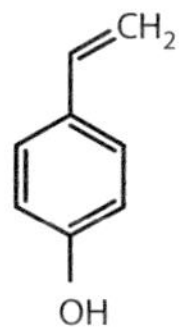

4-Vinylphenol
(Kanisawa, 1993, 1994; Perez-Silva, 2006)

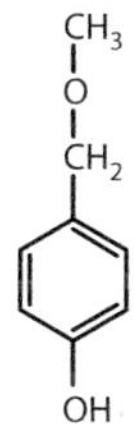

4-Hydroxy benzyl methyl ether
(Kanisawa, 1993, 1994)

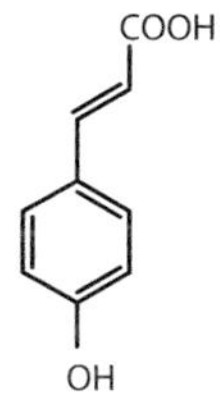

4-Hydroxycinnamic acid
(Kanisawa, 1993, 1994)

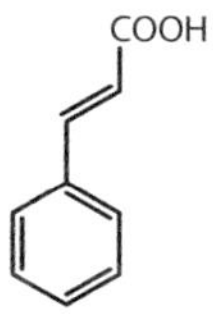

Cinnamic acid
(Kanisawa, 1993, 1994)

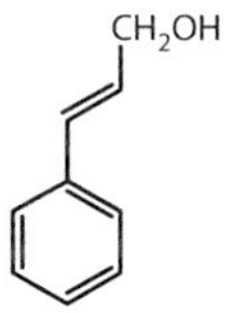

Cinnamyl alcohol
(Kanisawa, 1993, 1994; Perez-Silva, 2006)

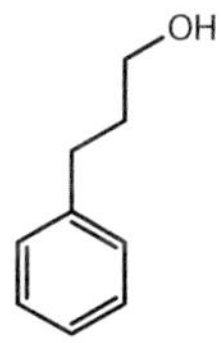

3-Phenylpropanol
(Kanisawa, 1993, 1994; Perez-Silva, 2006)

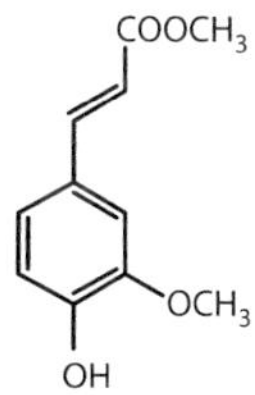

Ferulic acid methyl ester
(Kanisawa, 1993)

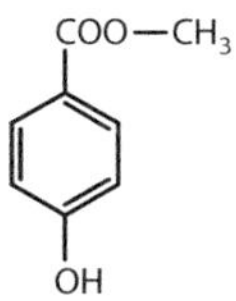

4-Hydroxybenzoic acid methyl ester
(Kanisawa, 1993)

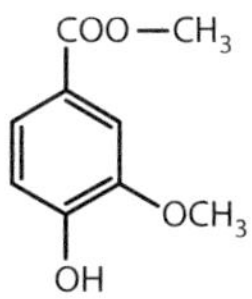

Vanillic acid methyl ester
(Kanisawa, 1993)

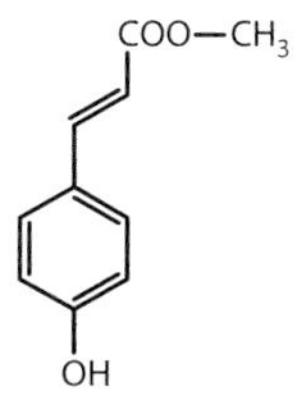

4-Hydroxycinnamic acid methyl ester
(Kanisawa, 1993)

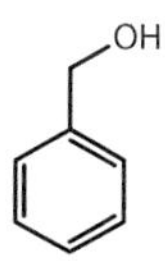

Benzyl alcohol (phenylmethanol)
(Kanisawa, 1993; Perez-Silva, 2006)

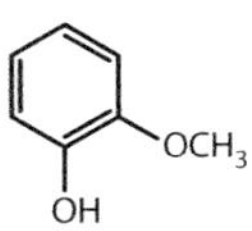

Guaiacol
(Dignum, 2002)

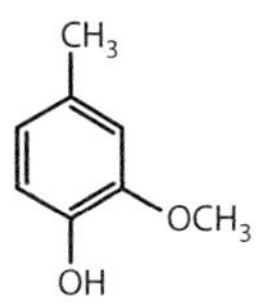

4-Creosol
(Kanisawa, 1993, 1994; Dignum, 2004; Perez-Silva, 2006)

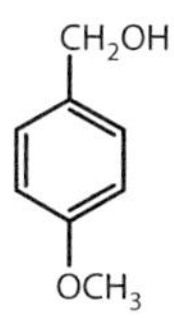

Anisyl alcohol
(Perez-Silva, 2006)

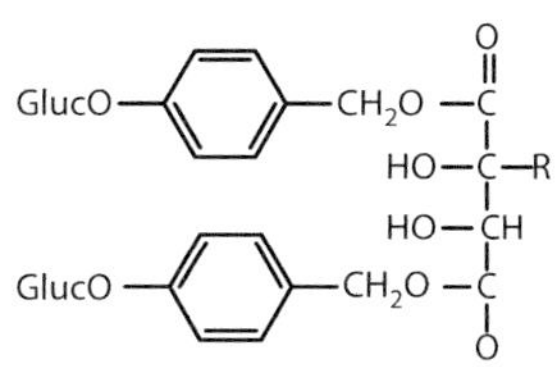

R = —CH(CH_3)$_2$: glucoside A
R = —CH(CH_2CH_3)(CH_3) : glucoside B

(Kanisawa, 1993, 1994; Dignum, 2004)

FIGURE 12.5 Chemical structures of the different aglycones found as glucosides in the green Vanilla beans according to different authors.

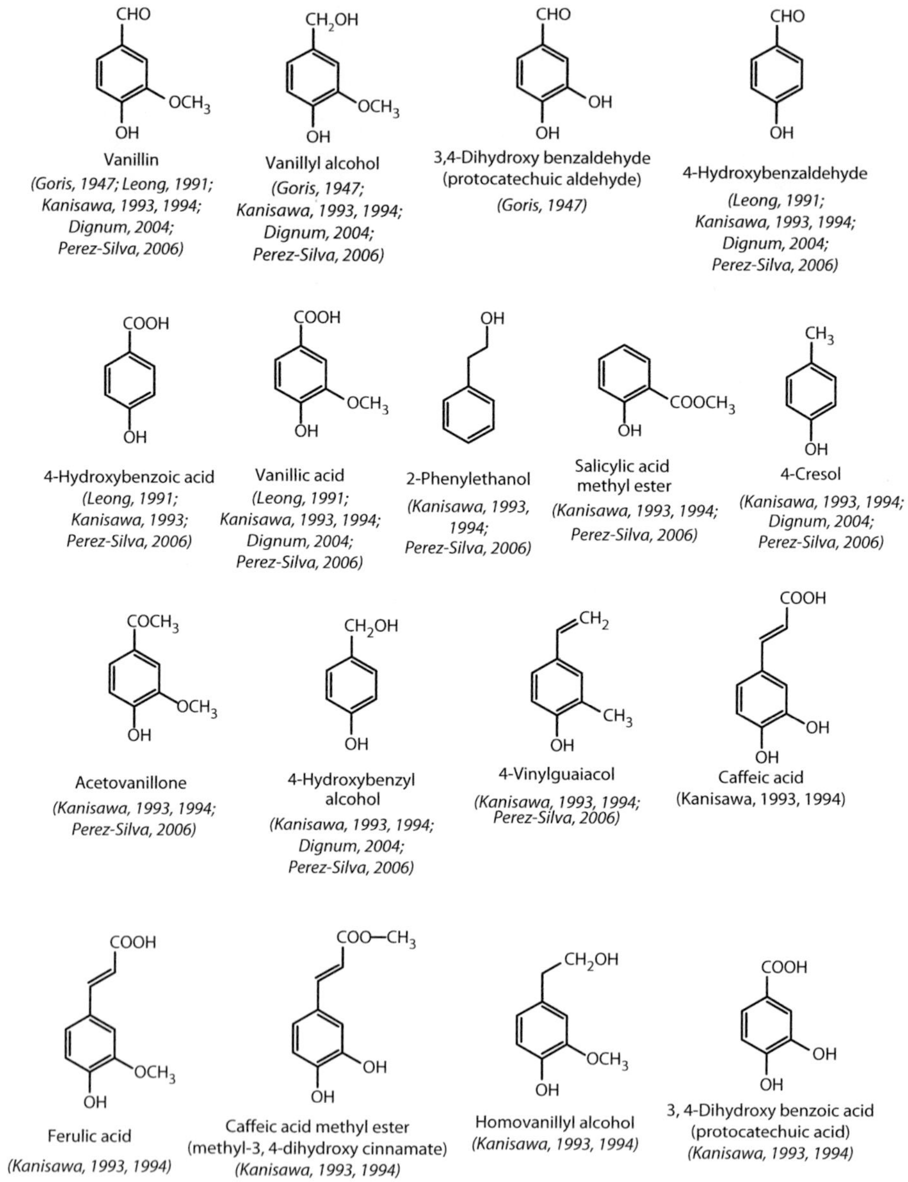

FIGURE 12.5 Continued.

Kanisawa (1993) identified the glucosides of 31 molecules (Figure 12.5) in the green vanilla bean (*V. planifolia* origin Indonesia), including the vanillin, *p*-hydroxybenzaldehyde, *p*-hydroxybenzoic acid, vanillic acid, and vanillyl alcohol already identified by the previous authors. Compounds such as 2-phenyl ethanol, salicylic acid methyl ester, *p*-cresol, and acetovanillone—aroma active compounds—were also found in glucosylated form.

This author also isolated and identified *p*-hydroxybenzyl alcohol and two diglucosides, bis[4-(β-D-glucopyranosyloxy)-benzyl] 2-isopropyl tartrate (glucoside A), and bis[4-(β-D-glucopyranosyloxy)-benzyl] 2-(1-methyl-propyl) tartrate (glucoside B), which are nevertheless only intermediaries in the biosynthetic pathway of the phenolic compounds.

In a later publication, however, Kanisawa et al. (1994) did not confirm the identification of the glucosides of 4 methyl esters: ferulic acid methyl ester, *p*-hydroxybenzoic acid methyl ester, vanillic acid methyl ester, and *p*-hydroxycinnamic acid methyl ester. More surprisingly, nor did they confirm the presence of glucosides of *p*-hydroxybenzoic acid and benzyl alcohol in this publication.

In a recently published study, Dignum et al. (2004) looked for different glucosides in the green bean (*V. planifolia* origin Indonesia). The different glucosides identified are those of vanillin, vanillic acid, *p*-hydroxybenzaldehyde, *p*-cresol, creosol, vanillyl alcohol, and the two diglucosides A and B (Figure 12.5). In a previous study (Dignum et al., 2002), these authors also found guaiacol glucoside, a compound that they were subsequently unable to isolate.

Finally, Perez-Silva (2006) identified 17 glucosides in green vanilla beans from Mexico, including the glucoside of anisyl alcohol (Figure 12.5), which was reported for the first time.

However, it is important to note that the presence of many of the aforementioned molecules must be confirmed, as only the glucosides of vanillin, *p*-hydroxybenzaldehyde, *p*-hydroxybenzoic acid, vanillic acid, and the glucosides A and B have been formally isolated and identified. Most of the other glucosides mentioned were first hydrolyzed by β-glucosidase action, and the free aglycones were identified. This kind of approach is likely to produce artifacts due to the reactivity of aglycones after their release.

FORMATION OF AROMA COMPOUNDS DURING CURING

GLUCOSIDES

As mentioned in Chapter 11, the best-known reaction in the development of the aromatic quality of vanilla during curing is the hydrolysis of the different glucosides that are precursors of the aforementioned aroma compounds. This is the fundamental reaction in the aroma development of vanilla, as it allows an aglycone that may have olfactory properties to be released from a glucoside that has none.

This reaction can be achieved either chemically under acidic conditions or enzymatically using a β-glucosidase (or more generally, with a glycosidase, depending on the nature of the sugar linked to the aglycone). In the case of glucovanillin, despite some debate (see Chapter 11), it is now acknowledged that this hydrolysis is the result of one or several β-glucosidases that are endogenous to the fruit.

This idea was first put forward by Miller in 1754 (quoted by Janot [1954]). Lecomte (1901) was then able to prove the existence of *ferments hydratant et oxydant*—probably the origin of the term fermentation, which is still often incorrectly associated with vanilla curing—and studied their role in the development of the vanilla aroma.

After the glucovanillin (and other glucosides) had been isolated and the glucosidase activity in the raw extracts of the green bean had been measured, it seemed clear that the main reaction involved in the aromatic development of vanilla was the hydrolysis of the glucosylated compounds by a glucosidase (Arana, 1943): Lecomte's *ferment hydratant* (1901). Many authors subsequently measured this enzymatic activity as part of their studies on the transformation process, the physiology of the fruit, and so on, but no research has been published on the purification and characterization of this glucosidase.

Kanisawa et al. (1994) mentioned the existence of two β-glucosidases in green vanilla beans: one is very specific to glucovanillin and *p*-hydroxybenzaldehyde glucoside, and the other has a much broader spectrum of activity. These results were obtained after precipitation using ammonium sulfate and cation-exchange chromatography. However, the experimental results are not included in the publication.

Odoux et al. (2003) purified and characterized a β-glucosidase of vanilla. An enzyme was isolated, with a native molecular weight of 200 kDa, an optimal pH of 6.5, an optimal temperature of 40°C, a K_m of 1.1 mM with pNPG and 20 mM with glucovanillin, and a V_m of around 5 μkatal mg^{-1} protein with the two substrates.

Hanum (1997) obtained a K_m value of 0.38 mM with *p*NPG and Dignum et al. (2004) obtained a K_m value of 3.3 mM from a raw enzyme extract of green beans. The latter also studied the kinetic parameters of the raw enzyme extract with different glucosides.

The findings show that the enzyme has greater affinity for the glucosides of vanillin, ferulic acid, and *p*-hydroxybenzaldehyde than for the glucosides of vanillic acid, guaiacol and creosol, and no activity for the glucosides of *p*-cresol and 2-phenyl ethanol.

A previous study (Dignum et al., 2002), based on monitoring aroma compounds during curing under laboratory conditions for a batch of vanilla from Indonesia, nevertheless, appeared to show a reduction in the glucoside of guaiacol (at least of the glucoside identified as such), which seemed less evident for the glucoside of vanillic acid. The results obtained for other compounds (aglycone form only) led these authors to conclude that the formation of the minor compounds, such as *p*-cresol and 2-phenyl ethanol, appears to occur via a chemical process rather than by an enzymatic one, where the lack of enzyme activity on the glucosides of these two compounds seems to confirm.

Research monitoring aroma compounds, during traditional curing in Mexico, Perez-Silva (2006) observed that of the 17 glucosides identified, 11 are only slightly hydrolyzed, if at all, which seems to confirm the findings of Dignum et al. (2002). These 11 compounds are the glucosides of creosol, 4-vinyl guaiacol, methyl salicylate, *p*-hydroxybenzoic acid, *p*-cresol, anisyl alcohol, 2-phenyl ethanol, phenyl propanol, benzyl alcohol, and cinnamyl alcohol (Figure 12.5).

To sum up, the action of a β-glucosidase on a certain number of aroma precursor glucosides is clearly established in the development of vanilla aroma during curing, but is far less evident for certain others, even though their corresponding aglycones may be present (see below).

It must nevertheless be stressed that given the very low concentrations of these compounds, the natural variability of these concentrations and their reactivity (especially the free aglycones), and so on, monitoring during curing is very difficult to implement and requires special attention regarding the sample size in order to avoid the sometimes erratic biases and fluctuations observed. This implies specific studies that are often lacking.

Approaches that consist in studying the kinetic parameters of the different aroma precursor glucosides on purified β-glucosidase(s) are undoubtedly more demonstrative for establishing once and for all whether these glucosides are substrates of the enzyme, but clearly imply obtaining glucosides and purified enzyme, which represents a good deal of work in itself.

It is also possible to work on raw enzyme extracts rather than on purified enzymes in order to simplify the task and to be sure to have all the β-glucosidase activities present in the green bean (if several exist). It should nevertheless be remembered that the enzyme(s) may be put into contact with products that affect the kinetic parameters (activators/inhibitors), which would otherwise not come in contact in the green bean or during curing.

From experience, we observed during the development of our enzymatic test that extracting the enzyme with low buffer/beans ratios (i.e., 4:1 v/w) led to rapid and considerable losses in β-glucosidase activity (even using a protective agent, such as antiprotease, antioxidants, etc.), while very high buffer/beans ratios (2000:1 v/w) made it possible to achieve high stability for over 4 h of storage at ambient temperature without any protective agent in the extraction buffer (unpublished results). These observations suggest the presence of compounds that inhibit β-glucosidase activity in moderately diluted green vanilla extracts.

Other Compounds

In most cases, the reaction mechanisms that lead to the appearance and/or disappearance of the other aroma compounds during curing remain to be determined.

For example, aldehydes are not found in green vanilla beans (Perez-Silva, 2006), but appear in advanced stages of curing when their content peaks, and then disappear with only traces found in cured vanilla. Some of these aldehydes could be products of the autooxidation of oleic acid and linoleic acid (found in large quantities in green vanilla).

Aliphatic acids on the contrary, of which linoleic acid is the major representative, are found in the green beans and their concentration tends to decrease during curing. Only acetic acid is not found in the green beans, but appears at a concentration of around 700 ppm after the first heat treatment, then decreases and stabilizes at around 200 ppm. The formation of acetic acid could be of microbial origin (Perez-Silva, 2006).

Aliphatic alcohols also appear after heat treatment, whereas the esters and hydrocarbons are already found in the green beans and their concentrations gradually decrease during curing.

As previously mentioned, among the compounds from the phenylpropanoid pathway, a certain number of aglycones are present despite the (apparent) lack of hydrolysis of the corresponding glucoside (as with *p*-cresol), or are present in far higher proportions than their concentration in glucosylated form suggested (as with vanillic acid), or are even present when the glucosylated form is absent (as with guaiacol). In fact, although the glucosylated forms are known to be relatively chemically unreactive, the aglycone released after hydrolysis may on the contrary become the substrate for chemical and/or enzymatic reactions, and thus be oxidized, reduced, decarboxylated, methylated, and so forth, and interconversions of one phenolic into another are therefore possible and even common.

For example, concentrations of vanillic acid higher than those expected can be explained by the oxidation of vanillin, and vanillic acid may also undergo decarboxylation into guaiacol (Perez-Silva, 2006), which could also explain the lack of glucoside of guaiacol. A solution of vanillin kept at 80°C at pH 5 in the presence of oxygen for several days also reveals the presence, in addition to the aforementioned molecules, of 2-methoxyhydroquinone, which can form quinones and semiquinones, which may lead to dimers (Perez-Silva, 2006). Different molecules of this kind, including divanillin, have been identified in cured vanilla (Figure 12.4) and are considered to contribute positively to the flavor of the product (Gatfield et al., 2006; Schwarz and Hofmann, 2009). Gatfield et al. (2006) suggest that the formation of divanillin is the result of the action of a peroxidase, a form of which has recently been purified (Marquez et al., 2008). This kind of reaction can also lead to the formation of large polymers that are at least partly responsible for the brown color of the vanilla.

These different reactions, which all start from vanillin, partly explain the considerable losses of this compound during curing (Odoux, 2000; Gatfield et al., 2007) that were already mentioned in Chapter 11.

Other reactions with no direct relationship to aroma formation may also explain these losses of vanillin, such as sublimation, coevaporation with water (Frenkel and Havkin-Frenkel, 2006), or the formation of a covalent bond with the lignin of the bean (Gatfield et al., 2007).

REFERENCES

Adedeji, J., Hartman, T.G., and Ho, C.-T. 1993. Flavor characterization of different varieties of vanilla beans. *Perfumer and Flavorist* 18:25–33.

Arana, F.E. 1943. Action of a β-glucosidase in the curing of vanilla. *Food Research* 8:343–351.

Dignum, M.J.W., van der Heijden, R., Kerler, J., Winkel, C., and Verpoorte, R. 2004. Identification of glucosides in green beans of *Vanilla planifolia* Andrews and kinetics of vanilla β-glucosidase. *Food Chemistry* 85:199–205.

Dignum, M.J.W., Kerler, J., and Verpoorte, R. 2001. Vanilla production: Technological, chemical and biosynthetic aspects. *Food Reviews International* 17:199–219.

Dignum, M.J.W., Kerler, J., and Verpoorte, R. 2002. Vanilla curing under laboratory conditions. *Food Chemistry* 79:165–171.

Ehlers, D. and Pfister, M. 1997. Compounds of vanillon (*Vanilla pompona* Schiede). *Journal of Essential Oil Research* 9:427–431.

Frenkel, C. and Havkin-Frenkel, D. 2006. The physics and chemistry of vanillin. *Perfumer & Flavorist* 31:28–36.

Galetto, W.G. and Hoffman, P.G. 1978. Some benzyl ethers present in the extract of vanilla (*Vanilla planifolia*). *Journal of Agricultural and Food Chemistry* 26:195–197.

Gassenmeier, K., Riesen, B., and Magyar, B. 2008. Commercial quality and analytical parameters of cured vanilla beans (*Vanilla planifolia*) from different origins from the 2006–2007 crop. *Flavour and Fragrance Journal* 23:194–201.

Gatfield, I., Hilmer, J.M., Weber, B., Hammerschmidt, F., Reiβ, I., Poutot, G., and Bertram, H.-J. 2007. Chemical and biochemical changes occurring during the traditional Madagascan vanilla curing process. *Perfumer & Flavorist* 32:20–28.

Gatfield, I., Reiβ, I., Krammer, G., Schmidt, C.O., Kindel, G., and Bertram, H.-J. 2006. Novel taste-active component of fermented vanilla beans. *Perfumer & Flavorist* 31:18–20.

Goris, A. 1924. Sur la composition chimique des fruits verts de vanille et le mode de formation du parfum de la vanille. *Comptes Rendus de l'Académie des Sciences* 179:70–72.

Goris, A. 1947. Formation du parfum de la vanille. *Industrie de la Parfumerie* 2:4–12.

Hanum, T. 1997. Changes in vanillin and activity of β-glucosidase and oxidases during post harvest processing of vanilla bean (*Vanilla planifolia*). *Bulletin Teknologik dan Industri Pengan* 8:46–52.

Hoffman, P., Harmon, A., Ford, P., Zapf, M., Weber, A., King, S., Grypa, R., Philander, E., Gonzalez, L., and Lentz, K. 2005. Analytical approaches to vanilla quality and authentication. In: D. Havkin-Frenkel, ed. *Vanilla, Proceedings of the First International Congress.* Allured Publishing, Carol Stream, pp. 41–49.

ISO 5565-1. 1999. Vanilla [*Vanilla fragrans* (Salisbury) Ames]. Part 1: Specifications, 6pp.

ISO 5565-2. 1999. Vanilla [*Vanilla fragrans* (Salisbury) Ames]. Part 2: Test methods, 11pp.

Janot, M.M. 1954. Formation du parfum de la vanille. In: G. Bouriquet, ed. *Le vanillier et la vanille dans le monde.*Paul Lechevalier, Paris, pp. 540–558.

Kanisawa, T. 1993. Flavor development in vanilla beans. *Kouryou* 180:113–123.

Kanisawa, T., Tokoro, K., and Kawahara, S. 1994. Flavour development in the beans of *Vanilla planifolia.* In: K. Kurihara, N. Suzuki, H. Ogawa, eds. *Olfaction Taste XI, Proceedings of the International Symposium.* Springer, Tokyo, pp. 268–270.

Kiridena, W., Poole, S., and Poole, C.F. 1994. Identification of 5-(hydroxymethyl)-2-furfural in vanilla extracts. *Journal of Planar Chromatography* 7:273–277.

Klimes, I. and Lamparsky, D. 1976. Vanilla volatiles—A comprehensive analysis. *International Flavours and Food Additives* 7:272–273, 291.

Lecomte, H. 1901. Sur la formation du parfum de la vanille. *Compte Rendu des Séances de l'Académie des Sciences de Paris* T133:745–748.

Lecomte, H. 1913. Formation de la vanilline dans la vanille. *L'Agriculture Pratique des Pays Chauds* 13:3–14.

Leong, G. 1991. *Contribution à l'étude des hétérosides des gousses de vanille vertes.* Th Sciences, Marseille III, 153pp.

Leong, G., Archavlis, A., and Derbesy, M. 1989a. Research on the glucosides fraction of the vanilla bean. *Journal of Essential Oil Research* 1:33–41.

Leong, G., Uzio, R., and Derbesy, M. 1989b. Synthesis, identification and determination of glucosides present in green vanilla beans (*Vanilla fragrans* Andrews). *Flavour and Fragrance Journal* 4:163–167.

Marquez, O., Waliszewski, K.N., Oliart, R.M., and Pardio, V.T. 2008. Purification and characterization of cell wall-bound peroxidase from vanilla bean. *LWT-Food Science and Technology* 41:1372–1379.

Nakazawa, Y., Wada, R., Izumitani, M., Fernandez, I.R., Vierma, D.F., Naito, T., Suyama, and Nakanishi, T. 1981. Studies on the vanilla flavours for food processing III. Characteristics of natural vanilla flavours from several beans produced in the some countries of the world. *Japanese Journal of Dairy and Food Science* 30:116–123.

Odoux, E. 2000. Changes in vanillin and glucovanillin concentrations during the various stages of the process traditionally used for curing *Vanilla fragans* beans in Reunion. *Fruit* 55(2):119–125.

Odoux, E., Chauwin, A., and Brillouet, J.M. 2003. Purification and characterization of vanilla bean (*Vanilla planifolia* Andrews) β-D-glucosidase. *Journal of Agricultural and Food Chemistry* 51:3168–3173.

Perez-Silva, A. 2006. *Contribution à l'étude de la genèse des composés d'arôme au cours du procédé mexicain de transformation de la vanille* (*Vanilla planifolia* G. Jackson).Th Sciences, Montpellier II:, 136pp.

Perez-Silva, A., Odoux, E., Brat, P., Ribeyre, F., Rodriguez-Jimenez, G., Robles-Olvera, V., Garcia-Alvarado, M.A., and Günata, Z. 2006. GC-MS and GC-olfactometry analysis of aroma compounds in a representative organic aroma extract from cured vanilla (*Vanilla planifolia* G. Jackson) beans. *Food Chemistry* 99:728–735.

Ramaroson-Raonizafinimanana, B., Gaydou, E.M., and Bombarda, I. 1997. Hydrocarbons from three *Vanilla* bean species: *V. fragrans*, *V. madagascariensis*, and *V. tahitensis*. *Journal of Agricultural and Food Chemistry* 45:2542–2545.

Ranadive, A. 2006. Chemistry and biochemistry of vanilla flavour. *Perfumer and Flavorist* 31:38–44.

Saltron, F., Langella, C., and Guerere, M. 2002. Evaluation de la qualité de la vanille malgache. Récolte 2000. *Annales des Falsifications, de l'Expertise Chimique et Toxicologique* 95:79–105.

Schulte-Elte, K.H., Gautschi, F., Renold, W., Hauser, A., Fankhauser, P., Limacher, J., and Ohloff, G. 1978. Vitispiranes, important constituents of vanilla aroma. *Helvetica Chimica Acta* 61:1125–1133.

Schwarz, B. and Hofmann, T. 2009. Identification of novel orosensory active molecules in cured vanilla beans (*Vanilla planifolia*). *Journal of Agricultural and Food Chemistry* 57:3729–3737.

Sostaric, T., Boyce, M.C., and Spickett, E.E. 2000. Analysis of the volatile compounds in vanilla extracts and flavorings by solid-phase microextraction and gas chromatography. *Journal of Agricultural and Food Chemistry* 48:5802–5807.

Vidal, J.P., Fort, J.J., Gaultier, P., and Richard, H. 1989. Vanilla aroma extraction by dense carbon dioxide. *Sciences des Aliments* 9:89–100.

Werkhoff, P. and Güntert, M. 1997. Identification of some esters compounds in Bourbon vanilla beans. *Lebensmittel - Wissenschaft und - Technologie* 30:429–431.

FIGURE 1.4 Representative genera of subfamily Vanilloideae, the "vanilloid orchids." (a) *Pogonia ophioglossoides* from the United States; (b) *Pseudovanilla foliata* from Queensland, Australia; (c) *Epistephium elatum* from Ecuador; (d) *Erythrorchis cassythoides* from New South Wales, Australia; (e) *Clematepistephium smilacifolium* vine and leaf with reticulate venation from New Caledonia; and (f) *Eriaxis rigida* from New Caledonia.

FIGURE 1.5 Representative species of *Vanilla*. (a) *Vanilla phaeantha*; (b) *Vanilla kinabaluensis*; (c) *Vanilla aphylla*; (d) *Vanilla mexicana*; (e) *Vanilla mexicana* in fruit with seeds that are visible; and (f) *Vanilla odorata*.

FIGURE 5.2 Flowers of Indian species of *Vanilla*: (a) and (b) *V. andamanica* with varying label colors, (c) *V. pilifera* showing indication of insect visits, and (d) *V. aphylla*.

FIGURE 5.11 Germination of cryopreserved shoots.

FIGURE 6.3 *Eulaema* sp. (*jicote*) bees, probable natural pollinator of *V. pompona*.

FIGURE 7.1 Chlorotic (left) and necrotic (right) flecks induced by CymMV on vanilla leaves.

FIGURE 7.2 Potyvirus symptoms on leaves of *V. planifolia* (a = BYMV-Réunion, b = BCMV-Madagascar, c = necrotic strain of WMV-Tonga) and *V. tahitensis* (d = WMV-FP, e = DsMV-FP) and on stem (f = WMV-FP).

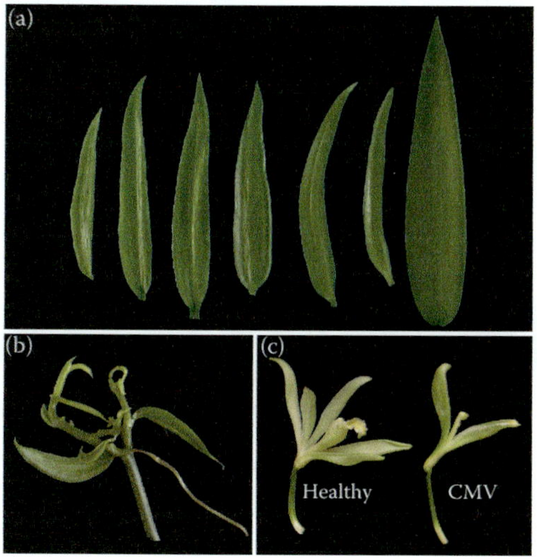

FIGURE 7.3 Symptoms caused by CMV infection in *V. tahitensis*: (a) young leaves embossed and deformed (healthy leaf on the right); (b) proliferation on infected young shoot; and (c) abnormal flower (right) on CMV-infected vine.

FIGURE 7.6 Mosaic on leaf of *V. pompona* infected by ORSV.

FIGURE 7.7 (a) Enations and necrosis on *V. planifolia* leaves infected with (b) enveloped virus-like particles visible under electron microscope.

FIGURE 8.1 Symptoms of Fusarium rot on vanilla vines: (a) dark brown lesion on stem internode with a chlorotic zone; (b) advanced necrotic stem internode and root; (c) numerous aerial roots are produced at the nodes but rot after reaching the soil.

FIGURE 8.4 Symptom of (a) Anthracnose on leaf; (b) sclerotium rot (brown water-soaked leaf tissue and white mycelium and *sclerotia* on soil; (c) Phytophthora stem rot on naturally infected shoot (Inset: symptom after artificial inoculation). (From Andriyani, N. et al. *Jurnal Biologi Indonesia*, 5, 227–234, 2008. With permission.)

FIGURE 9.1 Damage caused by the Angraecum scale: (a) on a vanilla plant; (b) detail of damage on a vanilla leaf.

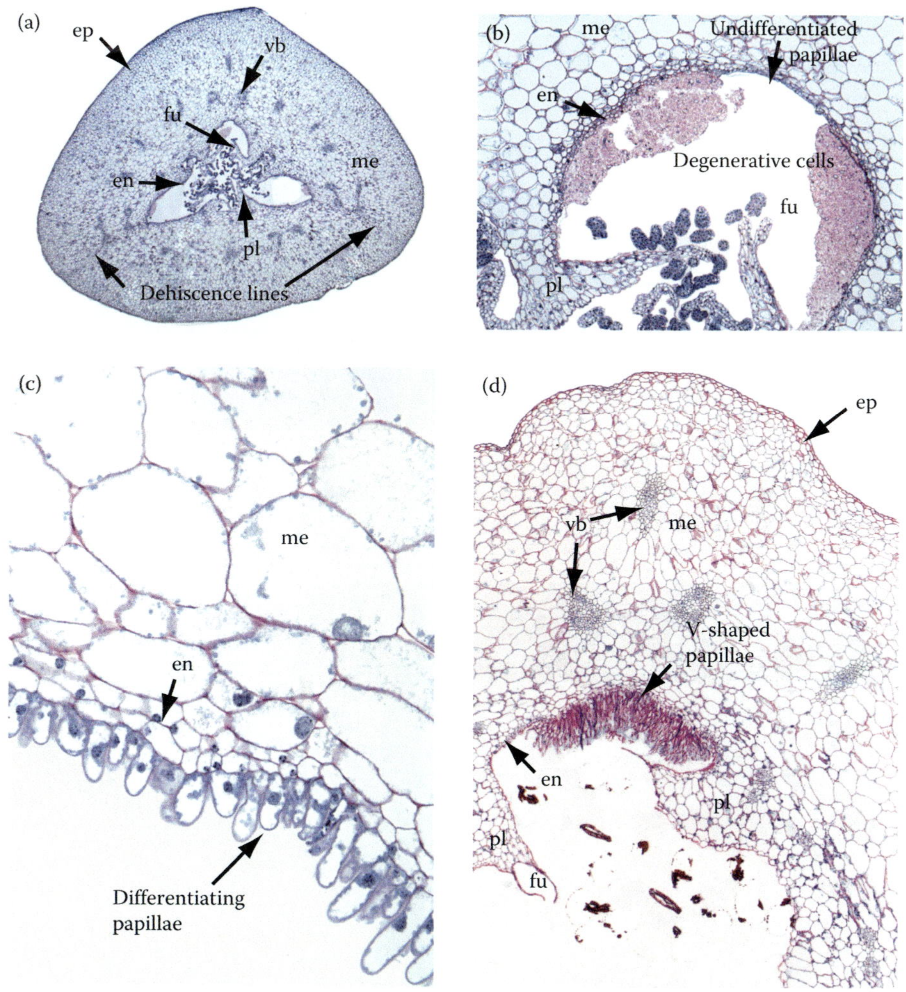

FIGURE 10.5 From the flower to the mature bean. Cross sections (3 μm) of a vanilla bean embedded in Technovit 7100 resin, at different stages after staining with periodic acid-Schiff (PAS)–Naphthol Blue Black: (a) 9 days after pollination (dap); (b) 14 dap; (c) 60 dap; (d) 8 months after pollination. (Data from Odoux et al., 2003. *Annals of Botany* 92: 437–444.) The walls and the storage sugars are stained in pink; the proteins in blue. en, Endocarp; ep, epicarp; fu, funicle; me, mesocarp; pl, placenta; s, seed; vb, vascular bundle.

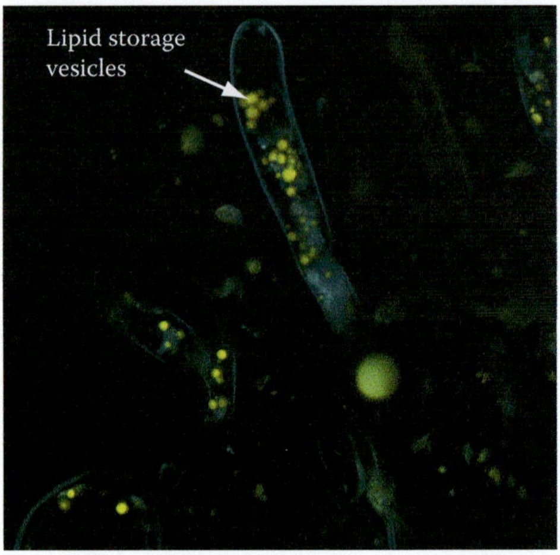

FIGURE 10.6 Visualization of lipid storage in the papillae of a mature vanilla bean after Nile Red staining (imaging with confocal microscope Zeiss 510 Meta, laser 488 nm and 405 nm, yellow: Nile Red staining, blue: autofluorescence of walls and papillae).

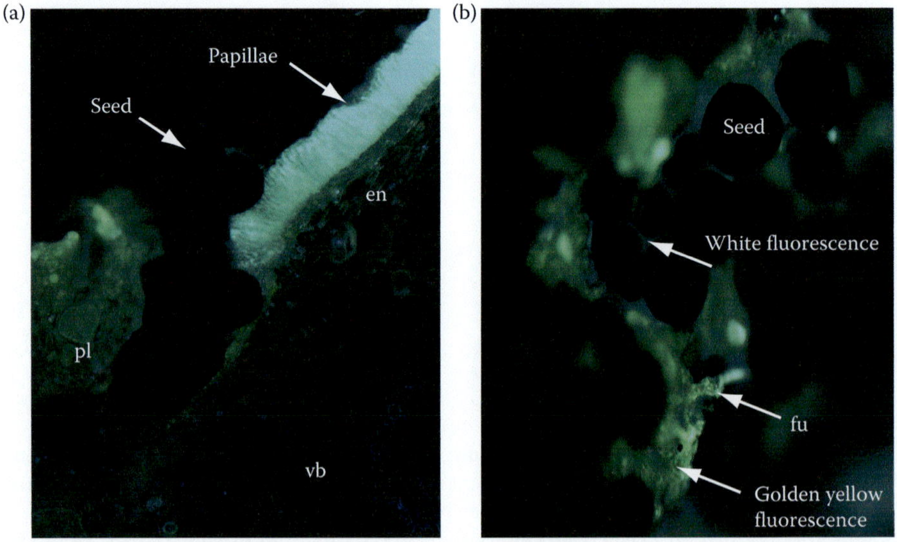

FIGURE 10.7 Fresh cross sections (100 μm) of mature vanilla bean (8 months after pollination) observed with epifluorescence microscope (Leica DM6000, filter A: 340–380 nm excitation, 425–800 emission). (a) A general view of placenta and papillae; (b) magnification of funicle, seeds, and "matrix." en, Endocarp; fu, funicle; pl, placenta; vb, vascular bundle.

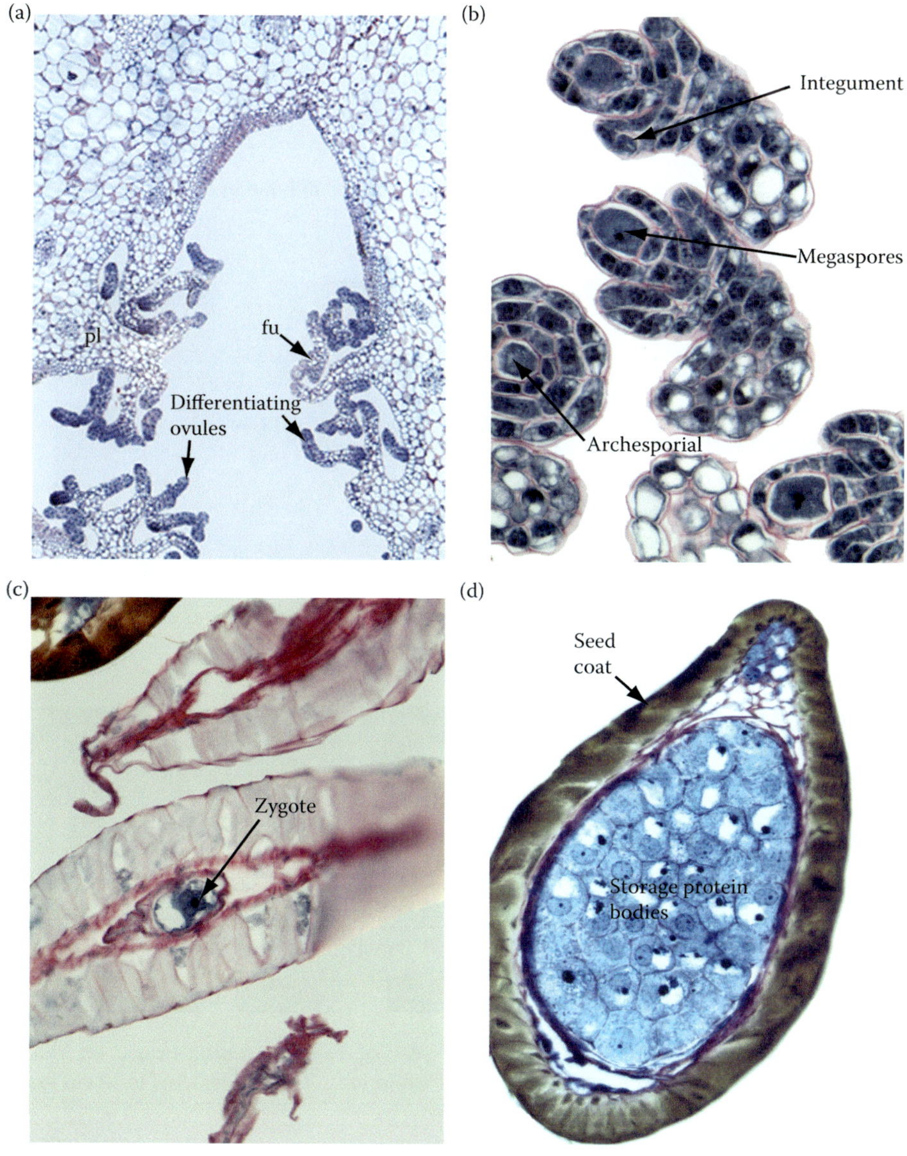

FIGURE 10.9 From the ovule to the seed: Cross sections (3 μm) of vanilla beans at different stages of development, embedded in Technovit 7100 resin after staining with PAS–Naphthol Blue Black. (a) 2 dap; (b) 15 dap; (c) 20 dap; (d) 200 dap. The walls and the storage sugars are stained in pink, the proteins in blue. fu: Funicle; pl: placenta.

FIGURE 13.1 Mature beans of Tahitian cultivar Haapape, still green or turning brown.

FIGURE 13.5 Color diversity of the labellum of Tahitian vanilla: (a) cultivar Rea rea; (b) cultivar Haapape.

FIGURE 13.6 Variation in size, shape, and color of Tahitian vanilla beans: (a) Tahiti, (b) Haapape, (c) Rea rea, (d) Parahurahu, and (e) Tahiti long. The yellow label is 10 cm long.

FIGURE 22.3 Adult of *T. confusus* Hsiao & Sailer (Hemiptera: Miridae) and the damages it causes in vanilla.

13 Morphological, Chemical, Sensory, and Genetic Specificities of Tahitian Vanilla

Sandra Lepers-Andrzejewski, Christel Brunschwig, François-Xavier Collard, and Michel Dron

CONTENTS

ABOUT THE ORIGIN OF VANILLA

The Tahitian vanilla plant, which is cultivated in French Polynesia since the nineteenth century, was described for the first time as a new vanilla species in 1933 by J.W. Moore: *Vanilla tahitensis*. This particular form of vanilla plant, found on trees, in a valley of Faaroa Bay (Raiatea, Society Islands, French Polynesia) is different from all the other cultivated vanilla forms known today. *V. tahitensis* differs from *Vanilla planifolia G.* Jacks. [syn. *Vanilla fragrans* (Salisbury) Ames] in particular by having more slender stems, narrower leaves, longer perianth segments, a lip that is shorter than the sepals, a longer column and shorter pods. For J.W. Moore, *V. tahitensis* is therefore a new vanilla species, endemical of French Polynesia.

However, the *Vanilla* genus, although amphitropical, is present as an indigenous species in the Pacific area only in Indonesia and Southeast Asia and is completely absent from the other Pacific islands except when cultivated (Florence and Guérin, 1996). Thus, its origin had to be searched among the various vanilla plants, which were introduced to French Polynesia.

There were at least three introductions of vanilla from various sources between 1848 and 1898 (Chevalier, 1946). The first one dates from 1848. Admiral Hamelin brought to Tahiti *V. planifolia* Andrews [syns. *Vanilla aromatica* Swartz and *V. fragrans* (Salisbury) Ames] from Manila in the Philippines (Costantin and Bois, 1915; Bouriquet, 1954; Florence and Guérin, 1996). Henry in 1924 noted that some horticultural tests were carried out in Tahiti between 1847 and November 1849; however, no details were given, and he indicated that the vanilla plant brought by Admiral Hamelin in 1848 did flower and produce pods. The seeds of these pods or the horticultural tests could be at the origin of the Tahitian vanilla.

The second introduction of vanilla occurred in 1850 (Costantin and Bois, 1915). French Polynesia received new plants from Paris, sent by Admiral Bonard. There were many *V. planifolia* plants coming from the botanical garden of Paris and many *V. pompona* plants from the Antillas (Florence and Guérin, 1996). These two species are still found in French Polynesia. But they are not really cultivated. The weather conditions do not always allow *V. planifolia* to flower. In addition, *V. pompona* is only considered as an ornamental plant in French Polynesia.

The third introduction of vanilla plant was by Major Pierre, in 1874, who brought new vanilla plants from Mexico (Costantin and Bois, 1915; Florence and Guérin, 1996). These plants were described as the best vanilla species from Mexico and Bourbon, with broader, thicker, and more round leaves than the vanillas cultivated in French Polynesia, which possess pointed and lanceolate leaves. According to these descriptions, it seems that this third introduction did also correspond to *V. planifolia*.

According to morphological and historical comparisons, many hypotheses on the origin of Tahitian vanilla were elaborated. According to Portères (1951), the morphology suggests that *V. tahitensis* was a clonal population resulting from segregations of natural crossings between *V. fragrans* (Salisbury) Ames (syn. *V. planifolia* Andrews) and *V. pompona* Schiede, which was probably already a hybrid of *Vanilla odorata* Presl. Portères (1953) suggests that this initial hybridization could have occurred in French Guiana or in Reunion Island between *V. fragrans* and a complex species (*V. pompona–V. odorata*). However, according to Florence and Guerin,

V. tahitensis does not show any intermediate traits that a hybrid between *V. planifolia* and *V. pompona* should exhibit. These authors consider *V. tahitensis* as synonymous of *V. planifolia*. They do see it as a segregation issued from a small number of introductions. The segregants have brought some variations resulting in traits specifying the Polynesian form.

At present, the origin of Tahitian vanilla is still unknown; nevertheless, morphological diversity exists among the vines encountered in French Polynesia. The most cultivated vanilla vines in French Polynesia are Tahiti and Haapape. The cultivar Haapape was first mentioned in 1914 (Bouriquet and Hibon, 1950). According to C. Henry (1924), Haapape seems to be an inbred of Tiarei, which was described in 1900 (Costantin and Bois, 1915). The cultivar Tiarei was reported as a case of gigantism (Henry, 1924). Comparing the morphological characters, color, and flavor of the beans, the Tiarei form cannot be the result of a cross between a Tahitian vanilla and a vine originating from Mexico. According to Bouriquet and Hibon (1950), the Tiarei cultivar is issued from a somatic mutation or is a spontaneous hybrid of *V. planifolia* and a Tahitian form.

BIOLOGY OF THE TAHITIAN VANILLA

Vanilla is mainly propagated by stem cuttings. Seed germination is difficult because it requests a symbiotic fungus: *Rhizoctonia*. Flowering necessitates approximately 18 months. However, this delay varies with the size of the vine. Cuttings of 20 nodes can flower within less than one year after plantation.

Flowering is induced by a decrease in the temperature (18–19°C) and by drier and sunnier weather during the fresh season. Two flowering periods often occur in a year: the main one is between July and September and another is possible in December–January. The ends of the hanging stems will then dry and 1–3 inflorescences from 10 to 15 flowers will appear. The flower opens at dawn and fades during the evening. On a single inflorescence only 1–2 flowers open each day.

There is no pollinating insect in French Polynesia and thus pollination, or "marriage," is hand-made early in the morning, to prevent pollen becoming too dry to adhere to the stigmata. The marriage is performed according to a common procedure developed in Reunion Island by Edmond Albius (Bouriquet, 1954, see also Chapter 17).

The bean is ripe after nine months. Green until then, the bean starts to yellow and then browns, starting from the heel. This indicates that the bean can be collected and cured.

CURING PROCESS

Thanks to original sensory and physical properties having been retained, the Tahitian vanilla has a unique flavor. In French Polynesia, Tahitian vanilla is nowadays considered as a bona fide source of cultural identity demonstrating the know-how of growers and curers. Polynesians developed a specific curing process based on harvesting pods when fully mature and flavorful, followed by a natural browning enabling the development of the aroma.

For most vanillas cultivated worldwide, notably *V. planifolia*, ripeness involves dehiscence of the beans. To avoid this phenomenon, clusters of beans are harvested before fully mature; and then ripeness is stopped by a treatment at high temperature. Tahitian vanilla beans rarely develop split ends and they are harvested when fully ripe, with complete aroma potential (Figure 13.1). Natural browning is achieved and allows a complete conversion of glycosylated aroma compounds into corresponding aglycones responsible for the vanilla flavor.

Traditional Polynesian curing involves water loss to concentrate the flavor and to enable a good conservation of the beans. In the course of the curing process, many biochemical reactions (enzymatic and nonenzymatic like oxidative reactions) occur in the vanilla pods, which result in the development of the flavor (Odoux et al., 2006, for more details see Chapters 11 and 12); such biochemical reactions are poorly studied in Tahitian vanilla.

Tahitian curing consists of three main steps that necessitate being very patient and meticulous (Larcher, 1989; Ranadive, 1994):

1. *Shade browning* After harvest, mature vanilla beans are exposed in the shade until uniform browning and initiation of sweet-scented flavor are attained. Then, they are washed and dried.
2. *Sun drying* Beans are exposed to sunshine for a few hours a day for several weeks. Their water content drops from 80% to the final moisture content desired (50–55%). When they are exposed to warm temperatures or the sun; water evaporates. Then, while still warm, beans are wrapped in cotton sheets called "faraoiti" and kept in wooden cases overnight to ensure water loss by sweating. The beans become flexible and glossy as their epidermis is increasingly covered with oil. They are massaged one by one, when necessary, to spread and homogenize the seeds lengthwise.
3. *Air drying (refining)* Beans are finally dried in the shade to homogenize batches and stabilize water content, enabling optimal shelf life.

The curing process allows the development of Tahitian vanilla flavor. The beans become sweet-smelling and rich anise flavored. Their physical properties are also intensified as they show a very attractive oily texture and appearance. The lipid components are involved both in these properties and aroma development because they restrict aroma loss from occurring during the process.

MORPHOLOGICAL SPECIFICITIES AND DIVERSITY

Tahitian vanilla is characterized by a more slender stem than *V. planifolia* or *V. pompona* with an average stem diameter of 8 mm and internodes of 13 cm length, sometimes very long, up to 19 cm. The leaves are also thinner (Figure 13.2), narrow–oval, lanceolate, gradually become very pointed, 3–7 times longer than their width, dark green in color, not very thick, not or very little in gutter toward the base, 16–25 cm in length, and 2–5 cm broad; the petiole is 15 mm in length on an average, and well canaliculate above.

FIGURE 13.1 **(See color insert following page 136.)** Mature beans of Tahitian cultivar Haapape, still green or turning brown.

FIGURE 13.2 Comparison of the size and shape of leaves of *V. pompona*, *V. planifolia*, and *V. tahitensis* cv Haapape (from left to right).

FIGURE 13.3 Comparison of flowers of *V. pompona*, *V. planifolia*, and *V. tahitensis* cv Haapape (from left to right).

The Tahitian vanilla flowers are assembled into an inflorescence. On an average, the inflorescences are 9.2 cm long, greenish, with some parts more yellow than those of *V. planifolia* (Figure 13.3). The sepals are narrow-oblanceolate, pointed at the base, attenuate-acute at the top, 9–16 mm broad, 55–77 mm long, and nonducted with the back. The petals are also 50–74 mm long, 7–13 mm wide, elliptic, and subacute. The labellum is 46–49 mm in length and funnel-shaped, shows small orange dotted lines on its inner face with a revolute margin, and is fimbriate-crenelated. The Tahitian vanilla fruits, aromatic, are generally indehiscent contrary to *V. planifolia* pods.

Today, the main Tahitian vanilla cultivars are Tahiti and Haapape. But local farmers distinguish about 14 cultivars, which differ by the shape, length, and width of the leaves, the size of the flowers, and the size and shape of the beans. The five main cultivated forms are described below and shown in Figures 13.4 through 13.6.

Tahiti

The stem is on an average 8 mm in diameter with internodes of 13 cm length. The leaves are not or very little in gutter toward the base and measure about 18 cm in length, and 3 cm in width with a petiole of 13 mm length. The sepals are 14 × 65 mm on an average, and the petals 11 × 64 mm. The labellum is white with short hairs and dotted lines of dark orange color. The fruit is clear green in color, trigonal and 10–19 cm long (14 cm on an average) and weighs on an average 10.5 g. The pods show little dehiscence with maturity.

Haapape

The stem is larger than "Tahiti," with, an average diameter of 10 mm and internodes of 16 cm length. The leaves present a gutter, measure about 17 cm in length and 3 cm in width with a petiole of 18 mm length. The flowering is less abundant than "Tahiti," but the larger flowers of Haapape are easier to marry. The sepals and petals are longer than those of "Tahiti": an average of about 13 × 70 mm and 10 × 68 mm respectively.

FIGURE 13.4 Plant and leaf shapes of the main Polynesian cultivars of vanilla.

FIGURE 13.5 **(See color insert following page 136.)** Color diversity of the labellum of Tahitian vanilla: (a) cultivar Rea rea; (b) cultivar Haapape.

FIGURE 13.6 **(See color insert following page 136.)** Variation in size, shape, and color of Tahitian vanilla beans: (a) Tahiti, (b) Haapape, (c) Rea rea, (d) Parahurahu, and (e) Tahiti long. The yellow label is 10 cm long.

The labellum is white, with long hairs and dotted lines of dark orange color. The fruits are of very dark color, rounded, blunt at the end and 13–23 cm long (average 18 cm) and weigh an average of 16.3 g. The beans are very rarely dehiscent with maturity.

Rea Rea

The stem is thinner, with, an average diameter of 7 mm and relatively short internodes of 13 cm length. The leaves, more oblong than lanceolate, are about 20 cm long, and 4 cm wide with a petiole of 16 mm length. The sepals are about 13 × 62 mm and the petals are 11 × 61 mm. The labellum is white with light orange color dotted lines. The fruits, trigonal, of green–light yellow color, and 11–17 cm long (average 14 cm) and weigh about 9.9 g. The beans are fairly dehiscent with maturity.

Parahurahu

The stem is also thin, measuring an average of 7 mm in diameter with internodes of 14 cm length. Leaves, not or very little in gutter toward the base and present a more apiculate end than pointed and are about 15 cm long and 3 cm wide, with a petiole of 16 mm length. The sepals are about 12 × 63 mm and the petals are about 10 × 61 mm. The labellum is white with white hairs and dotted lines of light orange color. The beans, of dark green color, present the shape of a bludgeon, are 10–18 cm long (14 cm on average) and weigh about 14.1 g. The beans are very rarely dehiscent with maturity.

Tahiti Long

The stem is thin, measuring about 7 mm in diameter with short internodes of 13 cm. The leaves are not or very little in gutter toward the base and are about 15 cm long and 3 cm wide with a petiole of 15 mm length. The flowers are smaller than those of the previous cultivars; the sepals are on average 11 × 59 mm and the petals are 9 × 58 mm. The labellum is white with long hairs and exhibits light yellow dotted lines. The end of the labellum has a thorny form. The fruits, of light green color, are rounded-off, thin, blunt at the end, and 11–24 cm long (average 17 cm) and weigh about 11.1 g. The beans are rarely dehiscent with maturity.

Growers distinguish many other morphological types than those described above. These types differ in the size and width of their leaves and the size, shape, and color of their fruits. These vanilla types are not really cultivated because their fruits do not obey the criteria defined by the regulations (for details, see Chapter 24), but they exhibit a real morphological diversity. Examples: Oviri, which has smaller leaves (14 cm × 3 cm) than "Tahiti" and shorter fruits (11 cm on average); Puroini, which possesses leaves with the shape of a heart; Potiti and "Tahiti" or "Haapape" have the same leave size and shape, but "Potiti" produces very small fruits (average 10 cm). Finally, a "bicolor" (variegated) vanilla plant whose leaves show light yellow stripes is also present in farming systems.

CHEMICAL COMPOSITION OF TAHITIAN VANILLA

Tahitian vanilla is very different from other vanillas cultivated in the world, particularly when compared to *V. planifolia* because of its subtle flavor and rich anise fragrance, and the glossy and flexible aspect of its beans.

To characterize the originality of Tahitian vanilla, methods were developed at the laboratory of the "Etablissement Vanille de Tahiti" to analyze its chemical composition, especially the molecules of interest–namely, aroma compounds and fatty acids.

AROMA COMPOSITION OF TAHITIAN VANILLA

An Original Aroma Composition

Thanks to their sensory properties, vanilla pods are mostly studied for their aroma compounds. The first comprehensive work on vanilla volatiles was carried out in 1976 by Klimes and Lamparsky, who identified, by gas chromatography (GC)–mass spectrometry (MS), 169 compounds in *V. planifolia*.

Then, several studies of Tahitian vanilla aroma composition highlighted its originality compared to *V. planifolia* by the presence of major phenolic compounds, including the widely mentioned anisyl compounds present in higher amounts, and less frequently the presence of minor compounds. The different studies of Tahitian vanilla describe the aroma as composed of the following molecules (see Table 13.1 for chemical structure) whose contents differ from *V. planifolia* as detailed later.

Ten major phenolic compounds are listed below:

a. Vanillin, *p*-hydroxybenzaldehyde, vanillic acid, and *p*-hydroxybenzoic acid (Fayet et al., 1987; Ranadive, 1992; Langella et al., 2003) usually quantified in *V. planifolia* to authenticate natural vanilla.
b. Four anisyl compounds characteristic of Tahitian vanilla:
   i. Anisyl alcohol and anisic acid, which are with vanillin the most concentrated compounds in Tahitian vanilla beans. They have been identified by GC–MS (Adedeji et al., 1993), confirming previous works (Shiota and Itoga, 1975; Lhuguenot, 1978; Tabacchi et al., 1978; Rey et al., 1980; Purseglove et al., 1981; Larcher, 1989; Hartman, 2003).
   ii. *p*-Anisaldehyde (Tabacchi et al., 1978) and methyl anisate (Shiota and Itoga, 1975; Adedeji et al., 1993; Lechat-Vahirua and Bessiere, 1998; Sostaric et al., 2000), which are less concentrated.
c. Other phenolic compounds in nonnegligible amounts: protocatechualdehyde, protocatechuic acid, syringic acid, and *p*-hydroxybenzyl alcohol (Fayet et al., 1987).

Minor components with various chemical functions:

d. Aromatic esters: anisyl acetate, anisyl formate, and methyl cinnamate (Shiota and Itoga, 1975; Adedeji et al., 1993; Sostaric et al., 2000; Hartman, 2003).
e. Other phenolic compounds (Shiota and Itoga, 1975; Scharrer, 2002).

TABLE 13.1
Major Aroma Molecules Identified and Quantified in Tahitian Vanilla

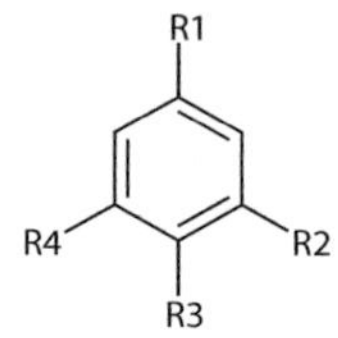

Family	Molecule	Name Used	R1	R2	R3	R4
Vanillyl	Vanillyl alcohol	Van_alc	CH_2OH	OCH_3	OH	H
	Vanillin	Van	COH	OCH_3	OH	H
	Vanillic acid	Van_ac	COOH	OCH_3	OH	H
p-Hydroxybenzyl	*p*-Hydroxybenzyl alcohol	Phb_alc	CH_2OH	H	OH	H
	p-Hydroxybenzaldehyde	Phb	COH	H	OH	H
	p-Hydroxybenzoic acid	Phb_ac	COOH	H	OH	H
	Methyl *p*-hydroxybenzoate	Me_paraben	$COOCH_3$	H	OH	H
Anisyl	Anisyl alcohol	Anis_alc	CH_2OH	H	OCH_3	H
	Anisaldehyde	Anis_ald	COH	H	OCH_3	H
	Anisic acid	Anis_ac	COOH	H	OCH_3	H
	Methyl anisate	Me_anis	$COOCH_3$	H	OCH_3	H
Protocatechuyl	Protocatechualdehyde	Pro_ald	COH	OH	OH	H
	Protocatechuic acid	Pro_ac	COOH	OH	OH	H
Isovanillyl	Isovanilline	Isovan	COH	OH	OCH_3	H
Others	Syringic acid	—	COOH	OCH_3	OH	OCH_3

f. Aldehydes, ketones, and lactones: 2,4-decadienal, 3-methylpentanal, 2-methyl-2-pentanone, 1-methoxy-2-propanone, isovaldehyde, and gamma nonalactone (Adedeji et al., 1993; Lechat Vahirua and Bessiere, 1998; Scharrer, 2002).
g. Monoterpenes (α-pinen, β-pinen, limonen, linalool, terpinen-4-ol, and α-terpineol (Scharrer, 2002).
h. Aliphatic esters (for instance, ethyl hexanoate, ethyl nonanoate, and ethyl decanoate) (Sostaric et al., 2000).
i. Heterocyclic compounds, aliphatic acids, and alcohols (Fayet et al., 1987).
j. Derivatives of anisyl compounds (Da Costa and Pantini, 2006).

The presence of piperonal (heliotropin) as a characteristic constituent of Tahitian vanilla has been a matter of debate for a long time. Definitively Joulain et al. (2007) showed that the supposed characteristic odorant constituent of Tahitian vanilla was *p*-anisaldehyde and that piperonal was present in Tahitian vanilla but only as a trace element as well as in *V. planifolia*, confirming previous works (Tabacchi et al., 1978; Ehlers et al., 1994; Lechat-Vahirua and Bessiere, 1998).

At the laboratory of the "Etablissement Vanille de Tahiti," the high flavored, original and pervasive flavor of the Tahitian vanilla was highlighted by the high-pressure liquid chromatography (HPLC) analysis of the major aroma phenolic compounds. The work focused on the role of the curing process in the development of a unique flavor, and on the aroma composition of some Polynesian cultivars.

More than 300 samples of Tahitian cured vanilla beans were analyzed from three successive annual harvests (2005, 2006, and 2007). Cured beans of Tahitian vanilla are rich in overall aroma compounds with an average of 4.6% of dry matter compared to 2.1–4.2% for *V. planifolia*. Moreover, the aroma composition of Tahitian vanilla is more different, with a smaller contribution of vanillin to the overall flavor (30% v. 80% for *V. planifolia*). Whereas its vanillin level is of less influence (1.2% as it reaches 1.7–3.5% for *V. planifolia*), Tahitian vanilla contains higher amounts of anisyl molecules (2.1% and less than 0.05% for *V. planifolia*) and *p*-hydroxybenzaldehyde. The most concentrated anisyl molecules are anisyl alcohol and anisic acid (respectively, 1.35% and 0.75% of dry matter); *p*-anisaldehyde and methyl anisate represent about 200 ppm each, whereas they are only at trace levels in *V. planifolia*.

Evolution of Aroma Compounds during the Curing Process

Many factors influence the flavor of Tahitian vanilla, such as genetic, agronomic, climatic, and transformation criteria. In fact, the highly valued flavor of Tahitian vanilla becomes more and more intense during the curing process. Nevertheless, if vanilla pods undergo a water loss when exposed to the sun, some aroma compounds are also evaporated and the batches of pods generate a pleasant vanilla flavor. Research on vanilla aroma in French Polynesia focuses on phenolic compounds that contribute to the overall flavor. Fourteen "aroma compounds" from an ethanolic Soxhlet extract were assessed by HPLC during the curing process (see Table 13.1).

In relation to the drying process, there is a subsequent loss of aromatic molecules. When beans are dried to 38% moisture content, they lose approximately one-third of their aroma potential, from 6.5% to 4.6% of dry weight. However, given that the loss of water is more important than the loss of aroma compounds, their concentration per gram of fresh weight actually increases during the curing period.

Differences in volatility between aroma compounds and oxidation reactions occurring in the beans (from alcohol to aldehyde and then to acid function) involve diverse changes of molecule concentrations (see Figure 13.7). The highly concentrated vanillin and anisyl alcohol undergo a significant decrease, whereas vanillic acid and anisaldehyde concentrations increase because of their production due to the oxidation of previous major compounds (respectively, vanillin and anisyl alcohol).

The moisture content has an influence on the flavor composition: drying favors accumulation of acid molecules (anisic acid and *p*-hydroxybenzoic acid) and decrease of the ratio of anisyl alcohol and vanillin (Figure 13.8). The odor-active compound *p*-anisaldehyde is a key component of the original Tahitian vanilla flavor, as its concentration doubles during the curing process. Since moisture content produces some effect on aroma quality, it is very important to provide details of moisture content when describing vanilla beans composition (Collard, 2007).

The curing process of Tahitian vanilla is a subtle compromise between optimal drying and conservation of aroma compounds. This results in an original aroma composition.

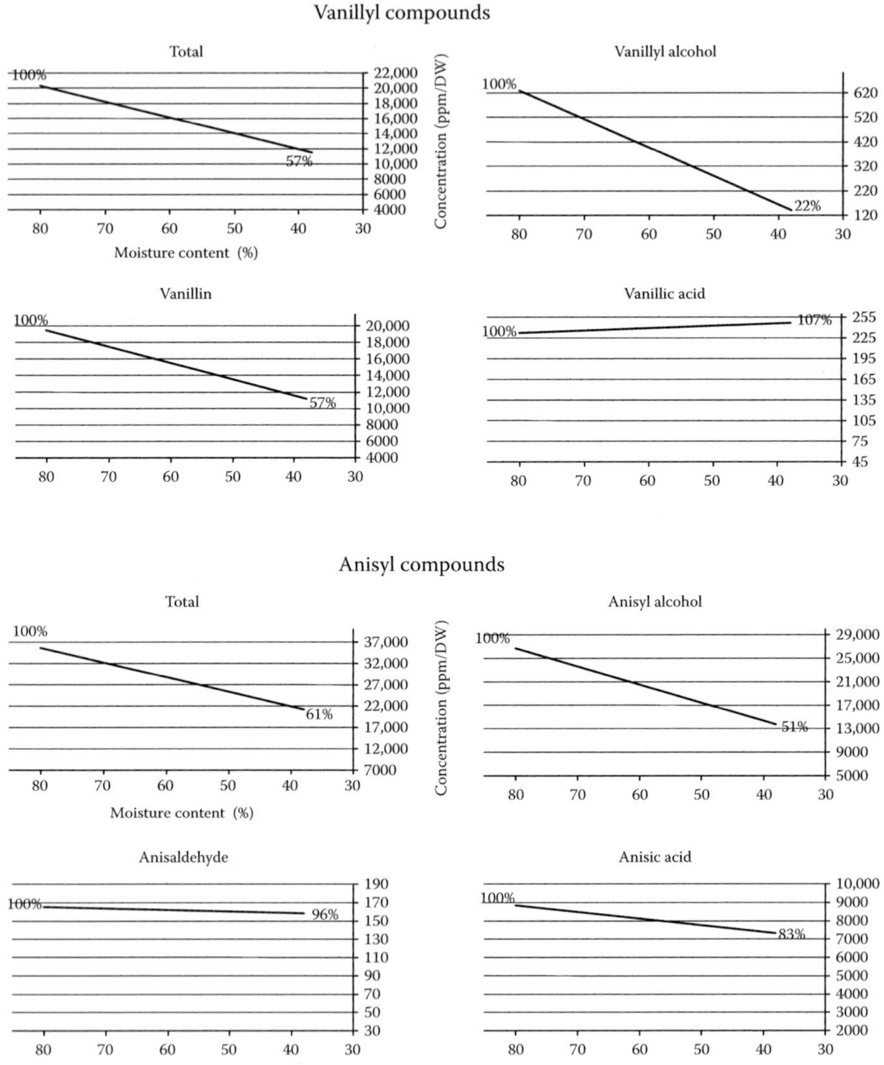

FIGURE 13.7 Evolution of major aroma compounds of Tahitian vanilla during the curing process. Variation in content of the compounds is indicated as a percentage of initial content.

Tahitian Vanilla Cultivars Show Various Aroma Compositions

Tahitian vanilla develops an exotic flavor compared to other vanillas. There is an intra-*tahitensis* diversity with various cultivars and nuanced flavors.

In Tahitian vanilla, there is a morphological diversity, which is confirmed by different chemical compositions. A study was carried out based on aroma quality of uncured beans of five cultivars of importance in French Polynesia (the most cultivated Tahiti and Haapape, Rea rea, Parahurahu, and Tahiti long).

Each cultivar develops its own aroma characteristics. Table 13.2 shows that the overall aroma amount differs from one cultivar to another and that some molecules

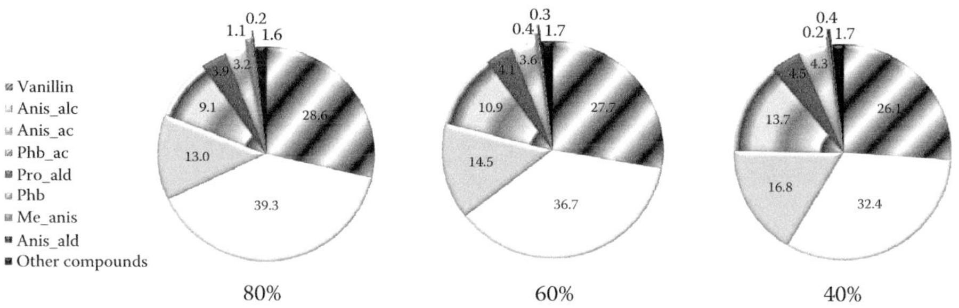

FIGURE 13.8 Aroma composition of Tahitian vanilla pods for different moisture contents (80%, 60%, and 40%). See Table 13.1 for complete names of molecules.

are discriminating as they contribute to structure groups by statistical analysis. Performing a factorial discriminant analysis shows very good correspondence between the observed classification obtained with aroma variables and naturally occurring cultivars (see Figure 13.9) (Collard et al., 2006).

Among Polynesian cultivars, Parahurahu is very specific with a significantly lower overall amount, a lower amount of vanillyl molecules and a higher level of anisic acid. The cultivar Rea rea demonstrates a remarkably lower amount of *p*-hydroxybenzaldehyde and a higher amount of vanillyl compounds, and it differs from the group formed by Tahiti, Haapape, and Tahiti long, all of which show quite

TABLE 13.2
Aroma Content (ppm) of Five Cultivars of Tahitian Vanilla (Uncured Beans) Quantified by HPLC at the "Etablissement Vanille de Tahiti"

Molecule*	Parahurahu	Rea rea	Tahiti long	Haapape	Tahiti
No. of samples	18	13	15	23	110
Van_alc	243^{a}	909^{e}	407^{b}	514^{c}	614^{d}
Van	806^{a}	17,718^{c}	11,732^{b}	13,337^{b}	20,184^{d}
Van_ac	87^{a}	316^{e}	177^{b}	197^{c}	227^{d}
Phb	2504^{c}	352^{a}	4221^{d}	2621^{c}	2081^{b}
Phb_ac	7101^{c}	4932^{a}	7275^{c}	7202^{c}	6007^{b}
Anis_alc	7156^{a}	13,108^{b}	16,397^{c}	20,158^{d}	21,117^{d}
Anis_ald	46^{a}	65^{b}	71^{b}	82^{b}	65^{b}
Anis_ac	16,026^{c}	4896^{a}	8103^{b}	8736^{b}	7907^{b}
Pro_ald	1101^{a}	1421^{b}	2585^{c}	2758^{c}	2628^{c}
Pro_ac	220^{b}	216^{b}	219^{b}	188^{ab}	169^{a}
Σ Van	1137^{a}	18,943^{c}	12,316^{b}	14,048^{b}	21,025^{c}
Σ Phb	9605^{c}	5284^{a}	11,496^{d}	9823^{c}	8088^{b}
Σ Anis	23,228^{b}	18,069^{a}	24,571^{b}	28,976^{c}	29,088^{c}
TOTAL	35,290^{a}	43,933^{b}	51,186^{c}	55,793^{d}	60,998^{e}

Averages with the same letter within rows are not significantly different ($p < 0.05$).
* See Table 13.1 for complete names of the molecules.

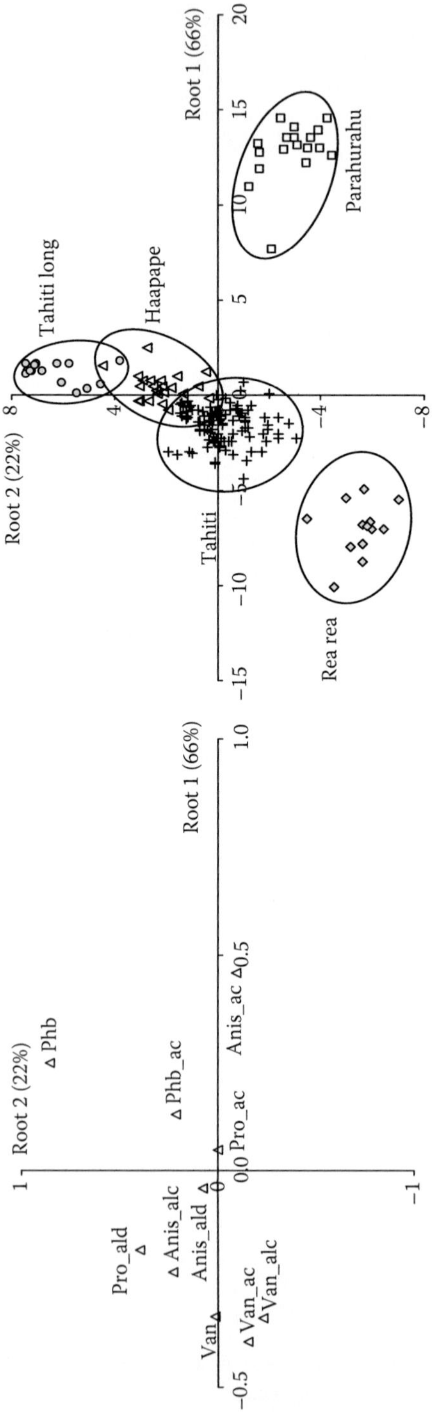

FIGURE 13.9 2D plot of five cultivars of uncured Tahitian vanilla beans in factorial discriminant analysis using aroma compounds (see Table 13.2) as variables.

similar aroma compositions. Beans of Tahiti long are characterized by a high amount of *p*-hydroxybenzaldehyde. Tahiti and Haapape are the cultivars that have the most similar aroma compositions, although Tahiti is richer in overall aroma and vanillyl compounds, and Haapape contains relatively more *p*-hydroxybenzyl and anisyl components.

The aroma composition can explain the original flavor of Tahitian vanilla and show slight differences between the main cultivars. As seen before, the originality of Tahitian vanilla is not only due to its flavor but also due to its oily texture.

Fatty Acid Composition of Tahitian Vanilla

The fatty acid composition has been a topic of interest to better understand the physical and sensory properties of Tahitian vanilla, which is oilier (see the section on Sensory Properties).

Some previous chemical studies of vanilla beans related to the presence of fatty acids or fatty acid derivatives. The major components identified were oleic or palmitic acid (Purseglove et al., 1981; Adedeji et al., 1993; Lechat-Vahirua and Bessiere, 1998). A comprehensive study of the lipidic fraction of three vanilla bean species *V. tahitensis*, *V. planifolia*, and *Vanilla madagascariensis* by GC revealed discriminating fatty acid compositions, with a predominance of linoleic acid among 31 fatty acids in every species (Ramaroson-Raonizafinimanana, 1988). Moreover, many additional lipidic components were identified: phytosterols (Ramaroson-Raonizafinimanana et al., 1998), hydrocarbons (Ramaroson-Raonizafinimanana et al., 1997), and two new product families: long-chain γ-pyrones (Ramaroson-Raonizafinimanana et al., 1999) and long-chain β-diketones (Ramaroson-Raonizafinimanana et al., 2000).

At the "Etablissement Vanille de Tahiti," work on the lipid extract aims at characterizing Tahitian vanilla and its different cultivars by the fatty acid composition. Derivatives of fatty acids are analyzed by HPLC. About 10 fatty acids were identified and the identification was confirmed by GC for a few of them (Brunschwig et al., 2007). Six common fatty acids coming from primary metabolites such as triglycerides were quantified: linolenic 18:3, linoleic 18:2, palmitic 16:0, oleic 18:1, stearic 18:0, and erucic 20:1 acids. Four very-long-chain monounsaturated fatty acids, rarely found in plants, were identified in vanilla beans: nervonic (24:1), ximenic (26:1), octacosen-19-oïc acid (28:1), and lumequeic (30:1) acids. These fatty acids derive from secondary metabolites (supposed β-diketones) and their composition may be much more variable than that of the primary metabolites (Ramaroson-Raonizafinimanana, 1988).

The fatty acid composition demonstrates that Tahitian vanilla is richer in fatty acids and that it could discriminate Polynesian cultivars. Cured Tahitian vanilla beans have an average fatty acid content higher than other vanillas in the world (an average of 2.5% of dry matter compared to 1.2–2.4%), which may explain the attractive glossy and oily aspect of the pods (Brunschwig et al., 2007). Tahitian vanilla beans contain mainly polyunsaturated fatty acids (with a preponderance of linoleic acid from 55% to 65%) and monounsaturated fatty acids (from 20% to 30%) with a remarkable content of monounsaturated long-chain fatty acids (5%–13%), and not

TABLE 13.3
Fatty Acid (FA) Composition (%) of Lipid Extract from Different Cultivars of Tahitian Vanilla (Uncured Beans)

	Parahurahu	Tahiti long	Haapape	Tahiti	*V. tahitensis**
No. of samples	8	6	8	8	1
16:0–palmitic	9.8[a]	10.4[a]	9.4[a]	9.6[a]	8.2
18:0 + 20:1 stearic + eructic	3.9[a]	4.4[b]	3.9[a]	4.7[b]	3.6
18:1ω9–oleic	15.7[a]	17.5[b]	15.2[a]	15.3[a]	12.9
18:2ω6–linoleic	53.9[a]	58.1[b]	63.7[d]	61.2[c]	67.4
18:3ω3–linolenic	3.7[b]	2.8[a]	2.5[a]	2.2[a]	1.3
Total common FA	87.0[a]	93.2[b]	94.7[c]	93.0[b]	93.3
24:1ω9–nervonic	6.6[c]	3.5[b]	2.4[a]	3.5[b]	4.5
26:1ω9–ximenic	2.0[c]	1.3[b]	0.9[a]	1.3[b]	1.0
28:1ω9–octasen-19oïc	2.1[b]	1.1[a]	1.0[a]	1.2[a]	0.6
30:1ω9–lumequeic	2.2[b]	0.9[a]	0.9[a]	1.0[a]	0.5
Total long-chain FA	12.9[c]	6.8[b]	5.2[a]	7.0[b]	6.7
Saturated FA	13.7[a]	14.8[a]	13.3[a]	14.3[a]	11.8
Monounsaturated FA	28.6[d]	24.3[c]	20.4[a]	22.3[b]	19.5
Polyunsaturated FA	57.6[a]	60.9[b]	66.2[d]	63.4[c]	68.6
Unsaturated FA	86.2[a]	85.2[a]	86.6[a]	85.7[a]	88.2
Total (ppm)	9147[a]	12,950[b]	16,978[b]	14,152[b]	—

Averages with the same letter within rows are not significantly different ($p < 0.05$).
* GC quantitation (Ramaroson-Raonizafinimanana, 1988).

much of saturated fatty acids (about 15%). These data consolidate the data of Ramaroson-Raonizafinimanana et al. (1988).

The cultivars from French Polynesia show quite different fatty acid compositions, as illustrated by Table 13.3. The intra-*tahitensis* chemodiversity is highlighted by a factorial discriminant analysis, showing that Parahurahu is very specific with a significantly higher content of monounsaturated long-chain fatty acids (especially nervonic acid) and linolenic acid (Figure 13.10).

The other cultivars have similar compositions with slight variations: Tahiti shows a greater amount of stearic acid, Tahiti long is distinguishable by its significantly higher oleic acid content, and Haapape is the oiliest among Polynesian cultivars.

The chemodiversity of Polynesian cultivars is underlined as well by aroma diversity as by fatty acids.

Conclusion

Tahitian vanilla beans have a specific aroma and fatty acid composition. The most original cultivar is Parahurahu, which contains a relatively high level of anisyl molecules and a low amount of vanillyl compounds, as well as a very high level of monounsaturated long-chain fatty acids.

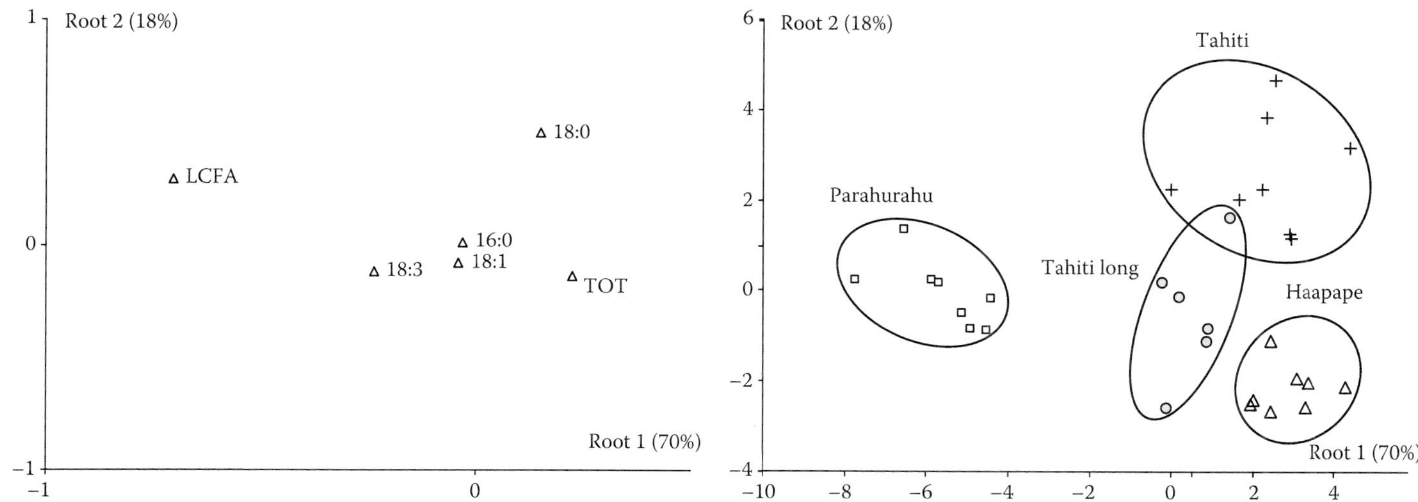

FIGURE 13.10 2D plot of four cultivars of uncured Tahitian vanilla beans analyzed in factorial discriminant analysis using fatty acids as variables (see Table 13.3).

The vanillas most cultivated in French Polynesia, namely Tahiti and Haapape, have homogeneous aroma and lipid characteristics. Their aroma composition and high fatty acid content explain, respectively, their subtle sensory properties and their attractive physical aspect.

SENSORY PROPERTIES

The sensory properties of Tahitian vanilla are unique. They are often analyzed either by sensory evaluation or by GC-O. These two techniques respectively (i) describe the overall flavor of vanilla and (ii) determine odor-active compounds contributing to the overall vanilla flavor.

Sensory Evaluation

The sensory properties of vanilla beans are primarily due to volatile constituents, but some nonvolatile compounds such as lipids may play a role in modifying the flavor perception (Purseglove et al., 1981). Their odor impact has not yet been well elucidated, but during the curing process, they fix some volatiles and restrict their release out of vanilla pods.

In sensory analysis, Tahitian vanilla is described as showing distinctive flowery, fragrant, perfumed, anise, almond, and cherry notes. The beans are also characterized by a shallow vanilla character as well as weaker phenolic, woody, balsamic, and smoky notes than *V. planifolia* (Ranadive, 1994, 2006; Petitdidier, 2005).

But there is no clear relationship between these sensory properties and the quantitative analytical parameters (as described in the section "Chemical Composition of Tahitian Vanilla"). For *V. planifolia*, high vanillin contents, which are indicators of quality, do not necessarily imply good sensory properties. In fact, due to a high odor threshold, some major compounds quantified in vanilla beans are not directly linked to the sensory properties of the extracts since minor compounds are involved in the overall vanilla flavor (Ranadive, 2006; Gassenmeier et al., 2008). That is why a GC-O analysis was performed.

GC-O Analysis

To describe the overall vanilla flavor, recent studies took into account the olfactory impact of compounds by GC-O and not only their concentration in the pods (Scharrer, 2002; Black, 2005; Perez-Silva et al., 2006). Compounds like anisaldehyde and anisyl alcohol account in great part for the characteristic flavor of Tahitian vanilla and are essential to the anise, sweet notes (www.flavornet.org). Some other compounds, such as aromatic anisyl esters (methyl anisate is present at 200 ppm in Tahitian vanilla), could have an olfactory impact on the overall aroma. Thus, a work on Tahitian vanilla enables the identification of 276 components. Among them, some minor specific anisyl molecules are scented and described as "anise, cherry, sweet" (Da Costa and Pantini, 2006).

Some phenolic compounds identified in Tahitian vanilla were found to be aroma-active by GC-O in *V. planifolia* beans (vanillin, vanillyl alcohol, *p*-hydroxybenzaldehyde, *p*-hydroxybenzyl alcohol, anisyl alcohol, and methyl cinnamate). Sweet, woody, balsamic, spicy, vanilla-like, and toasted notes were attributed to these compounds. Moreover, minor compounds present at less than 2 ppm have been found to contribute to the overall vanilla flavor, such as aldehydes, which were seen with green, oily, and herb-like floral notes, and aliphatic acids, which were described as having sour, buttery, and oily notes (Perez-Silva et al., 2006).

These studies have highlighted the specific anise flavor of Tahitian vanilla known as unique in the global market and considered as a luxury product.

GENETIC SPECIFICITY

Genetic Relationship between Tahitian Vanilla and Other Cultivated Vanilla Species

In order to determine if Tahitian vanilla was genetically different from the other cultivated vanilla species, a comparative study of the genetic diversity was carried out. This project involved the Bureau des Ressources Génétiques (BRG), the Centre de Cooperation Internationale en Recherche Agronomique pour le Développement (CIRAD), the Etablissement Vanille de Tahiti, the University of Reunion Island, and the Institut de Biotechnologie des Plantes (IBP) (Andrzejewski et al., 2006; Duval et al., 2006). Cultivated species *V. planifolia* (47 samples), *V. pompona* (6 samples), and Polynesian cultivars (6 samples) from Reunion Island, Central America, and French Polynesia, and 35 related species from Thailand, Brazil, Guiana, Cameroun, and botanical gardens were analyzed. The amplification fragment length polymorphism (AFLP) pattern of each sample was compared with other samples and the data were used to calculate dissimilarities between the samples according to the Sokal and Michener index. The structure of the genetic tree represented the species delimitation with a high level of accuracy (bootstraps of 100%). This study indicated that Tahitian vanilla is genetically very distinct from other cultivated or wild vanilla species.

Tahitian vanilla was also compared with several wild or cultivated species of *Vanilla* from Central America in collaboration with the Riverside University (Lubinsky et al., 2008). The putative origin of the Tahitian vanilla was assessed. Patterns of DNA sequences from the nuclear internal transcribed spacer (ITS) and chloroplast genomes were compared between Polynesian cultivars and samples collected in the tropical forest of Central America: *V. planifolia*, *V. odorata*, *V. insignis*, *V. pompona*, and a few other species. These data indicated that the two genetically closest species to Tahitian vanilla were *V. planifolia* and *V. odorata* and thus indicated that these two species are closely related to the ancestors of the Tahitian vanilla. This study favors the hypothesis of Porteres (1951) who assumed that Tahitian vanilla resulted from hybridization between *V. planifolia* and another hybrid involving *V. odorata*.

Polynesian Genetic Diversity

At the same time, the genetic diversity among Polynesian cultivars was also analyzed. The morphological diversity observed among Tahitian vanilla, which allowed

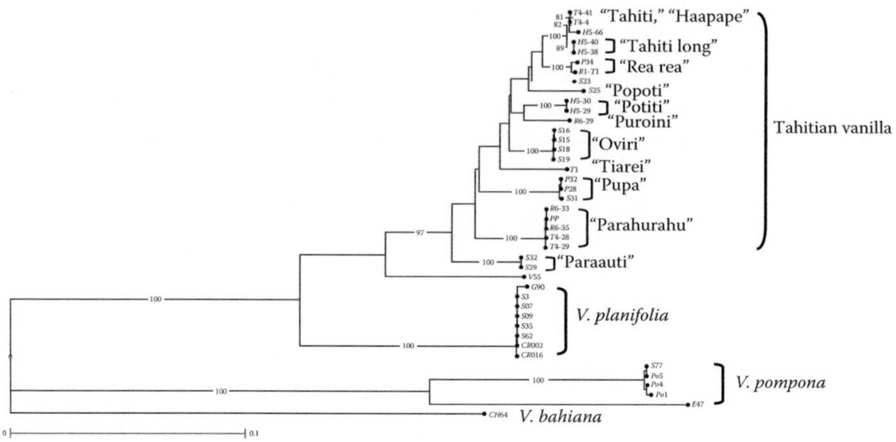

FIGURE 13.11 Diversity structure of 42 cultivated vanilla and *Vanilla bahiana* accessions and detailed relationships of the Polynesian accessions derived from neighbor joining trees using the distance matrix of Sokal and Michener (529 variables and 1000 bootstraps).

local growers to distinguish 14 cultivars, was related to genetic diversity. This genetic diversity was due to a variation at the level of the DNA sequences highlighted by the comparison of AFLP patterns between the different cultivars. AFLP patterns were defined by the presence or absence of each of the 529 tested AFLP markers. The comparison of these AFLP patterns indicated that in spite of the genetic diversity existing within Tahitian vanilla, all the accessions are gathered and form only one monophyletic group (Andrzejewski et al., 2006; Figure 13.11). Genetic diversity is weak (maximum distance between accession = 0.091) but coherent for a plant that is propagated by stem cutting and not by sexual reproduction. For some cultivars, many AFLP patterns were observed, indicating that diversity was underestimated by the growers and that new cultivars had to be defined.

Some others cultivars were not discriminated by their AFLP pattern; this was the case of Tahiti and Haapape. Their morphological differences were unambiguous; nevertheless, their genetic diversity has been found to be only the consequence of the number of copies of each AFLP marker. For this reason, cytogenetic analyses were developed.

As previously described for *V. planifolia* by Nair and Ravindran (1994), all the cells of a single root do not present the same chromosome value for Tahitian vanilla. We also observed these variations for the accessions of *V. planifolia* and *V. pompona*. Three ploidy levels were observed among the Polynesian accessions. The chromosome counts for numerous metaphase plates of root tips, and the genome size measurements by flow cytometry were concordant. Three groups were distinguished:

1. A diploid group (2×) with chromosome number ranging from 22 to 31 and with a genome size from 4.87 to 5.80 pg. These characteristics concerned the majority of the Polynesian cultivars: Tahiti (Figure 13.12), Rea rea, Parahurahu, Oviri, Paraauti, Potiti, Puroini, Pupa, Popoti, and Poura.

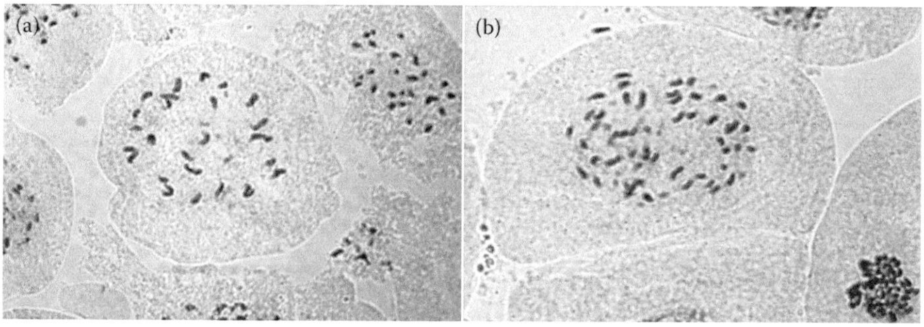

FIGURE 13.12 Metaphase plates of (a) cultivar Tahiti with 24 chromosomes and (b) cultivar Haapape with 52 chromosomes.

2. A triploid group (3×) for which the chromosome number varies from 33 to 41. This group contains two sterile vanilla plants called "Haapape sterile" showing abnormal flower physiology (absence of pollen and autoincompatibility).
3. A tetraploid group (4×) where the number of chromosomes and the genome size are twice as high as in the first group (43–56 chromosomes and genome size of 10.10–10.89 pg). The following are the cultivars: Haapape, Ofe ofe, Tiarei, and Tahiti long.

CONCLUSION

The originality of Tahitian vanilla is due to its morphology, its texture, and the aromatic content of its pods. These characteristics are indeed related to genotype. They are also influenced by geographical areas of cultivation. For example, beans of Tahitian vanilla produced and cured in French Polynesia are different from those produced in other countries. The rare and highly considered gourmet French Polynesian spice results from a combination of the factors the soil, the climate, and also the Polynesian people, who carefully produce and cure the vanilla beans.

ACKNOWLEDGMENTS

The authors thank Dr. S. Siljak-Yakovlev (Univ. Paris Sud) and Dr. S.C. Brown (CNRS Gif sur Yvette) for their help in cytogenetic analysis and for their interest and support to Tahitian vanilla genome studies, and J. Foster for his invaluable comments on the manuscript.

REFERENCES

Adedeji, J., T. G. Hartman, and H. Chi-Tang. 1993. Flavor characterization of different varieties of vanilla beans. *Perfumer & Flavorist* 18 (2):25–33.

Andrzejewski, S., S. Causse, C. Charon, M. Dron, M. F. Duval, S. Siljak-Yakovlev, and M. Wong. 2006. Diversité génétique des vanilliers cultivés en Polynésie française. Poster presented at the IVème *Colloque International sur les Plantes Aromatiques et Médicinales des Régions d'Outre-Mer*, Papeete, Tahiti.

Black, J. M. 2005. Vanilla bean volatile analysis origin & species. *Vanilla 3rd International Congress*. Veracruz, Mexico.

Bouriquet, G. 1954. *Le vanillier et la vanille dans le monde*. L'encyclopédie biologique no 46. Paul Lechevalier, Paris, 748 pp.

Bouriquet, G. and E. Hibon. 1950. Quelques vanilliers cultivés dans les établissements français d'Océanie. *L'agronomie tropicale* 1–2:54–61.

Brunschwig, C., F. X. Collard, M. Wong, J. P. Bianchini, and P. Raharivelomanana. 2007. The fatty acid composition of *Vanilla tahitensis*. Poster presented at the *Vanilla 3rd International Congress*, Jamesburg, New Jersey.

Chevalier, A. 1946. La systématique du genre *Vanilla* et l'amélioration des plants de vanille. *Revue de botanique appliquée et d'agriculture tropicale* 281–282:107–111.

Collard, F. X. 2007. Evolution of aroma compounds in *V. tahitensis* during the curing process. Poster presented at the *Vanilla 3rd International Congress*, Jamesburg, New Jersey.

Collard, F. X., C. Brunschwig, and M. Wong. 2006. Détermination de la composition aromatique de quelques variétés de *Vanilla tahitensis* par Chromatographie Liquide Haute Performance. Oral presentation at the *IVth CIPAM*, Papeete, Tahiti.

Costantin, J. and D. Bois. 1915. Sur trois types de vanilles commerciales de Tahiti. *Compte-rendus de l'académie des Sciences de Paris* 161:196–202.

Da Costa, N. C. and M. Pantini. 2006. The analysis of volatiles in Tahitian vanilla (*Vanilla tahitensis*) including novel compounds. *Flavour Science: Recent Advances and Trends* 43:161–164.

Duval, M. F., S. Bory, S. Andrzejewski, M. Grisoni, P. Besse, S. Causse, C. Charon, M. Dron, E. Odoux, and M. Wong. 2006. Diversité génétique des vanilliers dans leurs zones de dispersion secondaire. *Les actes du BRG* 6:181–196.

Ehlers, D., M. Pfister, and S. Bartholomae. 1994. Analysis of Tahiti vanilla by high-performance liquid chromatography. *Zeitschrift für Lebensmittel Untersuchung und Forschung* 199:38–42.

Fayet, B., C. Tisse, M. Guerere, and J. Estienne. 1987. Nouveaux critères analytiques dans l'étude des gousses de vanille. *Analusis* 15 (5):217–226.

Florence, J. and M. Guérin. 1996. A propos de la vanille à Tahiti. *L'orchidée* 13:84–87.

Gassenmeier, K., B. Riesen, and B. Magyar. 2008. Commercial quality and analytical parameters of cured vanilla beans (*Vanilla planifolia*) from different origins from the 2006–2007 crop. *Flavour and Fragrance Journal* 23:194–201.

Hartman, T. G. 2003. Composition of vanilla beans from different geographical regions. *First International Congress on the Future of the Vanilla Business*. Princeton, New Jersey.

Henry, C. 1924. Caractères généraux des différents vanilliers cultivés à Tahiti. *L'agronomie tropicale* 79:133–136.

Joulain, D., R. Laurent, J. Maason, J. C. Beolor, and H. Brevard. 2007. Heliotropin, heliotrope odor and Tahitian vanilla flavor: The end of a saga? *Natural Product Communications* 2 (3):305–308.

Klimes, I. and D. Lamparsky. 1976. Vanilla volatiles: A comprehensive analysis. *International Flavours and Food Additives* 7 (6):272–273, 291.

Langella, C., F. Saltron, and M. Guerere. 2003. Dosage et analyses isotopiques des composés phénoliques de la vanille: mesure de fidélité. *Ann. Fals. Exp. Chim.* 96 (962):73–86.

Larcher, J. 1989. Importance de la date de récolte sur la teneur en vanilline des gousses de *Vanilla tahitensis* J. W. Moore en Polynésie Française. *L'Agronomie Tropicale* 44 (2):143–146.

Lechat-Vahirua, I. and J. M. Bessiere. 1998. Plantes aromatiques de Polynésie Française: la vanille tahitienne. *Rivista Italiana* EPPOS 9: 569–578.

Lhuguenot, J. C. 1978. A propos de quelques particularités des extraits de vanille Tahiti (deuxième exposé). *Annales des Falsifications, de l'Expertise Chimique & Toxicologique* 71 (763):115–119.

Lubinsky, P., K. M. Cameron, M. C. Molina, M. Wong, S. Lepers-Andrzejewski, A. Gomez-Pompa, and S. C. Kim. 2008. Neotropical roots of a polynesian spice: The hybrid origin of Tahitian vanilla, *Vanilla tahitensis* (Orchidaceae). *American Journal of Botany* 95 (8):1040–1047.

Moore, J. 1933. New and critical plants from Raiatea. *Bernice P Bishop Museum Bulletin* 102:25.

Nair, R. R. and P. N. Ravindran. 1994. Somatic association of chromosomes and other mitotic abnormalities in *Vanilla planifolia* (Andrews). *Caryologia* 47 (1):65–73.

Odoux E, J. Escoute, and J. L. Verdeil. 2006. The relation between glucovanillin, beta-d-glucosidase activity and cellular compartmentation during the senescence, freezing and traditional curing of vanilla beans. *Annals of Applied Biology*, 149:43–52.

Perez-Silva, A., E. Odoux, P. Brat, F. Ribeyre, G. Rodriguez-Jimenes, V. Robles-Olvera, M. A. Garcia-Alvarado, and Z. Gunata. 2006. GC–MS and GC–olfactometry analysis of aroma compounds in a representative organic aroma extract from cured vanilla (*Vanilla planifolia* G. Jackson) beans. *Food Chemistry* 99:728–735.

Petitdidier, C. 2005. La vanille, reine des arômes chauds. *Arômes Ingrédients Additifs* 61: 38–45.

Portères, R. 1951. Nouvelles espèces de vanilliers et observations sur le vanillier de Tahiti. *Bulletin de la société Botanique de France* 98:126–127.

Portères, R. 1953. Le potentiel de variation clonale de *Vanilla tahitensis* J.W. Moore. *Agronomie Tropicale* 6:639–640.

Purseglove, J. W., E. G. Brown, C. L. Green, and S. R. J. Robbins. 1981. Vanilla. In Wrigley, D. R. a. G., ed. *Spices, Tropical Agriculture Series*. Longman Inc., London & New York, 644–735.

Ramaroson-Raonizafinimanana, B. 1988. Composition chimique de la fraction lipidique des gousses de trois espèces du genre Vanilla. PhD dissertation, Faculté des Sciences et Techniques de St-Jérôme, Aix-Marseille, France, 204pp.

Ramaroson-Raonizafinimanana, B., E. M. Gaydou, and I. Bombarda. 1997. Hydrocarbons from three Vanilla beans species: *V. fragrans*, *V. madagascariensis*, and *V. tahitensis*. *J. Agric. Food Chem.* 45:2542–2545.

Ramaroson-Raonizafinimanana, B., E. M. Gaydou, and I. Bombarda. 1998. 4-demethylsterols and triterpene alcohols from two vanilla bean species: *Vanilla fragrans* and *V. tahitensis*. *J. Am. Oil Chem. Soc.* 75:51–55.

Ramaroson-Raonizafinimanana, B., E. M. Gaydou, and I. Bombarda. 1999. Long chain g-pyrones in epicuticular wax of two vanilla bean species: *V. fragrans* and *V. tahitensis*. *J. Agric. Food Chem.* 47 (8):3202–3205.

Ramaroson-Raonizafinimanana, B., E. M. Gaydou, and I. Bombarda. 2000. Long chain aliphatic B-diketones from epicuticular wax of vanilla bean species. Synthesis of nervonoylacetone. *J. Agric. Food Chem.* 48 (10):4739–4743.

Ranadive, A. S. 1992. Vanillin and related flavor compounds in vanilla extracts made from beans of various global origins. *Journal of Agricultural and Food Chemistry* 40 (10):1922–1924.

Ranadive, A. S. 1994. Vanilla. Cultivation, curing, chemistry, technology and commercial products. In: Charalambous, G. ed. *Spices, Herbs and Edible Fungi*. Elsevier, Amsterdam, the Netherlands, 517–577.

Ranadive, A. S. 2006. Inside look: Chemistry and biochemistry of vanilla flavor. *Perfumer and Flavorist* 31:38–44.

Rey, S., F. Carbonel, and M. Van Doorn. 1980. Vanille: Composition aromatique des différentes espèces, *Vanilla planifolia* (Andrews), *Vanilla tahitensis* (Moore), *Vanilla pompona* (Shiede). *Annales des Falsifications, de l'Expertise Chimique & Toxicologique* 73 (788):421–431.

Scharrer, A. 2002. Vanille, Neues zur Authentizität. PhD dissertation, Fachbereich Chemische und Pharmazeutische Wissenschaften der Johann Wolfgang Goethe-Universitat, Frankfurt am Main, Germany, 114pp.

Shiota, H. and K. Itoga. 1975. The study of the aromatic components of vanilla beans (*V. planifolia* Andrews and *V. tahitensis* Moore). *Koryo (Japan)* 113:65–71.

Sostaric, T., M. C. Boyce, and E. E. Spickett. 2000. Analysis of the volatile components in vanilla extracts and flavorings by solid-phase microextraction and gas chromatography. *J. Agric. Food Chem.* 48 (12):5802–5807.

Tabacchi, R., G. Nicollier, and J. Garnéro. 1978. A propos de quelques particularités des extraits de vanille Tahiti (premier exposé). *Annales des Falsifications, de l'Expertise Chimique & Toxicologique* 71 (763):109–114.

14 Microbial Safety of Cured Vanilla Beans

Samira Sarter

CONTENTS

INTRODUCTION

Natural vanilla is an important ingredient in the food industry all over the world. The vanilla bean, the fruit of the climbing orchid *Vanilla planifolia*, is the main source for the commercial production of vanilla flavor, which consists of a mixture of vanillin and many other flavored compounds (Perez-Silva et al., 2006). However, the development of the vanilla flavor in harvested green beans is obtained only after a curing process comprising four stages in the traditional method; that is, killing, sweating, drying, and conditioning. As vanilla requires a humid tropical climate, the drying step is a key element to prevent microbial development, which might compromise its quality. The curing process, which takes several months, will then determine the vanilla bean quality regarding both commercial and safety standards. Over the past 20 years, the number of standards has grown rapidly due to the globalization and free trade. A very high value added product such as vanilla is very dependent on both the regulatory and private standards to be competitive on the international markets. In addition to the numerous technical regulations on food safety, plant protection, and labeling developed at the national and international levels, the private sector has increasingly established new standards covering the supply chain from farm to table. During these different stages of the vanilla beans preparation and storage, microbial hazards can occur at multiple points and then multiply or cross-contaminate other products once present. Thus, a farm-to-table approach is required to identify the hazards and the most effective points to control their occurrence. Since spices and food ingredients might be a source of contaminations (Scheuer and Gareis, 2002), this chapter focuses on the microbiological hazards associated with the processing and storage of cured vanilla beans and on the preventive measures to control those hazards.

MICROBIAL HAZARDS OF CURED VANILLA BEANS

Microbial contaminations (mainly molds and bacteria) of the vanilla beans can occur at the harvest and through the several steps of handling and processing. The curing process, besides enabling the biochemical process of aroma, is essential for the control of these microorganisms, which is beneficial for the preservation of the cured beans.

Curing of vanilla beans is a traditionally well-established process in the main countries of production (Madagascar, Mexico, Comores, Reunion, and Tahiti). It is laborious and takes up to six months, depending on the curing procedures adopted by different producing countries. In general, it is mainly based on the following four steps (Dignum et al., 2001):

- Killing the green beans by a thermal treatment (hot water, oven) to stop the vegetative development of the fresh beans.
- Sweating which takes place during the cycles of sunning and sweating, where the beans are spread out in the sun until they are hot, and then wrapped up in blankets and put into an airtight container overnight to maintain their warmth and moisture.
- Drying the beans slowly to prevent the microbial spoilage and to progressively allow the reactions for the development of aroma.
- Conditioning the dried beans in boxes for 3–4 months to obtain the desired aroma and flavor.

The first stage (killing) reduces the initial microbial load on the surface of the green beans. During curing, the major fungi found on vanilla beans are mainly black *Aspergillus* and green *Penicillium* strains (Röling et al., 2001). Several bacterial communities are present on the green beans, but after scalding (immersion in hot water 65–70°C for 2 min), this study has reported only the presence of *Bacillus* strains (*Bacillus subtilis*, *Bacillus firmus*, *Bacillus licheniformis*, *Bacillus pumilis*, *Bacillus smithii*) due to their thermophilic nature as being sporeforming microorganisms. For split beans, which have not been scalded, other genera were found such as *Xanthomonas*, *Cellulomonas*, *Vibrio*, and *Staphylococcus*. *B. subtilis* (including *B. licheniformis* as well) are food-spoilage organisms that have been isolated from other herbs and spices and that have also been implicated in food poisoning (te Giffel et al., 1996). Bourriquet (1954) has identified the presence of *Aspergillus niger*, *Penicillium lividum*, *Penicillium vanillae*, and *Penicillium rugulosum* on vanilla beans from Madagascar and Comores. Another study has shown that the main mold contaminations of vanilla beans consisted of *Penicillium glaucum*, *Trichoderma* sp., *Aspergillus oryzae*, *Aspergillus amstelodami*, and *Xantochpinirous* sp. at a total concentration of 8.4×10^4/g (Bachman et al., 1995).

The second key step is the drying step as the moisture content is a major factor in the preservation of cured vanilla beans because low moisture content is essential to prevent microbial growth. The beans should be well aerated while drying to avoid bacterial fermentation, which leads to undesirable flavor, as creosoted vanilla. When properly cured, the water content of beans must be sufficiently low to prevent the growth and activity of microorganisms. Actually, combination of low water with high phenolic content of which vanillin is the major compound, provide an inhibitory effect of the spoilage in cured beans (Havkin-Frenkel and Frenkel, 2006). According to ISO 5565-1:1999 (ISO, 1999), the maximum moisture specification in cured beans is 38% for classes 1 and 2, 30% for class 3, and 25% for class 4. The water content in cured vanilla beans correspond to water activity (a_w) values of, respectively, 0.89, 0.86, and 0.84 (J.M. Méot, pers. comm.). The presence of appropriate water content ranging from 25% to 30% has been reported to result in the desirable texture and appearance of the cured beans (Sreedhar et al., 2007).

Immature beans have been reported to be more susceptible to mold infestation by *Penicillium* and *Aspergillus* species (Sasikumar et al., 1992). As *Aspergillus* and *Penicillium* are, with *Fusarium*, the most important genera containing toxigenic isolates, prevention measures based on Good Agricultural Practices and Good Manufacturing Practices should be observed to mitigate preharvest and postharvest contaminations by mycotoxins (FDA, 2009). Mycotoxins are toxic secondary metabolites produced by fungi under specific conditions related to the physiological (microbial growth) and environmental (pH, temperature, a_w, preservatives) conditions. The toxigenic fungi may be eliminated during processing, while their corresponding extracellular mycotoxin remains in the food. The most important mycotoxins for human health risk are described in Table 14.1 (FAO, 2001). Among the fungi reported in vanilla beans, *A. niger* does not have the ability to produce aflatoxins (Schuster et al., 2002). However, this specie has been reported to produce ochratoxin A (Blumenthal, 2004). *A. oryzae* does not produce aflatoxins despite its very close taxonomic relatedness to *Aspergillus flavus* group, a major group that produces aflatoxins in food (Blumenthal, 2004). *A. oryzae* might produce other mycotoxin such as 3-nitropropionic acid. *P. lividum* has been reported to produce citrinin and penicillic acid (Frisvad, 1986). However, the risk, if any, of mycotoxin contamination of vanilla beans needs a detailed assessment (Codex Alimentarius Commission (CAC), 1999). A few research papers have reported the contamination of vanilla beans by mycotoxins. Analyzing a total of 681 samples of spices for the natural

TABLE 14.1
Molds and Mycotoxins of Worldwide Importance

Mold Species	Mycotoxins Produced
Aspergillus parasiticus	Aflatoxins B_1, B_2, G_1, G_2
Aspergillus flavus	Aflatoxins B_1, B_2
Fusarium sporotrichioides	T-2 toxin
Fusarium graminearum	Deoxynivalenol (or nivalenol), Zearalenone
Fusarium moniliforme (Fusarium verticillioides)	Fumonisin B_1
Penicillium verrucosum	Ochratoxin A
Aspergillus ochraceus	Ochratoxin A

Source: From FAO. 2001. *Manual on the Application of the HACCP System in Mycotoxin Prevention and Control.* FAO, Rome. With permission.

occurrence of ochratoxin A (OTA) and ochratoxin B (OTB), Scheuer and Gareis (2002) found one sample of vanilla extract that was positive for OTB at 156 ng/g.

Immaturity has been linked also to a poor yield of vanillin in the cured beans (Saltron et al., 2002, 2003). Cured beans contain about 2.5% of vanillin when properly prepared. Molds attack the bean from the fruit base, where the vanillin concentration is the least (Bourriquet, 1954). It should be stressed that vanillin has been reported to have antimicrobial properties against different food-related microorganisms (Nakazawa et al., 1982) and some pathogenic, indicator, or spoilage microorganisms: *Escherichia coli*, *Enterobacter aerogenes*, *Pseudomonas aeruginosa*, *Salmonella enterica*, *Candida albicans*, *Lactobacillus casei*, *Penicillium expansum*, and *Saccharomyces cerevisiae* (Rupasinghe et al., 2006). *In vitro* studies have shown that vanillin is effective in molds growth inhibition. The addition of 1000 ppm of natural vanillin for instance inhibited *Aspergillus ochraceus* growth for more than two months at 25°C, while growth of *A. flavus*, *A. niger*, and *Aspergillus parasiticus* was inhibited by 1500 ppm (Lopez-Malo et al., 1995, 1997). Vanillin has been proposed by these authors for hurdle technology in combination with other antimicrobial factors such as reduced pH and a_w to prevent mold spoilage in fruit purées. Another study using different food-spoilage molds and yeasts have suggested that the aldehyde moiety plays a key role in the antifungal activity of vanillin (Fitzgerald et al., 2005). These authors (Fitzgerald et al., 2004) have also investigated the mode of action of vanillin against *Escherichia coli*, *Lactobacillus plantarum*, and *Listeria innocua*. The antimicrobial activity of vanillin was dependent on the time of exposure, concentration, and the target organism. The *in vitro* minimum inhibitory concentrations of vanillin were 15, 75, and 35 mmol/L for the three tested strains. This inhibitory action of vanillin was found bacteriostatic rather than bactericidal for all. This study demonstrated that the vanillin disturbs the integrity of the cytoplasmic membrane, leading to the loss of ion gradients, pH homeostasis, and inhibition of respiratory activity of the tested bacteria.

The potential hazards for cured vanilla beans can thus be categorized as follows:

- Microbial spoilage due to bacterial fermentation or molds development
- Contaminations from microbial toxins (mycotoxins from molds)
- Bacterial contaminations related to poor hygienic conditions

To prevent these microbial hazards of cured vanilla beans, the relevant factors that should be controlled are the maturity of the green beans at harvest, the a_w of the processed beans, and the overall postharvest hygienic conditions. To preserve the quality of the beans during the storage that could last several months or up to one year, other parameters such as temperature, humidity, gas composition, and type of packaging should also be taken into consideration.

MICROBIAL HAZARDS MANAGEMENT AND CONTROL

Control measures aim to prevent, eliminate, or reduce food safety hazards to a tolerable level by either controlling initial levels of a hazardous agent, or preventing an increase in its levels, or reducing its levels. An important role of Hazard Analysis

Critical Control Point (HACCP) is to help the food producer and processor build safety into processes through identification of key or critical control measures that prevent, eliminate, or reduce hazards to acceptable levels.

The control of microbiological hazards of cured vanilla beans is particularly dependent on effective drying and the subsequent prevention of postprocess contamination and/or growth. The most effective microbial control measure in vanilla beans processing is to dry the commodity such that a_w is very low to support the microbial development, and in particular, the mold growth and/or prevent mycotoxin production from toxinogenic species. a_w is a measure of the water that is available to microorganisms. To prevent the growth of most molds, a_w needs to be ≤ 0.70. Each toxigenic mold has its own minimum a_w for growth and mycotoxin production and these translate into moisture contents for each commodity. These moisture contents are termed "safe" and would be the critical limit for the control measure. Most spoilage bacteria cannot grow at an $a_w \leq 0.91$. *Staphylococcus aureus* has, however, been found to grow at a_w as low as 0.84. Since molds are able to grow over a wide range of pH values, moisture conditions, and wider temperature range than bacteria, they are more susceptible for contaminating vanilla beans. A critical control point could be placed at the end of the drying process and one critical limit would be the water content or a_w of the beans, which would be easy to monitor in an HACCP system.

Poor hygienic conditions are a major source of contamination of the beans during the whole process. Since the majority of microorganisms are located onto the surface of the beans, the education of handlers is a priority. Hands as well as contaminated gloves and blankets can serve as vectors for the transmission of microorganisms. As reported in the literature, the hands of food workers are of major importance in the transfer of contaminants from person to person, from person to surfaces or vice versa, and from person to food (Guzewich and Ross, 1999). Good housekeeping procedures are necessary to minimize the levels of insects and fungi in storage facilities. Moldy beans should be discarded immediately because they lead to off-odors and lead to cross-contaminations between different lots of the vanilla beans.

From this point of view, the HACCP system is a widely recognized food safety management. It is a systematic approach in identifying, evaluating, and controlling food safety hazards. It is based on a preventive approach from the primary production till the end product sold on the market rather than relying solely on conventional inspections by regulatory agencies. Once an HACCP plan has been developed and introduced into a food operation it must be maintained on a continuous basis. The seven principles are: (1) identifying any hazards that must be prevented, eliminated, or reduced to acceptable levels; (2) identifying the critical control points (CCPs) at the step(s) at which control is essential to prevent or eliminate a hazard or to reduce it to acceptable levels; (3) establishing critical limits at CCPs, which separate acceptability from unacceptability for the prevention, elimination, or reduction of identified hazards; (4) establishing and implementing effective monitoring procedures at CCPs; (5) establishing corrective actions when monitoring indicates that a CCP is not under control; (6) establishing procedures, which shall be carried out regularly, to verify that the measures outlined are working effectively; and (7) establishing documents and records commensurate with the nature and size of the food business to demonstrate the effective application of the measures outlined. CCP, as defined in the Food Code, means a point at which loss of control may result in an unacceptable health risk (FDA, 2009). These principles have international acceptance and their application have been described by the food hygiene code of the Codex Alimentarius Commission (CAC, 2003), the National Advisory Committee on Microbiological Criteria for Foods (NACMCF, 1992), the ISO 22000 (ISO, 2005), the Food Code of the Food and Drug Administration (FDA, 2009), and the European regulation No 852/2004 (JO, 2004). Since 2006, this new European regulation on the hygiene of foodstuffs has been applied and it is also applicable to foreign operators who produce export food to the European Union. The new hygiene rules take particular account of the following principles:

- Primary responsibility for food safety borne by the food business operator.
- Food safety ensured throughout the food chain, starting with primary production.
- General implementation of procedures based on the HACCP principles.
- Application of basic common hygiene requirements, possibly further specified for certain categories of food.
- Development of guides to good practice for hygiene or for the application of HACCP principles as a valuable instrument to aid food business operators at all levels of the food chain to comply with the new rules.

The Recommended International Code of Practice—General Principles of Food Hygiene (CAC, 2003) is a basis for the implementation of Good Hygienic Practices as it gives the basic rules for the hygienic handling, storage, processing, distribution, and final preparation of food along the production chain. This code contains the Annex on HACCP system and guidelines for its application. In addition, the Code of Hygienic Practice for spices and dried aromatic plants (CAC, 1995) is more specific to our purpose and includes the minimum requirements of hygiene for harvesting, postharvest technology (curing, bleaching, drying, cleaning, grading, packing, transportation, and storage, including microbial and insect disinfestation), processing establishment, processing technology, packaging, and storage of processed products.

CONCLUSION

Standard specifications are important criteria determining the market value of a high value-added product highly demanded on the international market such as natural vanilla beans. During the postharvest preparation and processing of vanilla beans, microbiological hazards can occur at any stage and could compromise both safety and quality standards of cured vanilla beans. Key common factors for the beans preservation include (1) the maturity of the harvested beans to optimize the vanillin yield as having antimicrobial activity against molds and bacteria, (2) the killing and drying steps of curing to control the microbial development on the beans, (3) the hygienic conditions during the whole postharvest process to control the occurrence of contaminations and cross-contaminations between different batches of beans, and

(4) the conditioning step and postprocess storage to control the microbial growth and activity (production of mycotoxins from molds). To help producers and processors implementing these controls along the processing chain, it is necessary to build HACCP system based on sound Good Hygienic Practices (GHP), Good Agricultural Practices (GAP), and Good Manufacturing Practices (GMP) achieving an integrated approach to food safety management. Nevertheless, the most important restrictions for the implementation of HACCP throughout the whole supply chain of vanilla production might be organizational, since the actors chain involve farmers, collectors, processors, and exporters that do not necessarily comply with a vertical coordination, which has been reported to be more efficient for this purpose (Hobbs and Kerr, 1992). In this context, HACCP might face the problem of small operators with limited resources in developing countries.

REFERENCES

Bachman, S., Pietka, M., and Zegota, H. 1995. Studies on some microbiological and chemical aspects of irradiated vanilla beans. *Journal of Radioanalytical and Nuclear Chemistry* 189 (1):71–76.

Blumenthal, C.Z. 2004. Production of toxic metabolites *Aspergillus niger*, *Aspergillus oryzae* and *Trichoderma reesei*: Justification of mycotoxin testing in food grade enzyme preparation derived from the three fungi. *Regulatory Toxicology and Pharmacology* 39:214–228.

Bourriquet, G. 1954. Altérations et défauts de la vanille préparée. In: G. Bourriquet ed. *Le vanillier et la vanille dans le monde*. Paul Lechevalier, Paris, pp. 623–644.

CAC (Codex Alimentarius Commission). 1995. Code of hygienic practice for spices and dried aromatic plants. CAC/RCP 42-1995. FAO, Rome.

CAC (Codex Alimentarius Commission). 1999. Principles and guidelines for the conduct of microbiological risk assessment. CAC/GL 30-1999. FAO, Rome.

CAC (Codex Alimentarius Commission). 2003. Recommended International Code of Practice-General Principles of Food Hygiene. CAC/RCP 1-1969, Rev.4-2003. FAO, Rome.

Dignum, M.J.W., Kerler, J., and Verpoorte, R. 2001. Vanilla production: Technological, chemical, and biosynthetic aspects. *Food Reviews International* 17 (2):199–219.

FAO. 2001. *Manual on the Application of the HACCP System in Mycotoxin Prevention and Control*. FAO, Rome.

FDA (Food and Drug Administration). 2009. Food Code. FDA, Washington, DC 20204. Available at http://www.fda.gov/Food/FoodSafety/RetailFoodProtection/FoodCode/FoodCode2009/default.htm

Fitzgerald, D.J., Stratford, M., Gasson, M.J., et al. 2004. Mode of antimicrobial action of vanillin against *Escherichia coli*, *Lactobacillus plantarum* and *Listeria innocua*. *Journal of Applied Microbiology* 97 (1):104–113.

Fitzgerald, D.J., Stratford, M., Gasson, M.J., and Narbad, A. 2005. Structure–function analysis of the vanillin molecule and its antifungal properties. *Journal of Agricultural and Food Chemistry* 53:1769–1775.

Frisvad, J.C. 1986. Taxonomic approaches to mycotoxin identification (taxonomic indication of mycotoxin content in foods). In: R.J. Cole, ed. *Modern Methods in the Analysis and Structural Elucidation of Mycotoxins*. Academic Press, London.

Guzewich, J. and Ross, M.P. 1999. Evaluation of risks related to microbiological contamination of ready-to-eat foods by food preparation workers and the effectiveness of interventions to minimize those risks. *Available at:* http://vm.cfsan.fda.gov/;ear/rterisk.html.

Havkin-Frenkel, D. and Frenkel, C. 2006. Postharvest handling and storage of cured vanilla beans. *Stewart Postharvest Review* 2 (4):1–9.

Hobbs, J. and Kerr, W. A. 1992. Costs of monitoring food safety and vertical coordination in agribusiness: What can be learned from the British food safety Act of 1990? *Agribusiness* 8 (6):575–584.

ISO. 1999. ISO 5565-1:1999–Vanilla [*Vanilla fragrans* (Salisbury) Ames]—Part 1: Specification.

ISO. 2005. ISO 22000–Systèmes de management de la sécurité des denrées alimentaires–Exigences pour tout organisme appartenant à la chaîne alimentaire.

JO. 2004. Règlement (CE) No. 852/2004 du Parlement Européen et du Conseil du 29 avril 2004 relatif à l'hygiène des denrées alimentaires, Journal Officiel de l'Union Européenne Vol. L 139.

Lopez-Malo, A., Alzamora, S.M., and Argaiz, A. 1995. Effect of natural vanillin on germination time and radial growth of moulds in fruit-based agar systems. *Food Microbiology* 12:213–219.

Lopez-Malo, A., Alzamora, S.M., and Argaiz, A. 1997. Effect of vanillin concentration, pH and incubation temperature on *Aspergillus flavus*, *Aspergillus niger*, *Aspergillus ochraceus* and *Aspergillus parasiticus* growth. *Food Microbiology* 14 (2):117–124.

NACMCF. 1992. National Advisory Committee on Microbiological Criteria for Foods. Hazard analysis and critical control point system. *International Journal of Food Microbiology* 16:1–23.

Nakazawa, Y., Wada, R., Izumitani, M., Nakamura, H., and Chikuma, K. 1982. Studies on the vanilla flavours in food processing. VI. The antimicrobial action of vanilla compounds. *Japanese Journal of Dairy and Food Science* 31 (3):99–105.

Perez-Silva, A., Odoux, E., Brat, P., Ribeyre, F., Rodriguez-Jimenez, G., Robles-Olvera, V., Garcia-Alvarado, M.A., and Günata, Z. 2006. GC-MS and GC-olfactometry analysis of aroma compounds in a representative organic aroma extract from cured vanilla (*Vanilla planifolia* G. Jackson) beans. *Food Chemistry* 99:728–735.

Röling, W.F.M., Kerler, J., Braster, M., Apriyantono, A., Stam, H., and Van Verseveld, H.M., 2001. Microorganisms with a taste for vanilla: Microbial ecology of traditional Indonesian curing. *Applied and Environmental Microbiology* 67 (5):1995–2003.

Rupasinghe, H., Boulter-Bitzer, J., Ahn, T., and Odumeru, J. 2006. Vanillin inhibits pathogenic and spoilage microorganisms *in vitro* and aerobic microbial growth in fresh-cut apples. *Food Research International* 39 (5):575–580.

Saltron, F., Langella, C., and Guerere, M. 2002. Assessment of the quality of Malagasy vanilla: harvest 2000. *Annales des Falsifications de l'Expertise Chimique et Toxicologique (France)* 95 (958):79–105.

Saltron, F., Langella, C., and Guerere, M. 2003. Assessment of Malagasy vanilla bean quality: Harvest 2001. *Annales des Falsifications de l'Expertise Chimique et Toxicologique (France)* 95 (961):419–440.

Sasikumar, B., Rema, J., and Ravindran, P.N. 1992. Vanilla. *Indian Cocoa, Arecanut and Spices Journal* 16 (1):6–10.

Scheuer, R. and Gareis, M. 2002. Occurrence of ochratoxin A and B in spices. *Mycotoxin Research* 18 (2):62–66.

Schuster, E., Dunn-Coleman, N., Frisvad, J.C., and van Dijck, P.W.M. 2002. On the safety of *Aspergillus niger*—a review. *Applied Microbiology and Biotechnology* 59:426–435.

Sreedhar, R.V., Roohie, K., Venkatachalam, L., Narayan, M.S., and Bhagyalakshmi, N. 2007. Specific pretreatments reduce curing period of vanilla (*Vanilla planifolia*) beans. *Journal of Agricultural and Food Chemistry* 55 (8):2947–2955.

te Giffel, M.C., Beumer, R.R., Leijendekkers, S., and Rombouts, F.M. 1996. Incidence of *Bacillus cereus* and *Bacillus subtilis* in foods in the Netherlands. *Food Microbiology* 13:53–58.

15 Authentication of Vanilla Products

Jens-Michael Hilmer, Franz-Josef Hammerschmidt, and Gerd Lösing

CONTENTS

INTRODUCTION

Vanilla belongs to the most valued as well as the most expensive spices of the world. It represents the most important aromatic flavor, whose production constitutes a multimillion dollar a year for the industry (Krueger and Krueger, 1983). Worldwide production of vanilla observed is 2000 metric tons per year. Vanilla extracts are used extensively in chocolate and baked products, but most commonly in ice cream. The main flavor-active ingredient of vanilla extracts is vanillin. Industrial chemical synthesis of vanillin started more than 130 years ago (Tiemann and Haarmann, 1874). It is one of the most widely used flavoring components currently used in the flavor industry. The total consumption of vanillin is estimated to be at 12,000 metric tons per year (Eurofins Newsletter, 2003), about 82% for flavor purposes, 5% for fragrances and cosmetics, and 13% for pharmaceutical intermediates. The demand for it far outweighs its natural supply. Bensaid (Eurofins Newsletter, 2003) states that only 0.33% of the consumed vanillin originate from vanilla beans. Natural vanillin can be 100 times more expensive than vanillin from synthetic origin—in some years even more, depending on the harvest (Eurofins Food Newsletter, 2007). This price difference in combination with the "biotrend" and the demand for natural products has induced the flavor industry to develop alternative sources of natural vanillin, for example, based on the biotransformation of natural compounds.

Therefore, adulteration of vanilla extracts is a major problem in the commercial market for this product category. Consequently, there are plenty of governmental regulations concerning the authenticity of vanilla extracts to detect adulterations and to avoid cases of fraud.

In order to protect food manufacturers and consumers from adulteration of vanilla products, it is necessary to have powerful analytical tools available to prove authenticity of the respective vanilla products. Depending on the nature of the expected fraudulent activities as well as the specific product categories affected (such as vanilla beans, vanilla extracts, vanilla flavors, or vanilla food products), different strategies to detect these adulterations are applied.

There are numerous possibilities known about how the legal and sensory status of a vanilla product can be affected:

- Addition of non-vanilla-derived compounds to impart flavor; this includes:
    - nature-identical/artificial vanillin from synthetic origin
    - artificial ethyl vanillin
    - tonka bean extract
    - coumarin
- Adulteration of vanilla beans: addition of iron particles such as nails or even elementary mercury in order to increase the weight of the very precious vanilla beans (this was especially a topic, when vanilla bean prices were very high and reached several hundred dollars a kilogram some years ago).
- Adulteration by incorrect botanical origin: there are more than 100 different vanilla species known, and only three of them have a commercial relevance: *Vanilla planifolia* Jacks./Andrews, *Vanilla tahitiensis* J.W. Moore, and *Vanilla pompona* Schiede. For example, Bourbon Vanilla is related to *V. planifolia* species.
- Adulteration by incorrect geographical origin: since the geographical origin of vanilla beans is relevant for regulatory reasons [e.g., to claim "Bourbon Vanilla" is only allowed when using vanilla extracts derived from vanilla grown in Madagascar, Reunion Island (formerly called "Ile Bourbon"), the Comoros, Seychelles, or Mauritius].
- Correct concentration indication of *x*-fold vanilla extracts: The amount of vanilla beans used for the production of vanilla extracts is relevant, for example, in the US market. This is expressed in the "Standard of Identity" (FDA Code of Federal Regulations). A total of 100 g of vanilla beans used for the production of 1 kg vanilla extract gives a so-called onefold extract.

The analytical methods that could be applied to indicate an authentic vanilla product can be based on the identification of single ingredients as well as on the combination of ingredients typically present in vanilla. As an example, the main flavor-active compound present in vanilla, vanillin, can be monitored specifically with methods such as GC, HPLC, IRMS, or specific NMR methods (see below). In addition, the presence of typical by-products in vanilla, such as vanillic acid, *p*-hydroxybenzaldehyde, or *p*-hydroxybenzoic acid, is determined. The ratio of these components is also used to judge the authenticity of vanilla and vanilla products. Also, the absence of compounds, which are usually not present in vanilla, such as (artificial) ethyl vanillin or vanillin derived from chemical synthesis can be used to assess the quality. In order to detect adulterations indicated, various specific analytical methods were developed and presented in the following sections.

Both consumers as well as food manufacturers can be protected from fraudulent activities by using state-of-the-art comprehensive analytical techniques for the evaluation of the authenticity of food raw materials as well as food products.

HPLC ANALYSIS

The HPLC analysis is nowadays the routinely applied instrumental technique to analyze flavor compounds in vanilla products or their extracts. HPLC analysis is used to identify and quantify relevant compounds as well as in the further evaluation of authenticity or adulteration. Other techniques such as GC analyses have been published (Mosandl and Scharrer, 2001) but not been established in quality control and authenticity checks. An authenticity check based on the enantio selective analysis of the volatile fraction of vanilla extracts has been investigated (Mosandl and Hener, 2001). Among others, γ-nonalactone is one of the minor chiral compounds in vanilla and was extracted and separated into the *R*- and *S*-enantiomer. The low enantiomeric excess of *R*-γ-nonalactone of 45–63% was judged to be insignificantly high enough to provide an unequivocal proof for adulteration with racemic γ-nonalactone. HPLC methods established for vanilla are robust, rapid, and well suited for routine analysis. Sample preparation of the HPLC analysis is done by extraction in case of the beans, or simply dilution for extracts. The effects of the sample preparation techniques on the analytical results have been investigated (Ehlers et al., 1999). These results indicate that the ratios of typical vanilla ingredients such as vanillin, vanillic acid, *p*-hydroxybenzaldehyde, and *p*-hydroxybenzoic acid can vary depending not only on the raw material but also on the sample preparation method used.

Different HPLC methods have been developed and published and finally being adopted as official methods (Taylor, 1993; ISO, 1999) or even being part of the food law in France (Arrêté du juin 11, 1987).

The stationary phases are dominantly reversed-phase materials, mostly RP-18 materials but also RP-8 and others such as alkyl halogen-modified silica gel (Taylor, 1993).

The methods applying reversed-phase separation typically run with methanol/water mobile phases in isocratic mode or more often with gradient elution at acidic pH values and finally using UV- or DAD-detection (diode array detector). The quantification can be based on internal standards, and also on external calibration as most of the detected compounds are available as pure chemicals. Run times are around 20 min; however, with special columns with selected stationary phases, the run time can be reduced to less than 5 min (Tracy et al., 2008).

Typical parameters for HPLC analysis of vanilla extracts are given in Table 15.1.

The compounds that have been identified are also subject to publications and are described elsewhere in this book (see Chapter 12). Typical compounds that are detected with these analyses are vanillin, vanillic acid, *p*-hydroxybenzaldehyde, and *p*-hydroxy benzoic acid. These analytes are seen as impact compounds and represent a typical profile (Figure 15.1). A simple authenticity check can be performed based on the presence of untypical compounds such as ethyl vanillin but also on the absence of one or more of these impact compounds. Other minor compounds that are detected

TABLE 15.1
Typical Parameters for HPLC Analysis

Instrument	Agilent 1100
Column	Lichrospher 100 RP18 (5 μm) 124 × 4 mm
Injection volume	2 μL
Eluent/gradient	Acetonitrile/water resp. phosphate buffer
Detection	Diode array detector

by HPLC, such as anisaldehyde, may serve as indicators to distinguish *V. planifolia* from *V. tahitiensis* (Ehlers et al., 1994; Oberdieck, 1998).

Their quantitative distribution is typical for the kind of vanilla product and the extraction technique. Different studies have investigated the profile resulting from these impact compounds in vanilla extracts and found to be specific and therefore suitable to evaluate the authenticity of a vanilla extract (Fayet et al., 1987; Juergens, 1981). To transform the profile of these compounds into a numerical scale, the ratios of their concentration were calculated and limits were suggested. Five ratios of vanilla compounds have been introduced into French law to evaluate the authenticity (République Française, 1988) of vanilla products. In 2003, the ratios have been amended by the Direction Générale de la Concurrence, de la Consommation et de la Répression des Fraudes (DGCCRF) to reflect findings over a longer period of time and additional information from recent crops at that time (*Note d'information*, 2003). The limits are shown in Table 15.2. This evaluation is suitable only for alcoholic extracts of cured whole vanilla beans. Changes in the solvent composition or the use of alternative extraction agents such as CO_2 lead to ratios that do not necessarily comply with the limits given in Table 15.2 (Quirin and Gerard, 1998). Also the source of the vanilla beans used to prepare the extracts influence the profile and herewith the ratios. Further studies have partly disproved the conclusion of fraud when a ratio

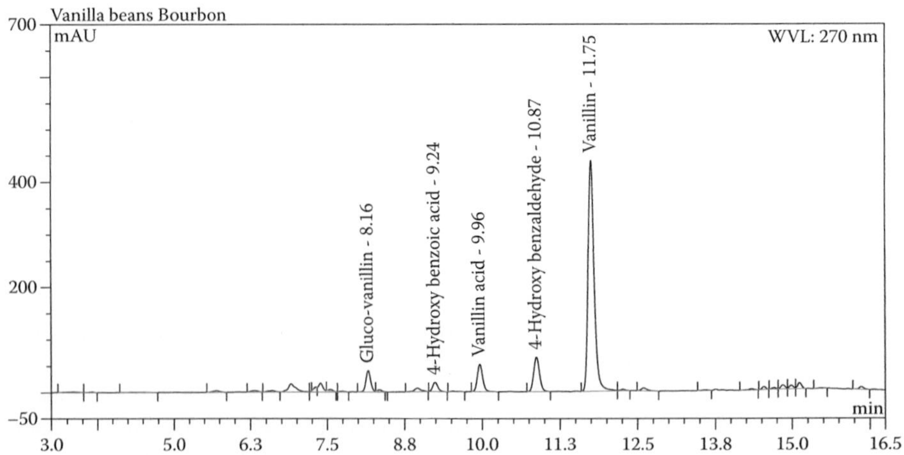

FIGURE 15.1 HPLC chromatogram of vanilla extract compounds.

TABLE 15.2
Component Ratios Used to Assess the Authenticity of Vanilla Products

Ratio	Ranges according to DGCCRF[a] 1988	Ranges according to DGCCRF[b] 2003
Vanillin/*p*-hydroxybenzaldehyde	10–20	10–20
Vanillin/*p*-hydroxybenzoic acid	53–110	40–110
Vanillin/vanillic acid	15–29	12–29
p-hydroxybenzoic acid/ *p*-hydroxybenzaldehyde	0.15–0.35	0.15–0.35
Vanillic acid/*p*-hydroxybenzaldehyde	0.53–1.00	0.53–1.50

[a] République Française (1988).
[b] *Note d'information* (2003).

is out of the ranges given (John and Jamin, 2004; Littmann-Nienstedt and Ehlers, 2005; Gassenmeier et al., 2008) and the suitability of the so-called ratios has been argued. In an information letter, the International Organization of the Flavor Industry (IOFI) has expressed their reservation toward the applicability of the ratios and overestimation of their validity (IOFI Information Letter, 2000) despite the fact that these ratios are sometimes erroneously cited as "IOFI values." The questionable validity of the ratios has been applied not only to vanilla beans and extracts but also to vanilla flavors and even flavored food. An interpretation of ratios for flavors and food is far beyond of what has seriously been investigated at the scientific level. Thus, the so-called ratios are a precheck for the authenticity of vanilla beans and extracts thereof, in case the conditions of extraction are well known to the evaluator. Other techniques such as stable isotope ratio mass spectrometry (IRMS) or quantitative site-specific nuclear magnetic resonance spectroscopy provide much clearer indication of adulteration of vanilla products (Kempe and Kohnen, 1999).

IRMS

A very advanced analytical method for the authentification of vanilla extracts and vanilla flavors is stable isotope ratio analysis (SIRA). It was developed in the 1970s and became the most important tool for authenticity testing of nonchiral compounds. This analysis can be performed either by IRMS, mainly combined with a gas chromatographic separation, or by quantitative NMR measurement of the natural abundance at individual atomic sites (Schmidt et al., 2007).

In nature, the main bioelements such as hydrogen, carbon, oxygen, and nitrogen occur as mixtures of isotopes. The natural abundances of the stable isotopes have global average values. Owing to physical processes, (bio)chemical reactions such as photosynthesis, geographic parameters, and climate, their relative ratio can vary. Always the "light" isotopes (^{1}H, ^{12}C, and ^{16}O) are by far dominant compared with the abundance of the "heavy" ones (^{2}H, ^{13}C, and ^{18}O). The products formed in plants (or animals) and the ingredients of the extracts or the food prepared from them obtain a characteristic isotope ratio, which allows to correlate it to the photosynthetic pathways, and the climatic and geographic conditions.

Compounds with an aromatic ring—for example, vanillin—are generally synthesized in plants through the shikimic acid pathway from erythrose-4-phosphate and phosphoenol pyruvate. The formation of the products of this pathway is accompanied by a corresponding depletion of ^{13}C relative to the primary plant products, the carbohydrates.

These carbohydrates are produced during photosynthesis of plants utilizing CO_2 and water. The primary step is enzymatically catalyzed, with $^{13}CO_2$ reacting somewhat more slowly than $^{12}CO_2$. This phenomenon is named the "kinetic isotope effect." However, the ^{13}C deficit is not identical for all plants.

Three major photosynthetic pathways for the CO_2 fixation are known for plants: C3, C4, and CAM. The so-called C3 plants (e.g., wheat, barley, sugar beet, and most trees) use the ribulosebisphosphate-carboxylase reaction, the Calvin pathway, while C4 plants (e.g., sugarcane, maize, sorghum, and millet) use the phosphoenolpyruvate-carboxylase reaction, the Hatch–Slack pathway. The *CAM* plants such as succulents, orchids (e.g., *V. planifolia*), and some tropical grasses have the Crassulacean acid metabolism. Each group shows different values of the $^{13}C/^{12}C$, $^{2}H/^{1}H$, and $^{18}O/^{16}O$ ratios for their metabolites as can be shown by IRMS (e.g., carbohydrates, fatty acids, isoprenoids, amino acids, phenylpropanes, etc.). By analyzing these ratios, it is possible to distinguish between compounds produced by plants during the biochemical pathways and those produced by synthesis.

The changes in the isotope ratio caused by these effects are very small. So they are not indicated in the atom% scale. These minimal changes are compared with the values of international standards and expressed in per million (‰) as a variance from standard, the so-called delta value (Schmidt, 2003):

$$\delta^{13}C\ (‰) = \left(\frac{[^{13}C_{sample}]/[^{12}C_{sample}]}{[^{13}C_{standard}]/[^{12}C_{standard}]} - 1\right) \times 1000$$

The standard for the $^{13}C/^{12}C$ ratio is $\delta^{13}C$: V-PDB (Vienna-PeeDee Belemnite).

The analysis of $\delta^{13}C$ values for C3 plants (the so-called "light plants") give values between −30‰ and −24‰, for C4 plants (the so-called "heavy plants") between −16‰ and −10‰. The CAM plants have $\delta^{13}C$ values from −10‰ to −30‰, making a differentiation from the other two types very difficult.

The authenticity proof of flavor substances by determination of their average $\delta^{13}C$ value implies the combustion at about 1000°C of the purified organic samples in order to convert them into CO_2. An online coupling of gas chromatography and IRMS via a combustion interface reduced the sample amount and made measurements easier. Later on, an online coupling of gas chromatography and IRMS via a pyrolysis interface (about 1300°C) was developed, primarily for ^{18}O analysis, allowing the simultaneous online measurement of $\delta^{18}O$ and $\delta^{13}C$ values. This interface also affords the determination of $\delta^{2}H$ when applying the pyrolysis at about 1450°C. Hener et al. (1998) compared the $\delta^{13}C$ values of different vanillin samples measured via CO_2 and CO and showed that the values are, in most cases, in good agreement. Nowadays, mainly the online coupling of gas chromatography with combustion or reductive pyrolysis to isotope ratio mass spectrometry (GC-C-IRMS or GC-P-IRMS) is used for the determination of the δ-values of carbon ^{13}C, hydrogen ^{2}H, and oxygen ^{18}O.

V. planifolia belongs to the CAM plants, and vanillin ex-beans shows $\delta^{13}C$ values in the range from −16.8‰ (*V. tahitiensis*) to −22.0‰ (*V. planifolia*) (see Table 15.3).

Apart from the extraction of vanilla beans, vanillin can also be produced by chemical synthesis or biotechnological pathways from different sources (lignin, guaiacol, eugenol, curcumin, and ferulic acid).

TABLE 15.3
$\delta^{13}C$-, $\delta^{2}H$-, and $\delta^{18}O$-Values of Vanillin from Various Origins

Origin of Vanillin	$\delta^{13}C_{V\text{-}PDB}$ (‰)	$\delta^{2}H_{V\text{-}SMOW}$ (‰)	$\delta^{18}O_{V\text{-}SMOW}$ (‰)
Ex-beans (Bourbon) *V. planifolia*	−21.5 to −19.2[a]		6.7–12.4[b]
Ex-beans (Tahiti) *V. tahitiensis*	−19.7 to −15.9[a]		
Ex-beans	−20.4 to −20.2[c]	−115 bis −52[d]	12.2–14.0[c]
	>−21.5[e]		8.1–10.7[f]
	−21.5 to −16.8[d]		
	−22.0 to −19.0[g]		
Ex-guaiacol	−36.2 to −29.0[h]	−23 bis −17[i]	−3.1 to −2.5[f]
	−26.1 to −24.9[i]		
Ex-eugenol	−31.7 to −29.9[h]	−87[j]	11.8–13.3[c]
			0.3[k]
Ex-lignin	−28.7 to −26.5[d]	−204 bis −170[i]	6.1–6.8[f]
		−195 to −178[b]	6.0–9.8[l]
Ex-ferulic acid Ex-rice bran (biotechnology)	−37 to −36[m]	−168 to −165[n]	12.4–13.2[c]
	−36.4 to −33.5[n]		10.7–11.2[n]

[a] Scharrer and Mosandl (2002).
[b] Gatfield et al. (2007).
[c] Bensaid et al. (2002).
[d] Schmidt et al. (2007).
[e] IOFI (1989).
[f] Kempe and Kohnen (1999).
[g] Koziet (1997).
[h] Hoffman and Salb (1979).
[i] Culp and Noakes (1992).
[j] Hoffman (1997).
[k] Brenninkmeijer and Mook (1982).
[l] Werner (1998).
[m] *Note d'information* (2003).
[n] Krammer et al. (2000).

The $\delta^{13}C$ values of vanillin derived from other sources than vanilla beans or its extracts are in the range from –24.9 to –37.0‰. Therefore, vanillin in vanilla extracts or derived from vanilla beans can be analytically differentiated from other sources by its $\delta^{13}C$ value. It can be used to indicate the authenticity of vanilla products based on vanilla or its extracts.

According to a recommendation of IOFI, the authenticity of vanilla flavors is estimated by the determination of the $\delta^{13}C$ value for vanillin. $\delta^{13}C$ values in the region of –21.0‰ ± 0.5‰ are considered to indicate the plant origin of the analyzed vanillin ex-vanilla beans (IOFI Information Letter, 1989). The variances resulting from the applied isolation techniques are currently under investigation through ring tests performed by a dedicated analytical working group of the German Chemical Society (GDCh).

As long as the vanillin is derived from biotechnological processes such as by fermentation starting from natural ferulic acid, it can be considered as natural and subsequently applied in natural flavors. Owing to the fact, that the $\delta^{13}C$ values of vanillin derived from natural ferulic acid by fermentation are very low, this quality can usually be differentiated from synthetic vanillin qualities that do not meet the naturalness requirements.

In order to detect an adulteration of vanilla bean extract with vanillin from other sources, it is possible to apply multielement GC-IRMS; this means, to also determine the ratios $^{2}H/^{1}H$ and $^{18}O/^{16}O$ and to evaluate this tridimensional dataset. The standard for these two ratios is Vienna-Standard Mean Ocean Water (V-SMOW). In a recent paper, Bensaid et al. (2002) show that the sp_2 oxygen atom in the aldehydic position can chemically exchange with water in industrial or laboratory procedures. Therefore, the authors used the $\delta^{18}O$ value of guaiacol, which is formed by the degradation of the vanillin molecule. However, they state that in practice, oxygen isotope ratios do not play an important role in authentication procedures.

The degradation of the vanillin molecule was already used by Krueger and Krueger (1983), when the instrumentation for the multielement IRMS was not yet well developed. Fraudulent adulterators had learned to imitate the $\delta^{13}C$ value of natural vanillin by blending it with "nature-identical" vanillin, with products artificially enriched in ^{13}C concentration (Krueger and Krueger, 1983, 1985).

IRMS measures the average $\delta^{13}C$ value of the whole vanillin molecule. Since it is relatively simple to enrich its methyl or carbonyl group in ^{13}C, synthetic vanillin blended with a ^{13}C-enriched fraction may be mistakenly taken as natural.

To check for [methyl-^{13}C]-vanillin, the methyl carbon is removed as CH_3I in refluxing HI. Then, IRMS is performed on the CH_3I (Krueger and Krueger, 1983). For analyzing [carbonyl-^{13}C]-vanillin, the molecule has to be oxidized to vanillic acid. After decarboxylation, the CO_2 formed is analyzed by IRMS (Krueger and Krueger, 1985).

NMR

Presently, the checking for enriched portions is more easily analyzed by quantitative NMR measurements of the abundance of ^{2}H or ^{13}C in the different positions of the molecule. This method was named site-specific natural isotope fractionation nuclear magnetic resonance spectroscopy (SNIF-NMR®) and it is based on the fact that the distribution of ^{2}H or ^{13}C at the different sites of the molecule is not statistical and depends on the origin of the particular compound.

SNIF-^{2}H-NMR spectroscopy is a powerful tool and allows a deeper insight into the biochemical mechanisms by the determination of isotope contents at specific molecule sites. In this respect IRMS, which usually needs to burn the sample before analysis, is not suited to the measurement of an "intramolecular" isotope distribution. Basic research has shown that deuterium is far from being randomly distributed in organic molecules (Martin and Martin, 1981). Therefore, the D/H-ratios of the different positions of the molecule have to be determined.

The relative sensitivity of NMR for ^{13}C is theoretically about 100 times higher than that for deuterium. But due to the fact that the kinetic isotope effects for carbon are much smaller than those for deuterium and due to shorter relaxation time for deuterium, quantitative ^{2}H-NMR became the best tool for the authenticity control in the flavor field.

The first application of this method to the vanillin molecule appeared in 1983 (Toulemonde et al., 1983). Now SNIF-^{2}H-NMR® is accepted as the official method by AOAC (No. 2006.05) for the authentication of vanillin. SNIF-^{2}H-NMR combined with ^{13}C-IRMS also allows the characterization of biotechnological vanillin derived from ferulic acid, eugenol/isoeugenol, or curcumin.

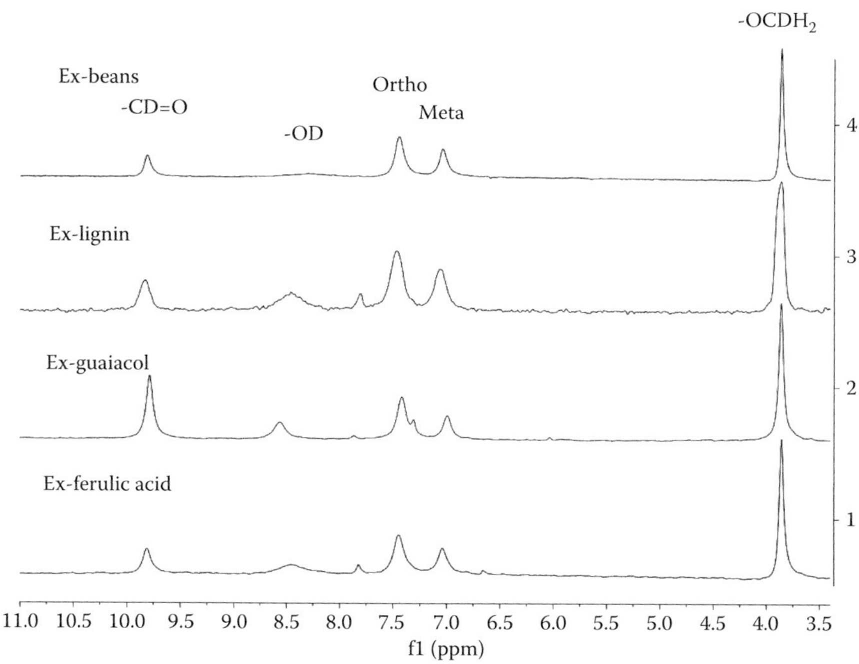

FIGURE 15.2 ^{2}H-NMR spectra of vanillin ex-vanilla beans, vanillin from synthetic origin (ex-lignin, ex-guaiacol), and vanillin from biotechnological origin ex-ferulic acid.

TABLE 15.4
^{2}H-NMR Results for Vanillin Samples of Various Origins (D/H) (ppm)

	Position in Molecule			
Sample	**D–C = O**	**Ortho**	**Meta**	**OCH_2D**
Ex-beans[a]	130.8 ± 3.1	157.3 ± 3.0	196.4 ± 2.5	126.6 ± 1.7
Ex-beans[b]	128.3	156.3	180.4	126.4
Ex-lignin[a]	119.9 ± 6.4	132.1 ± 2.6	168.8 ± 5.9	105.9 ± 1.4
Ex-guaiacol[a]	315.2 ± 56.9	138.8 ± 6.7	143.8 ± 5.3	139.1 ± 8.4

[a] Remaud et al. (1997).
[b] John and Jamin (2004).

The vanillin molecule has six monodeuterated isotopomers. Five signals can be seen in the ^{2}H-NMR-spectra of the three different vanillin qualities in Figure 15.2, the two deuteriums in the ortho-position fall into one signal. Table 15.4 shows that vanillin synthesized from guaiacol (2-methoxyphenol), by introducing an aldehyde group, is clearly separated from the other variants by a high degree of deuterium enrichment on the carbonyl function. Vanillin from lignin exhibits deuterium depletion at all positions compared with natural vanillin derived from the vanilla bean. Experimental values for the hydroxyl group are not included in Table 15.4, since they can easily be falsified by H/D exchange processes.

Also, the development of quantitative ^{13}C-NMR is ongoing. Therefore, first characteristic ^{13}C-distributions in vanillin qualities of different origins have already been obtained (Caer et al., 1991; Tenailleau et al., 2004a, 2004b).

CONCLUSION

The analytical determination of typical vanilla (extract) ingredients by HPLC and the calculation of their ratios can be considered as a precheck for the authenticity assessment for vanilla. IRMS is a powerful tool to prove the authenticity of vanilla products. Especially since the online technique provides a coupling of GC to IRMS, these measurements have become available without cumbersome isolation steps. The multielement assay with datasets of δ^{13}C-, δ^{2}H-, and δ^{18}O-values should allow to detect trials of fraud.

When a sufficient amount of vanillin can be isolated, quantitative ^{2}H- and ^{13}C-NMR measurements are particularly efficient for distinguishing the different origins of vanillin.

REFERENCES

Arrêté du 11 juin 1987 relatif à la méthode officielle de dosage de la vanilline, de l'aldéhyde *p*-hydroxybenzoïque, de l'acide vanillique et de l'acide *p*-hydroxybenzoïque dans les gousses de vanille et les produits vanillés. *Journal Officiel de la République Française* du 22 juillet 1987:8171.

Bensaid, F.F., Wietzerbin, K., and Martin, G.J. 2002. Authentication of natural vanilla flavorings: Isotopic characterization using degradation of vanillin into guaiacol. *Journal of Agricultural and Food Chemistry* 50:6271–6275.

Brenninkmeijer, C.A.M. and Mook, W.G. 1982. A new method for the determination of the $^{18}O/^{16}O$ ratio in organic compounds. In: H.-L. Schmidt, H. Förstel, and K. Heinzinger, eds. *Stable Isotopes*. Amsterdam: Elsevier, 661–666.

Caer, V., Trierweiler, M., Martin, G.J., and Martin, M.L. 1991. Determination of site-specific carbon isotope ratios at natural abundance by carbon-13 nuclear magnetic resonance spectroscopy. *Analytical Chemistry* 63:2306–2313.

Culp, R.A. and Noakes, J.E. 1992. Determination of synthetic components in flavors by deuterium/hydrogen isotopic ratios. *Journal of Agricultural and Food Chemistry* 40:1892–1897.

Ehlers, D., Pfister, M., and Bartholomae, S. 1994. Analysis of Tahiti vanilla by high-performance liquid chromatography. *Zeitschrift für Lebensmittel-Untersuchung und -Forschung* 199:38–42.

Ehlers, D., Schäfer, P., Doliva, H., and Kirchhoff, J. 1999. Vanillingehalte und Inhaltsstoff-Verhältniszahlen von Vanilleschoten—Einfluß der Extraktionsbedingungen auf die ermittelten Werte—Diskussion von Richtwerten. *Deutsche Lebensmittel Rundschau* 95:123–129.

Eurofins Newsletter 13, October, 2003. http://www.eurofins.com/documents/newsletter/food-testing/200310/200310.pdf

Eurofins Newsletter 23, March, 2007. http://www.eurofins.com/documents/newsletter/food-testing/2007/23/en/Eurofins_Newsletter_No_23_March_2007_engl.pdf

Fayet, B., Tisse, C., Guerere, M., and Estienne, J. 1987. Nouveaux critères analytiques dans l'étude des gousses de vanille. *Analysis* 15:217–226.

FDA Code of Federal Regulations, Title 21, Food and Drugs, Part 169, Sections 3 and 175–182; (e.g., www.cfsan.fda.gov/~lrd/FCF169.html).

Gassenmeier, K., Riesen, B., and Magyar, B. 2008. Commercial quality and analytical parameters of cured vanilla beans (*Vanilla planifolia*) from different origins from the 2006–2007 crop. *Flavour and Fragrance Journal* 23:194–201.

Gatfield, I., Hilmer, J.-M., Weber, B., Hammerschmidt, F., Reiß, I., Poutot, G., Bertram H.-J., and Meier, D. 2007. Chemical and biochemical changes occurring during the traditional Madagascan vanilla curing process. *Perfumer and Flavorist* 32:20–28.

Hener, U., Brand, W.A., Hilkert, A.W., Juchelka, D., Mosandl, A., and Podebrad, F. 1998. Simultaneous on-line analysis of 18O/16O and 13C/12C rations of organic compounds using GC-pyrolysis-IRMS. *Zeitschrift für Lebensmittel-Untersuchung und -Forschung A* 206:230–232.

Hoffman, P.G. and Salb, M. 1979. Isolation and stable isotope ratio analysis of vanillin. *Journal of Agricultural and Food Chemistry* 27:352–355.

Hoffman, P.G. 1997. C.A.I.S./I.S.C. isotopic analyses. *Report of the Isotopic Studies Committee*.

IOFI. 1989. Information Letter 775.

IOFI. 2000. Information Letter IL1271. Authenticity of natural vanilla products, Brussels.

ISO. 1999. Standard 5565–2, Vanilla [*Vanilla fragrans* (Salisbury) Ames]—Part 2: Test Methods.

John, T.V. and Jamin, E. 2004. Chemical investigation and authenticity of Indian vanilla beans. *Journal of Agricultural and Food Chemistry* 52:7644–7650.

Juergens, U. 1981. Zur Hochdruckflüssigchromatographischen Analyse von Aromen: Untersuchung von Lebensmitteln mit Vanillegeschmack. *Deutsche Lebensmittel Rundschau* 77:93–96.

Kempe, K. and Kohnen, M. 1999. Deterioration of natural vanilla flavours in dairy products during processing. *Advances in Food Sciences* 21:48–53.

Koziet, J. 1997. Isotope ratio mass spectrometric method for the on-line determination of oxygen-18 in organic matter. *Journal of Mass Spectrometry* 32:103–108.

Krammer, G.E., Gatfield, I., Güntert, M., Hilmer, J.-M., Schmidt, C.O., Sommer, H., Werkhoff, P., and Kaulen, J. 2000. Understanding flavors using isotopic patterns. In: P. Schieberle and K.-H. Engel, eds. *Frontiers of Flavour Science*. Garching: Deutsche Forschungsanstalt für Lebensmittelchemie, 111–116.

Krueger, D.A. and Krueger, H.W. 1983. Carbon isotopes in vanillin and the detection of falsified "natural" vanillin. *Journal of Agricultural and Food Chemistry* 31:1265–1268.

Krueger, D.A. and Krueger, H.W. 1985. Detection of fraudulent vanillin labeled with ^{13}C in the carbonyl carbon. *Journal of Agricultural and Food Chemistry* 33:323–325.

Littmann-Nienstedt, S. and Ehlers, D. 2005. Verfahren zur Beurteilung von Produkten mit Vanille oder Vanillearomen mit einer modifizierten Verhältniszahl. *Deutsche Lebensmittel-Rundschau* 101:182–187.

Martin, G.J. and Martin, M.L. 1981. Deuterium labeling at the natural abundance level as studies by high field quantitative ^{2}H NMR. *Tetrahedron Letters* 22:3525–3528.

Mosandl, A. and Scharrer, A. 2001. Reinvestigation of vanillin contents and component ratios of vanilla extracts using high-performance liquid. *Deutsche Lebensmittel-Rundschau* 97:449–456.

Mosandl, A. and Hener, H. 2001 GC/IRMS-Multielementanalyse zur Authentizitätsbewertung von Vanille(extrakten). Forschungsprojekt AiF-FV 12062 N der Forschungsstelle Universität Frankfurt und dem Verband der Deutschen Essenzenindustrie e.V., Bonn. Project 1999–2001.

Note d'information No 2003-61,16th June 2003 regarding vanilla products, published by Direction Générale de la Concurrence, de la Consommation et de la Répression des Fraudes.

Oberdieck, R. 1998. Ein Beitrag zur Kenntnis und Analytik von Vanille. *Deutsche Lebensmittel Rundschau* 94:53–59.

Quirin, K.-W. and Gerard, D. 1998. CO_2-vanilla extracts—New multipurpose flavor ingredients. *European Food and Drink Review* autumn: 53–55.

Remaud, G.S., Martin, Y.-L., Martin, G.G., and Martin, G.J. 1997. Detection of sophisticated adulterations of natural vanilla flavors and extracts: Application of the SNIF-NMR method to vanillin and *p*-hydroxybenzaldehyde. *Journal of Agricultural and Food Chemistry* 45:859–866.

République Française. 1988. Ministère de l'économie, des finances et du budget—Direction Générale de la Concurrence, de la Consommation et de la Répression de Fraudes—Service de la Consommation de la qualité et de la sécurité, Sous Direction H. Bureau H2: Note de service N° 5387. Paris, October 17, 1988.

Scharrer, A. and Mosandl, A. 2002. Progress in the authenticity assessment of vanilla. 2. $\delta^{13}C_{V\,PDB}$ correlations and methodical optimizations. *Deutsche Lebensmittel Rundschau* 98:117–121.

Schmidt, H.L. 2003. Fundamentals and systematics of the non-statistical distribution of isotopes in natural compounds. *Naturwissenschaften* 90:537–552.

Schmidt, H.L., Werner, R.A., Rossmann, A., Mosandl, A., Schreier P. 2007. Stable isotope ratio analysis in quality control of flavourings. In: H. Ziegler, ed. *Flavourings*. Weinheim: Wiley-VCH, 602–663.

Taylor, S. 1993. Improved determination of vanillin and related phenolic components in vanilla (*Vanilla fragrans* (Salisb.) Ames) by high-performance liquid chromatography. *Flavour and Fragrance Journal* 8:281–287.

Tenailleau, E.J., Lancelin, P., Robins, R.J., and Akoka, S. 2004a. NMR Approach to the quantification of nonstatistical ^{13}C distribution in natural products: Vanillin. *Analytical Chemistry* 76:3818–3825.

Tenailleau, E.J., Lancelin, P., Robins, R.J., and Akoka, S. 2004b. Authentication of the origin of vanillin using quantitative natural abundance ^{13}C NMR. *Journal of Agricultural and Food Chemistry* 52:7782–7787.

Tiemann, F. and Haarmann, W. 1874. Ueber das Coniferin und seine Umwandlung in das aromatische Princip der Vanille. *Berichte der deutschen chemischen Gesellschaft* 7:608–623.

Toulemonde, B., Horman, I., Egli, H., and Derbesy, M. 1983. Food-related applications of high-resolution NMR. Part II. Differentiation between natural and synthetic vanillin samples using ^{2}H-NMR. *Helvetica Chimica Acta* 66:2342–2345.

Tracy, M., Xiadong Liu, and Pohl, C. 2008. Fast analysis of vanilla extracts on the acclaim RSLC 2 μm C18 column, LC GC Europe. *The Applications Book*. http://chromatographyonline.findanalytichem.com/lcgc/Articles/Fast-Analysis-of-Vanilla-Extracts-on-the-Acclaim-R/ArticleStandard/Article/detail/575600.

Werner, R.A. 1998. Entwicklung neuer Verfahren für die on-line $^{18}O/^{16}O$- und $^{2}H/^{1}H$-Isotopenverhältnismessung und ihre Anwendung zur Authentizitäts- und Herkunftsbestimmung bei Naturstoffen. PhD thesis, Technical University Munich, Munich, Germany.

16 Vanilla Use in Colonial Mexico and Traditional Totonac Vanilla Farming

Patricia Rain and Pesach Lubinsky

CONTENTS

MESOAMERICA

The tribes living in the northern regions of vanilla's natural range were the first to incorporate vanilla into their lives, perhaps as early as 2000–2500 years before the Spanish Conquest of Mexico in 1520. From Central Mexico to Costa Rica, there was a passion for pungent, aromatic fragrances. Vanilla had sacred and religious connotations as did corn and cacao. These were gifts bestowed upon them by the gods and were treated with reverence. Corn provided nourishment, cacao was a ceremonial drink, and vanilla was a fragrant incense. Vanilla beans were ground and mixed with copal (a dried resin from the Copalli tree with a pleasing pine-like odor) to perfume their temples. The native peoples were highly knowledgeable about the medicinal use of herbs, and may well have used ground vanilla bean for lung and stomach disorders as well as used the liquid from green beans as a poultice for drawing out insect venom and infections from wounds. Their medicinal skills far surpassed those of the Europeans at the time of their arrival in Mexico in the sixteenth century. While vanilla taken to Spain in the early 1500s was valued as a commodity, for the tribal peoples of the Americas vanilla was reverently considered as a sacramental herb.

During the pre-Classic period, approximately 1500 BC, extensive trade routes developed throughout Mesoamerica, and the bartering of goods between groups living in the various different climates and altitudes, brought about cultural exchanges including the sharing of important discoveries and spiritual beliefs. This is likely the time when vanilla was first used and traded among coastal tribes.

As the hunting and gathering nomadic lifestyle was slowly replaced by a more agrarian and settled existence, the population expanded, and adequate food production became increasingly important. Greater focus was placed on the nature gods, such as Tlaloc and Xipe Toltec, as it was believed that they controlled the rain, the sun, the winds, and the harvests. Pleasing the gods meant abundant food supplies.

It was during this time period that the Olmecs, the "mother culture" of Mexico, emerged. They lived in the humid forests and open savannahs along the Gulf Coast of southern Vera Cruz and Tabasco. In addition to being skilled artists and having highly developed commerce, the Olmecs made huge contributions to the development of what came to be the backbone of Mesoamerican diet—maize.

The Olmecs developed a technique that inadvertently changed the nutritional value of maize, which had not been particularly important until then, and subsequently, literally fueled the growth and development of all of Mexico. Instead of boiling dried maize kernels until they were soft, or pounding them into powder, corn was cooked with white lime or wood ashes and then left overnight in the cooking liquid. The transparent hulls (pericarp) of the treated kernels now slipped off easily, making it much easier to grind the corn into a smooth dough (known as *nixtamalli* in Nahuatl, *masa* in Spanish). This simple shift in preparation changed corn from a low-protein grain into a high-protein food, creating Mexico's staff of life.

The Olmec were quite possibly the first to also domesticate cacao, and with the domestication of cacao, the use of vanilla as a flavor followed at some point in this early history. They also created *atole*, a mildly fermented corn drink that was sometimes flavored with vanilla.

Over a period of time, the Olmeca culture was eclipsed by the Maya, who were brilliant artists, architects, and scientists, and who spawned a renaissance in pre-Hispanic Mexican culture. They also made major culinary advances, including perfecting the cultivation of cacao and creating the drink, *chocolatl*, which later became popular among Aztecs at the time of the Conquest. And it was during the time of the Maya, that vanilla was probably introduced as one of the flavorings for *chocolatl*.

However, vanilla was not always used in this adored beverage. There are at least 12 different *chocolatl* flavor versions, depending on the event and what was at hand, not unlike our flavor choices for coffee, milkshakes, and Italian sodas. While the Totonacs, an important group of people who emerged a little later in Mesoamerican history, were possibly the first to domesticate vanilla, their predecessors certainly enjoyed the fruits wild from the forest.

According to Totonac legend and popular belief, the Totonacs were the first people to identify and domesticate vanilla. Written history challenges this belief. The Totonacs who came from the Central Valley of Mexico probably knew nothing of vanilla until they came to the Gulf Coast. However, there is now strong evidence that there were also Totonacs living on the Coast and that the groups converged at some point in Mesoamerican history. What is known for certain is that vanilla has been central to the lives of Totonacs for at least hundreds of years, and it has become so enmeshed with their history and lore, that it belongs to them culturally (see the appendix).

The Maya called it *Zizbic*, the Zoques-Popolucas, *Tich Moya*, the Totonacs, *Xanat*, and much later, the Aztecs named it *Tlilxochitl*. Interestingly, *Xanat* (also spelled

Xanath and pronounced *cha-nat*) is the generic term for the flower in the Totonac language, which is curious, given its stature within their culture. It is also sometimes called caxixanath (catch-e-chanat), which means "hidden flower," as the orchids are neither showy nor prominent in the forest, and only last for a day.

Vanilla, in early times, was harvested from the forest when it was fully ripe and split open, filling the air with its distinctive fragrance. Instead of the 100–200 beans that are typical on hand-pollinated plants, there were eight or nine beans, pollinated at random by the forest insects. With such a seductive aroma, it is certainly understandable that the coastal and low mountain forest dwellers would be drawn to the seedpods and experiment with ways to fragrance their lives with its alluring perfume. If cacao was the food of the gods, vanilla was definitely the nectar that accompanied it!

VANILLA IN NEW SPAIN

There are several accounts of vanilla in New Spain during the 1500s. An Aztec herbal book was compiled in Nahuatl and written in Latin in 1552 by Martin de la Cruz and Juan Badiano. The manuscript entitled *Libellus de Medicinalibus Indorum Herbis* was finally translated in English and published in 1939 (*An Aztec Herbal: The Classic Codex of 1552*, William Gates). This is the earliest known document that specifically mentions vanilla. The book tells us that the Aztecs ground flowers with other aromatics and wore them around the neck as a medicinal charm or amulet. It is very likely that amulets such as these were first created by the Gulf Coast tribes who gathered vanilla. On the other end of the spectrum, vanilla was made into an ointment as a treatment for syphilis. As we know that vanilla flowers last only for a day, the flowers may have been dried and then brought by the *pochteca* traders to the Valley of Mexico, though there are anecdotal comments about Moctezuma growing vanilla in his extensive botanical gardens. The book also contains the first known illustration of the vanilla plant.

The Aztecs were talented herbalists as were most of the Mesoamerican peoples. Whether they independently identified vanilla's curative powers or learned of it from the *pochteca* traders, the rare and valued vanilla was important as a medicine as well as an aphrodisiac in their culture.

In 1529, Franciscan Fray Bernardino de Sahagun came from Spain with the assignment of converting Aztecs into Christianity. The enlightened friar first learned Nahuatl, the Aztec "Mother language," then taught young indigenous men to write their history in Nahautl using Spanish spelling. For the next three decades, they collectively recorded all aspects of the Aztec culture. The 12-volume history, written in both Nahuatl and Spanish went to Spain, but it was considered heretical by the Catholic Church, and much of the body of work was not published until 1829. However, the Aztec materials ultimately ended up in the Laurentian Library in Florence, Italy, where it became known among scholars as the Florentine Codex.

Fray Bernardino Sahagun mentioned the widespread use of hot chocolate among the colonizers. He said, "It is perfumed, fragrant, precious, good, and a medicine. It is toasted and mixed with cacao. I add vanilla to cacao and drink it like vanilla." (These were his words; what he meant by "drinking it like vanilla," is uncertain.)

Cacao and vanilla could both be purchased in the market place in New Spain, but only the wealthiest people and the clergy in New Spain could afford its use.

VANILLA AS MEDICINE IN NEW SPAIN

According to Fray Sahagun, chocolate, mixed with vanilla and two other aromatic herbs was a remedy for cough and a cure for spitting blood. On the basis of Sahagun's comments, researchers until now have wondered whether vanilla was used to treat lung diseases such as tuberculosis, which plagued the native peoples since prehistoric times. Recent medical discoveries have found that another dreaded disease that caused severe lung distress, may also have been what Fray Sahagun referred to in his notes as *cocoliztli* (Nahuatl for pests) one of the deadliest plagues in history. Doctor Francisco Hernandez was sent to New Spain by King Philip II in 1570 to study the colony's natural history, the medical usefulness of the herbs and other native treatments, and to do an anthropological history of the country. He spent seven years in the Valley of Mexico, learned Nahuatl, studied indigenous medicine, and wrote his observations in Latin, creating *The Natural History of New Spain* (*Rerum Medicarum Novae Hispaniae Thesaurus*) a six-volume collection describing over 3000 plants. The books were accompanied by 10 folio volumes of illustrative paintings by Mexican artists.

In his writings, Hernandez referred to vanilla by its Nahuatl name, *Tlilxochitl*, which means black pod for the vanilla bean. However, he erroneously translated it to mean black flower, which led to centuries of confusion over the increasingly popular bean. He also classified vanilla as *Araco aromatico*. Hernandez wrote about the medicinal properties of vanilla that he learned from the native physicians, "A decoction of vanilla beans steeped in water causes the urine to flow admirably; mixed with *mecaxuchitl*, vanilla beans cause abortion; they warm and strengthen the stomach; diminish flatulence; cook the humours and attenuate them; give strength and vigor to the mind; heal female troubles; and are said to be good against cold poisons and the bites of venomous animals."

In itself, vanilla sounds like a miracle cure, but Francisco Hernandez also talks about three flavorings that "excites the venereal appetite," the Viagra of the sixteenth century. Vanilla, cacao, and *macaxochitl* (a plant related to black pepper and known now as *acuyo*) were the top three contenders that would do the trick. In fact, centuries later, vanilla's reputation as an aphrodisiac still continues.

TRADITIONAL TOTONAC VANILLA FARMING

Until the late 1600s, vanilla only grew wild in the forest. Over a period of time it was domesticated by cutting the vines and planting them closer together in the forest to create ease in harvesting. Domesticating vanilla also made it easier for the insect pollinators to reach the flowers and production improved dramatically.

It was not until 1875 that some French men living on the Vera Cruz Coast brought from France the technique for artificially pollinating vanilla. Initially, they kept this information among themselves and vanilla processors, but the Totonacs accused them of stealing their beans because of their larger harvests, so the Totonacs were taught the technique as well. This innovation changed the Mexican industry completely, of course, and the Totonac growers became wealthier, though never as wealthy as the Euro-Mexicans who purchased their green beans and sold the dried beans to Europe and America.

The following explains the traditions, uses, and spiritual value of vanilla in the Totonac culture.

The Totonacs grow their own vanilla in a similar environment to the way it grows naturally in the forest. Their method and rituals remained virtually unchanged until the 1970s. Today, the Totonac farmers continue to grow vanilla in a similar fashion, but most of the rituals surrounding the cycle are no longer practiced.

Before any work began, *Kiwigilo*, the "old man of the forest," was consulted and permission was requested to clear the land. At this time, they also requested that no venomous snakes would come into the vanilla plantation.

It was customary to make a promise to *Kiwigilo* at the time of planting to ensure a good harvest. A ritual meal of fiery *mole* with turkey was prepared and given to the land as an offering for fertility and to ensure that the vanilla thrived. Small tortillas were prepared in the numbers of two, four, five, eight, ten, and twelve to use as ritual foods for the gods. These were traditional numbers that were syncretized with Catholicism to represent the 12 Apostles.

There were two altars in the traditional Totonac home. The main altar, the one most people saw upon entering the house, was the Catholic altar. It was—and still is—decorated with woven palm ornaments, pictures of saints, icons, flowers, and candles. In a prominent spot there would be offerings of food and drink.

The traditional Totonac altar was below the main altar and dedicated to *Kiwigilo*. It was usually hidden behind small decorative cut paper banners (*papel picado*). Fabrics decorated with ritual designs covered the altar. Idols and figures made of stone or iron—legitimate archeological pieces found by the Totonacs in the countryside during their work in the fields—decorated the altar. These idols were passed from fathers to sons over the generations. Offerings consisted of native plants and fruits, minerals, bones, animal skins, bird feathers, maize and seeds, earth, and water. Cigars made from locally grown tobacco were also placed on the altar. In the Catholic churches in the rural areas throughout Totonacapan, it is still common to see the patron saints of the town dressed in traditional Totonac dress with small offerings of water, *refino* (sugarcane alcohol), and maize kernels at the feet of the saints. Placed there at the beginning of the planting season, would guarantee a good harvest.

Next, the land was cut back and then burned to clear out the forest understory. Acahual were allowed to remain as they provided the necessary shade for the vanilla. Rapid-growing bushes that were easy to propagate were planted to serve as tutors for the vanilla; traditional bushes and small trees included *ramon*, *laurel*, *chaca*, *capulin*, *pata de vaca*, *balletilla (cachuapaxtle)*, *cojon de gato*, and *pichoco*. These bushes were chosen for their small leaves and year-round foliage so that the sun/shade balance would be maintained. Healthy vanilla vines (*esquejes*) 2–3 m in length were cut from the forest and planted individually or in rows of twos next to the tutors.

For the Totonacs, each of the many varieties of vanilla that grew in the region had religious significance. *Vanilla pompona*, called *Vainilla bastarda*, was considered the

"Queen of Vanilla," and was always planted at a key point on the plantation. As *V. pompona* is larger and hardier than *V. planifolia*, it was believed that it would protect the other plants, and that if bad spirits or harm to the family came through disease or curses, it would affect *V. pompona* first and be absorbed, leaving the family safe. *V. planifolia* was known as *Vainilla mestiza. Vainilla rayada* also known as *Vainilla rayo or Vainilla de taro* (bamboo vanilla) has a striped leaf and similar fragrance to *planifolia*. It was the vanilla that was always dedicated to the most important cult of fertility. There was also *Vainilla de puerco* (pig vanilla), *Vainilla de mono* (monkey vanilla) and *Vainilla oreja de burro* (donkey ear vanilla) *Vainilla de monte alto* (vanilla of the tall forest), and others, each with its own special story, most of which, unfortunately, have never been recorded and are probably lost forever.

Traditional plantations ranged in size from 10 to 30 ha. A compost of dead leaves and other forest matter was applied to the base of the plants in February, August, and December.

It took between three and four years for the first blooms to appear on the traditional plantation. March 18th was the *Fiesta de Fecundacion*, the celebration of fertility. Flowering commenced at this time and continued into early May. During pollination there were dietary restrictions. Beef and fish were prohibited as were some other foods, in order not to forfeit the setting of the flowers. They also abstained from sex during the period dedicated to pollination.

Once artificial pollination was discovered, it was always called "the marriage of vanilla." Because the pollination was done manually with a small stick, moving the pollen from the male anther and depositing it on the female stigma, it was similar to intercourse. The fact that it took between eight and nine months for the vanilla bean to develop, the entire cycle was not unlike that of human procreation. For this reason, the vanilla was perceived, in the Totonac vision of the world, as divinely tied to humankind.

According to Totonac belief and practice, vanilla orchids have both male and female plants. The males produce lots of flowers but the beans do not set, and quickly drop-off. As the plants essentially look the same, if the plants do not bear fruit over four years they were removed from the plantation, a practice continued today. It was recently shown that a chromosome alteration (triploidy) was responsible for sterility in some vines because of unviable pollen.

The stick used for pollination was—and still is—carefully prepared. Some farmers believed that the type of wood does not matter, but others believed that the heart of the *chaca* was the only wood to use. The tip was whittled with a knife or machete until it had a thin, chisel-like point. Plants high up and out of reach were pollinated because they were less likely to be stolen. Ropes or ladders were used to reach the blossoms. *V. pompona* was sometimes used for pollination. The flower was cut and carried to the cultivated vine; the pollination process was the same except that it used two flowers rather than just one. The pods from this cross were larger and heavier, but not as desirable. A day or two after pollination, the flowers were checked to make sure the beans had set. If not, additional flowers were pollinated to ensure maximum production.

Once the vanilla was pollinated, there was little that needed to be done for the plants until the time of harvest. This allowed time for caring for the family *milpa* (cultivated crops), hunting, and other necessary tasks. It was also a time of concern, especially if the rains did not return on time. Papantla has always had very inconsistent rainfall and consequently there has been great preoccupation with the rain cycles. As the soil is thin and much of the land sits on a great limestone shelf, the heavy rainfall is quickly absorbed into the land. If rain does not return by May, it can be disastrous for vanilla and food crops alike. Therefore, prayers and offerings were made for rain on a continual basis.

The beans were originally harvested when they were ripe and beginning to dry. Later, as techniques improved and standards were set, the beans were harvested when they were nearly ripe. The finest beans—those with the greatest amount of oil—were harvested in late January and early February. The majority of the vanilla was brought to the *casas de beneficio* and sold green. Depending on where the families lived, it was sometimes easier to keep the beans on the ranchos and dry them there. Also windfalls and vanilla that matured early were dried at home. These beans were not considered as premium quality. However, some of the Totonacs became *beneficiadores* and purchased beans from their neighbors, then dried them using the same methods as the Europeans and Mestizos. This was especially common in the *pueblos* tucked into the sierra that were not easily accessible to Papantla though a few of the *beneficiadores* in Papantla would travel to the ranchos to collect the vanilla.

In the initial stages of commercial vanilla production, vanilla was counted by the thousands. Later, it was sold green in lots of 100 pods. Then it was sold by the pound in rolls of 3–5 pounds each; 100 green beans were considered about 5 pounds. Now it is sold by the kilogram, both as green beans, and as dried.

Papantla was known as *Kachikin*, or "the city." Trips to *Kachikin* were planned for the delivery of beans or for festivals and holy days, and combined with picking up supplies unavailable outside of the town. Burros or mules were the primary method of transporting goods. When the vanilla was finally ready for shipment from Papantla, it was loaded into tin boxes and then into larger cedar boxes. The boxes were carried by the mules either through the hills to Tecolutla, where they were floated in flat-bottomed boats called *chalanes* or be taken by mule back into the mountains to the rail head at Tezuitlan, where the vanilla traveled to the port of Vera Cruz. The majority of the vanilla headed to the United States, but until the early years of the twentieth century, it also went to Europe.

A successful harvest and sale were always celebrated, usually with local fiestas. The successful transport of money back home was important as the sale. Robbery, assault, and homicide have always been a problem integrally tied in with producing vanilla, just as being "tied to the company store" has been a problem for small farmers everywhere. Many of the *beneficiadores* had stores and essentially kept the Totonacs in servitude. They would not tell the Totonacs the selling price for their vanilla, but instead would issue credits, which rarely covered the cost of supplies. These two problems were always in the forefront of the minds of the growers. However, as Spanish was not spoken fluently by most Totonacs, and even those who spoke Spanish were usually illiterate, they had no access to the current prices for vanilla, and so were forced to accept the pay or not sell at all. As vanilla often went through several hands until it reached the *beneficiadores* in Papantla, the price difference between what the farmer got and what the vanilla ultimately was sold for, would be significantly different.

The families who either dried their own vanilla or who worked in the *casas de beneficio* used the vanilla oil that ran off during the first stages of curing and drying to rub on their skin or to shine their hair. Vanilla was also used as an air freshener and as a perfume for clothing. Dried beans were tucked into hatbands along with flowers and feathers. A vanilla-flavored *aguardiente* was always prepared for baptisms, weddings, and other significant family events, a tradition that has been carried forward and followed today even as the other rituals have slipped away.

ACKNOWLEDGMENT

The coauthors wish to acknowledge the important contribution of Rocio Aguilera Madero for the preparation of this chapter through personal communications notably on Totonac vanilla farming, and through the information published in the newsletter *La Voz del Vainillero*, Union Agricola Regional de Productores de Vainilla, Papantla, Mexico.

REFERENCES

Bruman, H. 1948. The culture history of Mexican vanilla. *The Hispanic American Historical Review* 28:360–376.

Chenaut, V. 1995. *Aquellos Que Vuelan: Historia de los Pueblos Indigenas de Mexico, los Totonacos en el Siglo XIX.* Centro de Investigaciones y Estudios Superiores en Antropologia Social, Mexico City.

Chenaut, V., ed. 1996. *Procesos Rurales e Historia Regional (Sierra y Costa Totonacas de Vereacruz).* Centro de Investigaciones y Estudios Superiores en Antropologia Social, Mexico City.

Harvey, H.R. and Kelly, I. 1969. *The Totonac* (*Handbook of Middle American Studies*), University of Texas Press, Austin, Texas.

Kelly, I. and Palerm, A. 1950. *The Tajin Totonac, Part 1.* Smithsonian Institution, Washington, D.C.

Ossenbach, C. 2005. History of orchids in Central America. Part I: from Prehispanic times to the independence of the new republics. *Harvard Papers in Botany* 10:183–226.

APPENDIX: THE LEGEND OF VANILLA

This is a Totonac legend about the origin of vanilla in Mesoamerica. The time period of when this legend emerged is uncertain.

In early times, the Land of the Good and Resplendent Moon, was the kingdom of Totonacapan, ruled by the Totonacas. The palm-studded sands, verdant valleys, and shimmering hills and sierra in what is now known as Vera Cruz, were overseen from several locations. One was Papantla, place of the papan birds. Another was El Tajin, the thunderbolt, an ancient city built in honor of the deity, Hurakan/Tlaloc, god of the storms. It was here in this dense, tropical rainforest that vanilla was first cultivated and cured. It was here that the fragrance from the vanilla was so exquisite, that Papantla later became known as, The City That Perfumed the World.

There was a time, however, before the reign of Tenitzli III, when there was no vanilla. In this city, famous for its artists and sculptors, Tenitzli and his wife were blessed with a daughter so incredibly beautiful that they could not bear the thought of giving her away in marriage to a mere mortal. They dedicated her life as a pious offering to the cult of Tonoacayohua, the goddess of crops and subsistence, a powerful goddess who affected their very life and survival. Their daughter, Princess Tzacopontziza (Morning Star), devoted her time at the temple, bringing offerings of foods and flowers to the goddess.

It was during her trips from the forest, carrying flowers for the temple that the young prince Zkatan-Oxga (Young Deer) first caught sight of Morning Star and immediately fell under her spell. He knew that even allowing his eyes to remain upon her for a moment, gazing at her innocent beauty, could bring him death by beheading, but he was obsessed to have her as his wife and companion. The love in his heart for Morning Star outweighed the dangers of being captured and killed. Each morning, before Morning Star went into the forest in search of flowers and doves as offerings for the goddess, Young Deer would hide in the undergrowth and await the arrival of the beautiful princess.

One morning, when the low, dense clouds clung to the hills following the rain, Young Deer was so overcome with desire that he decided to capture Morning Star and flee with her to the sierra. As she passed close by, he leapt from the bushes, then taking her by the arm, ran with her, deep into the forest. Although Morning Star was startled by Young Deer's abrupt arrival and ardent passion, she too came under the spell of their star-crossed destiny and willingly followed.

Just as they reached the first mountains, a terrifying monster emerged from a cave, spewing fire, and forced the young lovers to retreat to the road. As they did, the priests of Tonacayohua appeared and blocked their path. Before Young Deer could utter a word, the priests struck him down and beheaded him. Swiftly, Morning Star met with the same terrible fate. Their hearts were cut from their bodies, still beating, taken to the temple and were placed on the stone altar as an offering to the goddess. Their bodies were then thrown into a deep ravine.

Not long after, on the exact site of their murders, the grasses where their blood had spilled began to dry and shrivel away as if their death was an omen of change. A few months later a bush sprang forth so quickly and prodigiously, that within a few days it had grown several feet and was covered with thick foliage. Shortly after, an emerald-green vine sprouted from the earth, its tendrils intertwining with the trunk and branches of the bush in a manner at once delicate and strong, much like an embrace. The tendrils were fragile and elegant, the leaves full and sensual. Everyone watched in amazement as, one morning, delicate yellow-green orchids appeared all over the vine like a young woman in love in repose, dreaming of her lover. As the orchids died, slender green pods developed, and over a period of time they released a perfume more splendid than the finest incense offered to the goddess.

It was then that priests and devotees of Tonacayohua realized that the blood of Young Deer and Morning Star was transformed into the strong bush and delicate orchid. The orchid and vine were designated as a sacred gift to the goddess and from that time on has been a divine offering from the Totonacas to their deity and to the world.

17 Vanilla's Debt to Reunion Island

Raoul Lucas

CONTENTS

Vanilla came to Bourbon Island, now known as Reunion Island, at the dawn of the nineteenth century, from Central America, on the other side of the planet. The genius of a 12-year-old black slave transformed the fruitless orchid into a pod-bearing source of opulence. Reunionese (inhabitants of Reunion) industriousness brought out the best of its fragrance. Toward the end of the century, it invested new shores; the adjacent islands of the Indian Ocean. International recognition was obtained through the quality label bestowed during 1964—Bourbon vanilla.

When vanilla was introduced at the beginning of the nineteenth century in Reunion Island, previously called Bourbon Island, it had come a long way from the other side of the planet and possessed a rich history (Lucas, 1990).

However, it was due to its acclimatization in Reunion, which fostered its commercial success. The Reunionese transformed this exotic orchid into a lucrative commodity before exporting it throughout the world.

By their technical know-how, their knack of inventiveness, and their native genius, the Reunionese were the architects of the fabulous adventure of vanilla; a saga which triggered off in Reunion Island and in other areas where the orchid was introduced, led to vast fortunes but also its inevitable controversies and tragedies.

FROM TLILXOCHITL TO VANILLA: FROM MEXICO TO REUNION

Originally, the orchid was an indigenous plant growing wild in the Mexican forests and the outlying expanses of Central America. It was better known to the Aztecs as tlilxochitl (which means black flower in Nahuatl). Biographers have trailed it back to Bernardino de Sohagun's book, *General History of New-Spain* (Sohagun, 1801).

The book is a major landmark not only because it is the oldest known reference on tlilxochitl, but also because it gives valuable information about the way it was used by the Aztecs.

Sohagun, a Spanish monk, landed in Mexico eight years after the conquest by the Spaniards. He relates that "after each meal, delicately-flavored cocoa beverages were served, like the one made from the bean and which is very tasty; the one made of honey and another made from tender tlilxochitl."

It is said that emperor Moctezuma offered this precious mixture to the illustrious Cortez. History does not reveal if Cortez was conquered by the cocoa drink flavored with tlilxochitl but we do know what happened afterward. Profusion of gold and precious plants were shipped to Europe, and among them was the tlilxochitl.

The Spaniards fell under the charm of its aroma and named the orchid vanilla, short for vaina, which means pod, no doubt because of the shape of the fruit.

Besides Spain, other European countries discovered vanilla and its commercialization started in France, where in 1692 it was marketed exclusively through a royal monopoly. Disseminated and largely naturalized, throughout the eighteenth century vanilla formed the topic of many studies (Lucas, 1990).

A century later, this already popular plant was brought to Reunion. Located 800 km east of Madagascar in the middle of the Indian Ocean, this mountainous island, 2500 square kilometers in area, was devoid of inhabitants when French settlers arrived in July 1665.

They were sent by the French East India Company founded in 1664 by Colbert. During the French colonial period, slavery contributed significantly to the agricultural expansion. Emancipation occurred in 1848, when the liberated slaves largely of African and Malagasy descent were replaced on the fields by Indian indentured labor (Lucas and Serviable, 2008).

The introduction of vanilla is in keeping with a long-time tradition of acclimatization carried by scientists, administrators, naval officers, or laymen; they were keen on enriching the vegetal heritage thereby contributing to the economic development of Reunion. We owe it to a local born navy officer who was the first one to introduce vanilla plants on the island.

On June 26, 1819, commander Philibert (Figure 17.1), the head of a naval squadron, stops in his native land, after he returns from a mission from "the south seas," presumably the warm Pacific shores and from Cayenne in French Guyana.

It is from La Gabrielle estate in Cayenne, formerly awarded to General Lafayette as a token from the Nation in return for services rendered, that samplings of various plants were taken including vanilla. It is a short, stubby big-pod specimen, probably *Vanilla pompona*. For Philibert the question does not arise: the introduction of vanilla in Reunion can be a source of prosperity. And even more! In a letter to the Colony's governor he writes: "this plant could be a cash crop traded in Asia. And the colonists just cannot miss making big money in cultivating it" (Billiard, 1822).

A year later, returning from another mission, this time from the Philippines, commander Philibert brings back a second lot of vanilla cuttings. The plant had been identified in the forests close to Manila by Perrotet the botanist, a member of the survey mission; he was also in the former expedition. The vanilla differed from the one brought from Cayenne. Its stem and leaves were smaller. Called "little vanilla" as compared to *V. pompona*, it was formally introduced in the Reunion Island on May 6, 1820 (Thomas, 1828).

FIGURE 17.1 The commander Philibert, the first to introduce vanilla plants (probably *V. pompona*) in Reunion Island in 1819. (Lithography from the *Album de La Réunion*, A. Roussin, courtesy of Océan Editions.)

The third introduction is to the credit of Marchant, a local colonial administrator. Benefiting from a trip to France, he pays a visit to the Paris Museum and obtains cuttings of Mexican vanilla. This greenhouse-grown variety stems from the plants brought over by the Marquis of Blenford, constituting the Charles Greville collection at Paddington (England). This particular species was introduced on September 25, 1822 and differed from the two precedents—in fact, it was *Vanilla planifolia*.

Nonetheless, all those samplings, from Cayenne or Paris, had in common the care shown in their selection and the vigilant concern displayed during their transfer. A. Delteil, author of a study on vanilla, describes in detail Perrotet's operations: "He took the stems which measured 4 or 5 meters long, rolled them into rings and laid them down horizontally in the boxes. He watered them moderately so as to prevent the liana's organic tissues drying up. The boxes were covered with a tight metallic fabric to protect them from the sun glare. They arrived in Reunion after two and half months, in perfect condition, some of them having even developed buds and tendrils" (Delteil, 1884).

In Reunion, the cuttings were subjected to a methodical treatment. Philibert who was accused by the governor for not having delivered the vanilla directly to the Royal Botanical Garden of Saint-Denis justified his choice: "As these different species were collected in various localities, distant from one another, I am convinced that to ensure the success of this plant, it is judicious to disseminate it in various parts of the island with different temperatures; so that if it does not succeed in a region, it may find more appropriate prevailing conditions in others. Indeed, if it comes out well in one spot, success is assured.

Therefore the decision which I've taken to distribute the cuttings to a number of inhabitants seems to be the best to ensure the acclimatization of this precious plant. Moreover, I've distributed them only to those whose agricultural talents are well-known, such as Mr Hubert: as you can see, this is an additional precaution which I've taken" (Focard, 1862).

Thus, vanilla was introduced in Reunion at three distinctive periods by different protagonists. It comes from three separate areas and is made of three different species. Acclimatization will be the result of various factors: the choice of soil and climatic conditions, judicious advice from the individuals having introduced the plant, the vigilance of the authorities, and never-ending care given by the islanders.

A SLAVE OF GENIUS

Introduced and cultivated in Reunion from 1819, vanilla is not yet a commercial produce as it does not bear any fruit. Natural fertilization of the flower is a haphazard operation because of its morphology (Figure 17.2). Indeed, the male and female organs are separated by a large membrane, which hinders their contact. Direct pollination is impossible and needs the intercourse of insects or of humming birds. In Central America, this essential part is played by small bees (melipone). Without external intervention, the precious orchid is doomed to fruitless sterility.

To solve such a deadlock, a way had to be found to fertilize the plant by artificial means. Two experiments were launched in Europe: the first in Liege in 1836 and the second in Paris in 1838.

But the methodology proposed, likewise the attempt some years later in Guadeloupe, was abandoned as being highly complex and with a poor record of success.

The most effective process of pollination was discovered in 1841 by a young Reunionese slave; his surname Edmond, slaves having no family names.

He was the property of a rich family of planters in Sainte-Suzanne, the Bellier-Beaumont. Born in 1829, Edmond lost his mother, a female slave on the

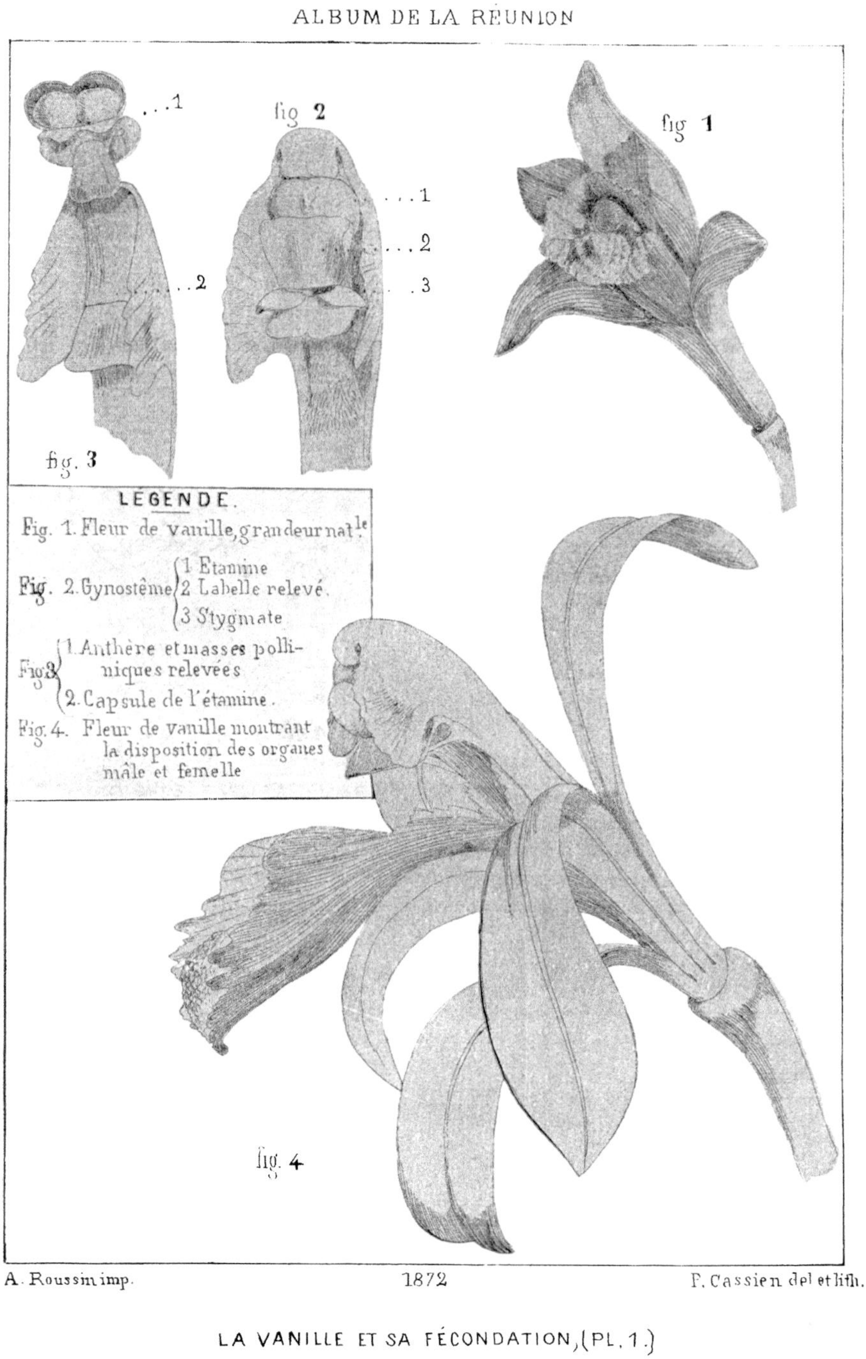

FIGURE 17.2 Morphology of a vanilla flower. (Lithography from the *Album de La Réunion*, A. Roussin, 1872, courtesy of Océan Editions.)

Bellier-Beaumont compound, at his birth. Although Edmond has no school education owing to his slave status, he develops a rare acumen for horticultural tasks from his master. Meziaire Lepervanche reported that Edmond "identifies all flowers by their botanical names and that on many occasions he attempted artificial caprification on his flowers which for some reason or other cannot be pollinated naturally" (Focard, 1862).

He conducted the same experiment on the vanilla flower, bringing together the male and female components. Edmond was only 12 years old and he discovered the process of artificial pollination in vanilla. A discovery which Fereol Bellier-Beaumont, Edmond's master, relates in these terms: "Walking around (with Edmond), I noticed on the only vanilla plant that I had, a glossy black pod. I showed my surprise and he told me that he had actually pollinated the flower. I refused to believe him and moved on. But two or three days later, I saw a second pod close to the first one. He repeated his assertion. I asked him how he did it. He performed his manipulation in front of me (...) and I had to acknowledge it when I saw the operation repeated each day and every time with the same success" (Focard, 1862).

A year later, in 1842, Fereol Bellier-Beaumont who did not wish to maintain secrecy, wrote an article in the local newspaper *Le Moniteur*. It did not pass unseen. The piece of good news spread like wild fire throughout the country and Bellier-Beaumont was beset with demands to borrow his young slave. Never in the whole history of Reunion, was a slave so fawned upon as young Edmond (Figure 17.3).

Planters approached him, "thinking rightly that it was so much simpler and so much safer to deal straight with the inventor" (Focard, 1862).

Edmond's discovery (Figure 17.4) sparked fortune for many planters; it provided a new agricultural activity and fostered the development of vanilla throughout the world. However, initially it did not bring any reward to the slave, not even freedom. He was set free seven years later, in 1848, and was named Albius, in Latin meaning white. Following his epoch-making discovery, every personal request remained unanswered, either from the authorities or from the vanilla planters.

But worse than ingratitude, and adding insult to injury, Edmond's discovery aroused jealousy and disputes. Many questioned his merit, pointing out that he was an "illiterate little nigger"; that the discovery was pure luck! Others even claimed copyright and the responsibility of the discovery, putting forward their status and their learning. Others conceded that Edmond Albius was the sole and legitimate inventor of the process of pollination of vanilla, adding that he was a white man.

In wretched poverty, abandoned by all, Albius died in Sainte-Suzanne on August 9, 1880 (Lucas, 1990).

THE CURING: THE REUNIONESE PROCESS

After naturalization and manual pollination, the Reunionese know-how was harnessed for the transformation of the vanilla pod. In fact, when it is picked, the pod is odorless. It is only after a long preparation, taking about several months, that its transformation takes place.

It acquires thereafter a dark chocolate tint and the delicious sweet fragrance, which is much appreciated.

FIGURE 17.3 Portrait of Edmond Albius, inventor of manual pollination of vanilla. (Lithography from the *Album de La Réunion*, A. Roussin, 1863, courtesy of Océan Editions.)

If numerous methods of pod preparation exist in the warm regions of Central America, they may be summarized into two main types—direct and indirect preparation. The first process allows the natural ripening of the pod by leaving it exposed alternately in the sunshine and in the shade. The second process aims at stopping the pod's natural evolution. It is said that the vanilla "is killed." Mexicans used both methods (Bourriquet, 1954). In Reunion Island, vanilla was prepared by desiccation in the shade and in the sun. But in the early part of the nineteenth century, these processes developed poor products, which could not withstand the long voyage by ship. And Reunion had heady ambitions for vanilla. This is why dedicated and creative planters tried to aim at excellence (Delteil, 1884).

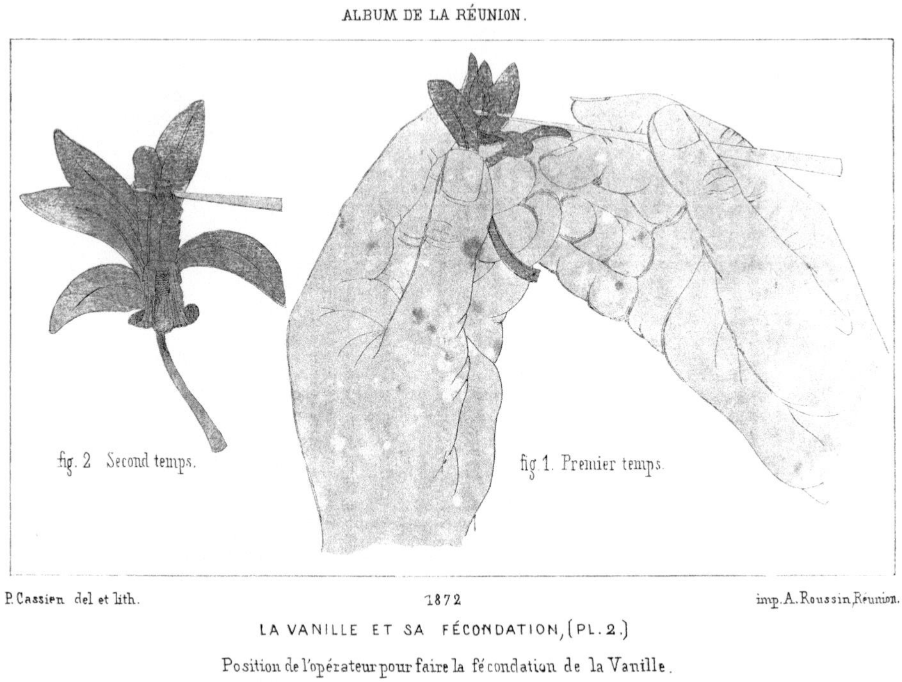

FIGURE 17.4 Manual pollination of vanilla. (Lithography from the *Album de La Réunion*, A. Roussin, 1872, courtesy of Océan Editions.)

In 1851, Ernest Loupy, an attorney by trade, a planter by family tradition, and animated by a keen sense of learned curiosity, decided to test a new method—the blanching in hot water, which he found in the *Encyclopédie méthodique* (1787 edition). The results were conclusive. The "boiling-water" process was born.

David de Floris, a retired navy official and a planter in Saint André improved on Loupy's process (Floris 1857). Vanilla was far from being something novel for him, and he was not indifferent to the high expectations which it conjured. It was de Floris himself who captained the vessel bringing Marchant's samples in 1822. The new process devised by the two planters of Saint-André, with pods picked at full maturity from healthy plants, gave excellent results. In 1857, de Floris edited a handbook on vanilla preparation, the first ever Reunionese handbook! The book met with large success and brought the process to the attention of one and all (Figure 17.5).

CONQUERING THE WORLD

If seven years after Edmond's discovery, Reunion exported its first vanilla production (about 50 kg), exports simply rocketed after the innovation brought about by Loupy and de Floris. From 267 kg in 1853, exports reached more than 3 tons in 1858 and 15 tons in less than five years later. At the end of the nineteenth century, vanilla

FIGURE 17.5 Drying and conditioning of vanilla beans in Reunion Island at the beginning of the twentieth century. (Post card from author's personal collection.)

was almost as lucrative as sugar. Never before, in the economic history of Reunion Island, a commodity had such a meteoric development and generated so much wealth (Figure 17.6). The 1867 and 1900 "Expositions Universelles" (International commercial fairs) established the success of Reunion vanilla. Local planters and curers carried off numerous prizes (Lucas, 1990).

For Marius and Ary Leblond, the famous local novelists, Creole civilization may be equated with vanilla. Summoning history, geography, and sensuality, the two authors waxed lyrical: "You have all heard about Reunion's vanilla which won the first prize at so many international fairs, no one can imagine how moving and lovely its manufacturing by skilful handiwork proves to be. Some words are sweet-scented spoonfuls of perfumes and souls. The words vanilla and vanilla-factory enable us to imagine and even to feel Saint-Joseph, Vincendo, Langevin, all those quaint hamlets with blessed names, cherishing like an innocent incense the fragrance of vanilla to bid us the sweetest of welcomes" (Leblond, 1931).

Confident in their craft, crowned with laurels awarded in numerous national and international events, the Reunionese were filled with self-confidence at the turn of the century (Figure 17.7). With great plans of success, they set out to conquer the Indian Ocean islands with their perfumed trunks of vanilla. In 1866, they introduced vanilla in Seychelles, which met with lightning success. In less than 10 years, it became the archipelago's main source of prosperity; at the dawn of the twentieth century, Seychelles' exports were almost equal to that of Reunion.

FIGURE 17.6 Vanilla curing in Reunion Island. (Post card from author's personal collection.)

FIGURE 17.7 Vanilla exporters at the beginning of the twentieth century. (Post card from author's personal collection.)

In 1873, the Reunionese moved on with their vanilla dreams to the Comoros. It was first introduced in Mayotte before spilling to the other three islands. As in Seychelles, it was so successful that it soon replaced sugarcane in the first decade of the twentieth century. Comoro's production before long toppled that of Reunion. An identical script was played in Madagascar where some Reunionese brought vanilla in 1880 to the Nosy Be islet before overrunning the north-west of Madagascar's mainland. Reunionese vanilla thrived in a favorable environment.

And a spectacular rush for vanilla was triggered off. Everyone wanted to produce it. Reunionese colonists and Malagasy peasants started vanilla on every bit of available land; and land was plentiful. In 1929, Madagascar exported 1032 tons and became the number one world producer. It still holds this rank until today (Lucas, 1990).

The pollination problem solved, the vanilla preparation mastered, new territories cultivated, the planters headed by Reunionese set out henceforth to settle a new challenge—the reform of the vanilla economy. In 1964, the Indian Ocean vanilla producers, Madagascar, Reunion, Seychelles, and Comoro agreed to form an organization. The major objectives were: promoting a quality product, stabilizing the market, and focusing a common policy. A label was created: Bourbon vanilla, thus, paying international tribute to Reunion's pioneering technological breakthroughs: those of Albius, Loupy, de Floris, and many others.

At the beginning of the nineteenth century, Reunion received the first vanilla saplings from faraway resorts, she greeted them ungrudgingly. The inhabitants did not just welcome another beautiful orchid to a land of beaches and blooms. Similar to a fairy tale of rags-to-riches, the Reunionese have turned a plain flowering bean plant into a world asset.

REFERENCES

Billiard, A. 1822. *Voyage aux Indes orientales*. Le Dantu, Paris.

Bourriquet, G. 1954. *Le vanillier et la vanille dans le monde*. Lechevallier, Paris.

Delteil, A. 1884. *La vanille, sa culture, sa préparation*. Challamel, Paris.

Floris, D. (de). 1857. *La culture du vanillier*. Lahuppe, La Réunion.

Focard, V. 1862. L'introduction et la fécondation du vanillier à l'île Bourbon. *Bulletin des Sciences et Arts*, 222–235.

Leblond, M.A. 1931. *L'île enchantée*. Librairie de la Revue Française, Paris.

Lucas, R. 1990. *La Réunion, Ile de vanille*. Océan-Editions, La Réunion.

Lucas, R. and Serviable, M. 2008. *Commandants et gouverneurs de l'Ile de la Réunion*. Océan-Editions, La Réunion.

Sohagun, B. (de). 1801. *Histoire générale des choses de la Nouvelle-Espagne*. Lesguilhez frères, Paris.

Thomas, P. 1828. *Essai de statistique de l'Ile Bourbon*. Bachelier libraire, Paris.

18 Recognizing the Quality and Origin of Vanilla from Reunion Island

Creating a PGI "Vanille de L'île de la Réunion"

Bertrand Côme

CONTENTS

STRONG HISTORICAL LINKS BETWEEN VANILLA AND REUNION ISLAND

Although the genus *Vanilla* is native to Mexico, Reunion Island can reasonably be considered its second home. Vanilla cultivation and processing have very strong links with the Reunion Island (formerly the *Ile Bourbon*). The vanilla plant was introduced on the island in three successive years (1819, 1820, and 1822) from Cayenne and Manila (Philibert and Perrotet) and from Mexico, by means of the *Jardin des Plantes* (botanical garden) in Paris (Marchant).

Grown in different parts of the island, the vanilla plants survived in the ideal climatic conditions and flowered, but failed to bear fruit (as Reunion Island lacked the pollinating insects found in Mexico). It was not until 1841 that a young slave, Edmond Albius aged 12, invented a simple and practical method for pollinating vanilla orchids, allowing vanilla cultivation to rapidly develop on the island (see Chapter 17).

With the first harvests, producers were faced with the problem of the dehiscence of *Vanilla planifolia* beans. Once again, they invented an ingenious method to prevent the beans from ripening, while activating the aroma production process (David de Floris in 1857). The foundation for the present-day vanilla cultivation and processing was initiated within a span of a few years.

Presently, the techniques that are used were disseminated worldwide under the impetus of the Reunionese producers, who exported their vanilla plants and know-how to the Comoros (1873) and then to Madagascar (1890).

WORLD-RENOWNED QUALITY

At the outset, Reunion Island producers strove to produce high-quality beans that rapidly gained an excellent reputation with the consumers. This reputation, based on selecting only the best beans and following very strict production processes, was built around the designation "Bourbon vanilla" (from the former name of the colony under the monarchy) from the very first exports in 1848, and has endured ever since.

In 1964, to face competition from synthetic vanillin and ensure the promotion of the natural product, Madagascar, the Comoros, and Reunion Island met in Saint Denis to form a cartel of Indian Ocean producer countries (which represented 85% of global production at that time). Its aims were to organize production, stabilize prices by establishing country-specific import quotas, promote the product in Europe and the United States, combat counterfeiting (references on labels), and design a logo identify the natural vanilla. To finance these initiatives, a contribution of $1.5 was charged for every kilogram exported.

To clearly identify the product, the signatories agreed on a common designation, "Bourbon vanilla" (whose reputation was already largely established). The term "Bourbon vanilla," thus, became generic and ceased to identify only Reunion Island as the origin of the product (this generated confusion for people from Reunion Island and many buyers), although its aroma differs considerably from one origin to another.

Finally, more recently, products of Indonesian origin have appeared on the world market under the designation "Bourbon-style vanilla," but the quality of these beans is not comparable with that of Reunion Island vanilla.

REUNIONESE PRODUCTION REDUCED TO A NICHE MARKET

Owing to its high cost price, vanilla produced in Reunion Island is no longer competitive on the international market of standard vanilla. However, it remains highly prized by certain connoisseurs looking for very special products. But the term "Bourbon vanilla" is no longer enough to enable Reunionese production to distinguish itself from competition. Many retailers play on the ambiguity of this term to hide the true origin of their product and thereby mislead consumers, who often wrongly assume that "Bourbon vanilla" originates in Reunion Island. Producers have, therefore, thought about ways to clearly identify their products.

To guarantee the origin of their products, their options were either an AOC ("Appelation d'Origine Contrôlée," i.e., a PDO: Protected Designation of Origin) or a PGI (Protected Geographical Indication).

The AOC is based far more on the concept of *terroir* (the assumption that the land on which a product is grown imparts a unique quality to that crop) than on the production process. Given the microclimates that exist in Reunion Island and the considerable differences in soil from one area to another, or even from one ravine to another, the AOC would have to be divided into several designations:

- An AOC for the region of Sainte Marie, Sainte Suzanne, and the hills above Saint André (nitisols, umbrisols, and andosols, between 2000 and 3000 mm of rainfall)
- An AOC for the coastal region of Saint André and Bras Panon (detrital cone of the Mât river and 2000 mm of rainfall)
- An AOC for the region of Sainte Rose and Saint Philippe (recent lava flows and between 3000 and 4500 mm of rainfall)

It would also be necessary to deal with these productions separately.

Given the low production volume at present and the resulting constraints, it seemed inappropriate to divide Reunionese production into three parts, even if the AOC is far better known by consumers than the PGI.

The PGI, a designation officially recognized throughout the European Union, provides a very precise definition of the production and processing conditions for a product, corresponding to an ancestral tradition (the concept of know-how) from a well-defined geographical area (the concept of *terroir*). This designation is therefore more suited to the characteristics of vanilla production in Reunion Island, and this is why the PGI was finally chosen.

THE PGI: A PREFERENCE SHARED BY ALL REUNIONESE PRODUCERS

The PGI approach is the result of a desire shared by all producers and processors in Reunion Island to promote their product by means of an unambiguous designation (unlike "Bourbon vanilla"). In 2000, they formed the *Association pour la Valorisation de la Vanille de l'Ile de la Reunion Island* (the association for the promotion of Reunion Island vanilla, A2VR), an organization to protect and manage the designation, with the aim of drawing up the specifications for labeled production and protecting the designation against any attempts at imitation or counterfeiting. The association currently includes producers (Provanille, the association of farmers from Saint Philippe), processors (UR 2) and distributors (Réunion Agricom). It remains open to all (companies and individual producers), provided the terms of the specifications are met. In 2007, A2VR represented 78% of all producers in Reunion Island and 65% of the volume processed on the island.

CHOOSING THE DESIGNATION

Vanilla originating in Reunion Island is traditionally known as "Bourbon vanilla." However, this designation has become the common term employed for all vanilla produced in the Indian Ocean (Madagascar, Comoros). Recovering the exclusive use of this term was therefore not a possibility.

A2VR thus decided to turn the page and to call its PGI production "Vanille de l'île de la Réunion" (i.e., "Vanilla from Reunion Island"). The origin thus clearly appears in the product name, and the only vanilla that is now authorized to use the geographical reference "Réunion" in its designation is that produced in Reunion Island under the PGI.

PRODUCTS COVERED BY THE PGI

The "Vanille de l'île de la Réunion" PGI applies only to vanilla beans from plants belonging to the species *V. planifolia* G. Jackson and currently found in Reunion Island. The by-products such as vanilla powder or extract are not covered by the PGI, nor are the beans from the hybrid plants or recently imported plants.

Among other specifications, PGI beans must meet those specifications under categories 1 and 2 of the NF ISO 5565-1 standard of March 2000, with the two other categories not applicable. The beans must be at least 14 cm long and split for no more than 3 cm. They must be whole, supple, and full, with a characteristic flavor and uniform color ranging from brown to dark chocolate brown. They may bear the producer's mark on their lower third. They are oily, supple in texture, and malleable. Labeled beans have a maximum moisture content of 38% at sale and a vanillin content of 2% (moist-weight basis). Finally, they must undergo maturation for at least seven months in order to develop their aroma.

DEFINING THE GEOGRAPHICAL PRODUCTION AREA OF THE PGI

The area in which vanilla is produced, processed, and conditioned is traditionally situated on the east coast of Reunion Island, due to the requirements of the plant and the soil and climate conditions found there, but also due to socioeconomic factors.

Climate Factors

To develop under favorable conditions, vanilla requires at least 2000 mm of rainfall per year, evenly spread throughout the year. Below this, the conditions are no longer met for guaranteeing quality production with no external inflow (irrigation). Altitude is also a factor in the development of the plant through nighttime temperatures. The conditions are favorable below 600 m altitude. These two criteria were thus selected for defining the PGI area.

Climate factors affect not only the plant but also the conditions for processing the beans. An atmosphere that is too dry will be detrimental to the vanilla drying process. High relative humidity is needed if the beans are to dry slowly, ensuring their flexibility and quality. These conditions are only met on the east coast of the island.

Socioeconomic Factors

The development of sugarcane farming in Reunion Island pushed vanilla back onto land that is difficult to reach or farm on the east coast—the foothills of the volcano (State forests) or edges of ravines (steep slopes). Its hardiness meant it nevertheless succeeded in adapting to these difficult conditions and thereby helped to develop woodland and fallow land and to deter plant pests. It plays an important role at the environmental level and fits perfectly into a policy of sustainable development. Farmed extensively in forest or fields, it occupies an area of around 250 ha, and creates economic activity in a region that is particularly hard hit by unemployment. It contributes to the developing social and economic life in eastern Reunion Island. These socioeconomic criteria are just as important as climate factors in the choice of the PGI area.

The PGI area, defined according to the above-mentioned criteria, thus, stretches along the whole of the east coast of the island (from Sainte Marie to Saint Joseph), up to 600 m altitude, excluding the west part of the island and the hills, which are unsuitable for traditional vanilla production (Figure 18.1). To benefit from the PGI, the beans must be produced and processed within the determined PGI area.

PROOF OF ORIGIN: A QUESTION OF TRACEABILITY

To guarantee consumers the origin of the PGI product, a traceability system has been set up for every stage of bean production and processing. These elements of traceability have been collated in a table that defines the entries to be made at each stage and their supports (Table 18.1).

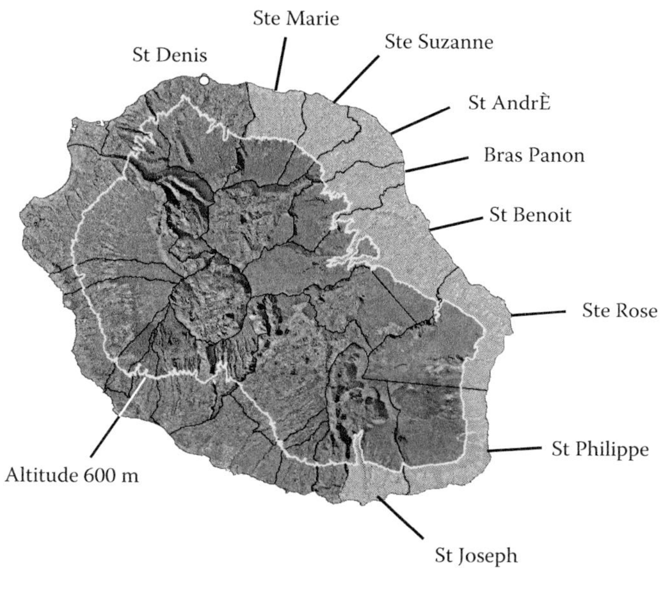

FIGURE 18.1 Map of Reunion Island showing in light gray the PGI area.

TABLE 18.1
Traceability System for All Stages of Bean Production and Processing

Production Stage	Entries	Identification Supports
Stage I: Adherence of producer	Producer number	Commitment agreement
Stage II: Geographical area	Cadastral number	Cadastral map
Stage III: Parcel selection	Parcel number	Commitment agreement Plot record
Stage IV: Plantation	Number of cuttings taken Origin of cuttings	Plot record
Stage V: Parcel maintenance	Date of operation Commercial specialty Quantity supplied	Plot record
Stage VI: Fertilization of vanilla plant	Date of supply Type of soil improvement Quantity supplied	Plot record
Stage VII: Pollination	Dates of beginning and end of pollination	Plot record
Stage VIII: Harvesting	Date of harvest Quantity of vanilla beans harvested	Plot record
Stage IX: Storage at plantation	Date of harvest Date of delivery to processing unit	Plot record Supply forms
Stage X: Identification of processing units	Processor number	Commitment agreement
Stage XI: Receiving vanilla beans	Number of collection sheet (n° of calendar week or n° of supplier) Names and numbers of producers Dates of supply Quantity of compliant and non-compliant vanilla beans received Number of supply forms	Supply form Collection sheet
Stage XII: Storing beans before killing stage	Date of supply Date of killing	Processing sheet
Stage XIII: Killing	Date of operation Exposure temperature for beans Time of beginning and end of exposure	Processing sheet
Stage XIV: Sweating	Dates of beginning and end of sweating Temperature	Processing sheet
Stage XV: Drying	Oven drying: date of beginning and end of drying and temperature Sun drying: date of beginning and end of drying	Processing sheet

continued

TABLE 18.1 (continued)
Traceability System for Every Stage of Bean Production and Processing

Production Stage	Entries	Identification Supports
	Shade drying: date of beginning and end of shade drying	
Stage XVI: Sorting beans	Number of processing sheet	Processing sheet
	Date of sorting	
	Quantity of compliant and noncompliant vanilla beans transferred	
	Number of maturing box	
Stage XVII: Aromatic maturing	Number of maturing box	Maturing sheet
	Dates of beginning and end of maturing	
	Dates of inspections	
	Quantity of beans downgraded	
Stage XVIII: Grading	Date of bean removal	Maturing sheet
	Quantity removed	
	Number of storage box	
Stage XIX: Storing vanilla beans	Number of storage box	Storage sheet
	Number of maturing box	
	Date of storage in box	
	Quantity of beans transferred	
Stage XX: Packaging	Number of production batch	Packaging sheet
	Dates of removal	
	Quantity of vanilla beans removed	
	Number of storage box	
	Quantity produced	
Stage XXI: Labeling	Legal references	Label
	Certification references	
	Best-before date	
	Quantity of certified vanilla beans	
	Number of production batch	

A plot record identifies each parcel farmed (referencing, date of plantation, etc.) and details the different operations carried out in the field: looping, pruning, weeding, treatments, pollination, thinning, and harvesting.

Monitoring sheets accompany each batch of vanilla during processing, with a batch being made up of all beans harvested in one calendar week or by all beans harvested by one single producer.

- A harvesting sheet identifying the producers at the origin of the batch
- A processing sheet detailing the operations applied to each batch: killing, sweating, drying, and sorting

- A maturing sheet monitoring the aromatic development of the beans
- A storage sheet monitoring stabilized batches awaiting sale
- A conditioning sheet monitoring the packaging of beans

Finally, stock records are kept, mentioning the quantities of incoming green vanilla and outgoing black vanilla for certifiable and noncertifiable batches.

METHOD FOR PRODUCING VANILLA UNDER THE PGI

Characteristics of Production

Producers must sign a commitment agreement with A2VR in which they undertake to comply with the specifications of the PGI. They must of course be situated in the geographical production area of the PGI (the east coast of the island). The plantation must only contain *V. planifolia* plants from referenced parcels or registered nurseries. The planting density must not exceed 5000 plants/ha. Three types of plantations are authorized: woodland planting, in fields, or under shade houses with no irrigation. Plot maintenance is carried out mechanically in crop rows and mechanically or chemically between rows. The vanilla plants may only be fertilized using organic matter (compost, natural organic substrates, etc.); chemical fertilizers of any kind are forbidden. Vines must be looped at least once a year. The pollination of flowers must be adapted to the robustness of the plant and should never exceed 15 flowers per inflorescence. Harvesting shall take place when the beans have reached optimum ripeness, in other words when the tip of the bean turns yellow. Finally, storage of beans after harvesting is limited to 72 h. Beans produced under the PGI must be at least 14 cm long and should be split for no more than 3 cm. Any beans that fail to meet these criteria are systematically downgraded.

Characteristics of Processing

Each processing unit must be referenced and situated within the geographical area of the PGI. The first stage of processing involves killing the beans on arrival to prevent them from splitting (the beans dehisce when ripe) and thereby losing their qualities. This operation involves immersing the beans in water heated to 65°C for 3 min (a technique known as the "Reunionese process"). After draining, the hot beans are placed in boxes lined with blankets for 24 h; this is the sweating phase, during which the beans lose some of their moisture. They then turn a chocolate brown color. Next comes the drying phase to ensure the beans have a moisture content that will enable them to be preserved. Drying is divided into three stages: first, discontinuous oven drying at 65°C (which is in fact an optional step), then sun drying between 5 and 15 days depending on the stage at which beans were harvested and their consistency, and finally shade drying on wire racks to ensure the beans dry slowly, which makes them supple and oily. During this drying phase, the beans are sorted at regular intervals according to dryness. Dry beans are put into maturing boxes and the others returned to the drying cycle. This maturing phase is very important for the development of the aromatic qualities. To benefit from the designation, PGI beans must be kept in maturing boxes for at least seven months. Regular inspections must be made to check the state of the beans (removing any that are moldy). Once the aromatic maturing is finished, the beans are graded and then packed in bundles before the final storage. They will be packaged as required for sale. Each package must be hermetically sealed and must have a label stating not only the references required by the consumer code (designation of the product, net quantity, producer references, production batch number), but also certification references (the PGI "Vanille de l'île de la Réunion" logo, the name and address of A2VR, and the name and address of the certifying organization).

STRICT CONTROLS TO GUARANTEE THE PRODUCT'S CERTIFICATION

The PGI is held by A2VR rather than its individual members, who are only users of the designation. The association must therefore make every effort to ensure that the specifications are met and thereby protect the designation.

The different stages of production and processing are set out in the specifications. Compliance with these specifications by the producers and processors that benefit from the label is monitored at several levels. An initial internal inspection is conducted by an A2VR technician, who goes on-site to meet the producers and guarantee compliance of plots and farming practices used. The technician also inspects each processing unit, checks the processing parameters, conducts the necessary tests (vanillin and moisture content, etc.), and inspects the bean monitoring sheets (harvesting, processing, maturing, storage, and conditioning), and stock accounting. This internal inspection, carried out by A2VR, is supervised by an external inspection conducted by an official certifying organization, in this case Organisme Certificateur Tropique Réunion Océan Indien (OCTROI), which carries out spot checks on producers, examining documentation, and plantations to ensure internal inspections are conducted satisfactorily. It is the certifying organization that maintains or denies certification of PGI products depending on the results of its annual inspections.

The "Vanille de l'île de la Réunion" PGI is indisputably a tool for the protection of traditional production in Reunion Island and guarantees consumers the origin and quality of beans sold. It is therefore essential to ensuring the safeguard of Reunionese production, even though the process remains difficult to implement.

REFERENCE

Floris, D. (de). 1857. La culture du vanillier. Lahuppe, La Réunion.

19 Vanilla Production in Indonesia

Robber Zaubin, Mesak Tombe, and Edward C.Y. Liew

CONTENTS

INTRODUCTION

The centers of vanilla production in Indonesia are the provinces of North Sumatera, Lampung, West Java, East Java, Bali, West Nusa Tenggara, East Nusa Tenggara, North Sulawesi, Central Sulawesi, and South Sulawesi, with increasing areas being developed, including those in Kalimantan, Maluku, and Papua (Figure 19.1). Most vanilla estates are run by smallholders and only small parts are of private entrepreneurs. In the international market Indonesian vanilla is known as Java Vanilla Beans.

The main constraints in the vanilla industry in Indonesia are low productivity and the relatively low bean quality (Hadipoentyanti et al., 2007). According to the Diratpagar (Directorate General of Estate Crops, 1995), vanilla production ranged from 125 to 876 kg/ha, and the quality had still to be improved. Production is affected, among others, by environmental conditions, types of vanilla grown, cultural practices, and occurrence of pests and diseases. Vanilla quality (vanillin content of bean) is commonly influenced by time of harvest, length of beans, and postharvest processes. This chapter attempts to provide some information on the low productivity and quality of vanilla beans in Indonesia, and on the results of research conducted to improve the situation.

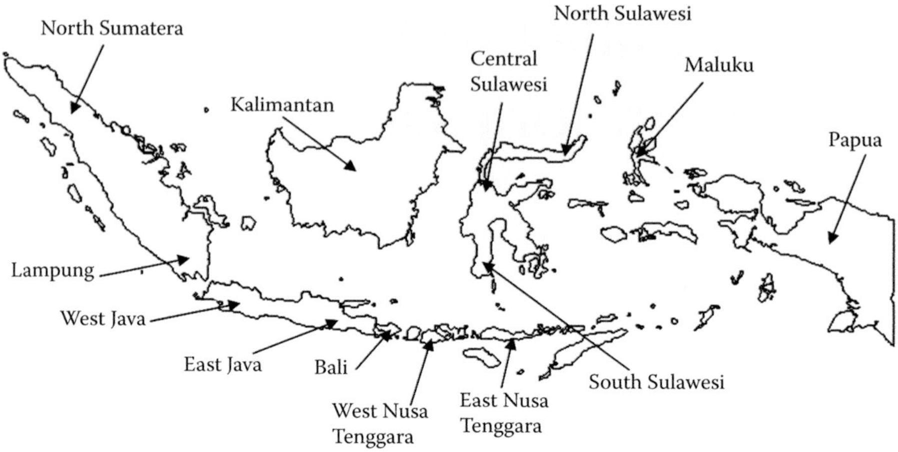

FIGURE 19.1 Map of Indonesia showing the main producing areas.

THE CURRENT SITUATION

Vanilla planifolia cuttings were introduced to Bogor (West Java) in 1819 by Marchal. The plants grew well, produced flowers but bore no fruit. It soon became obvious that self-pollination of the flowers was not possible in the absence of the natural pollinating bees in Java. In 1836, Morren succeeded in pollinating the flowers manually at the botanical garden at Luik (Belgium) and a few years later, the technique was improved for routine farm use by Edmond Albius on Réunion Island. In 1850, Theysman succeeded in producing vanilla beans in Java (Deinum, 1949), and since then plantations of vanilla spread all over Indonesia.

Prior to 1967, vanilla production in Indonesia had been centered in Java, such as Malang (East Java), Temanggung (Central Java), and Garut (West Java). Owing to the heavy damages caused by disease outbreaks and long periods of drought, vanilla production was developed outside Java, for example, Bali, Lampung, and North Sumatera in the 1970s, and since 1980 further expanded to include North Sulawesi, West Sumatera, Aceh, and East Nusa Tenggara (Anonymous, 1993).

Vanilla Production

Farmers in Indonesia are not specialized vanilla producers; they also grow miscellaneous crops, which function as the main source of income. Such crops include durian, mango, papaya, corn, cassava, sweet potato, peanut, and others planted as secondary crops in the dry season. As a smallholder commodity, vanilla is farmed in small areas scattered in regions with limited agronomic inputs. A common practice is the maintenance of farms only when the price of vanilla is high, and new farms are often initiated without much consideration given to understanding the technical requirements for vanilla cultivation.

Soil and climatic conditions in the production areas are often not optimal, in particular, sandy soil and prolonged dry season, while the plant material or genetic source of vanilla is mostly of inferior types. There are many types of *V. planifolia* grown in Indonesia, such as the Anggrek, Gisting, Cilawu, Malang, Ungaran Daun Tipis, Ungaran Daun Tebal, Bacan, Bandialit, and Bali. Although derived from the same species, these types of vanilla have varying production potentials (Asnawi and Nuryani, 1995). Production output therefore differs greatly from one farm to another, especially when multiple factors are involved.

Most vanilla farms use gamal (*Gliricidia maculata*) or lamtoro (*Leucaena glauca*) as supporting trees with planting spacing of about 1.0 m × 1.5 m and 1.5 m × 2.0 m, respectively. Cultural practices, such as weed control, pruning of support trees to control shade, fertilizer use, and pest and disease control are limited. Applications of fertilizers, insecticides, and fungicides are commonly ignored due to the high cost of input and the high risk of production loss due to diseases. Only during periods when the price of vanilla is relatively high are fertilizers applied. This further contributes to the inconsistency in production levels. The suggested dosage of fertilizer use is 50–100 g NPK (1:2:2)/plant/year for plants of less than two years old, and 100–200 g NPK (1:2:3)/plant/year for plants more than two years old, applied 50% at the beginning and 50% at the end of the rainy season. The high dosage of potash compared to nitrogen is supposed to increase stem rot tolerance (Zaubin et al., 1994). Where necessary, vanilla also needs foliar spray of nutrients every 1–2 weeks, with a concentration of 5–8 g NPK (1:1:1)/L of water. The nutrient solution is applied in the morning between 6 and 7 a.m. or late afternoon between 5 and 6 p.m. when the relative humidity is high (Ernawati, 1993). Gusmaini and Tarigan (1999) suggested the use of 0.4% of Gandasil D (NPKMg 14:12:14:15 plus trace elements) every four days. Vanilla farmers, however, prefer to use compost or manure, the application rates of which unfortunately are too low (3–5 kg/plant/year).

Vanilla is grown from sea level to the highlands (up to 1200 m asl). Although vanilla can grow in a wide range of temperatures, from 9°C to 38°C, Chalot (in Deinum, 1949) mentioned that the optimum temperature for vanilla is about 25°C, which occurs at elevations between 200 and 400 m asl. Temperatures lower than 18.6°C are considered very low and will affect the activity of enzymes responsible for the aroma and the development of vanilla beans (Zaubin and Wahid, 1995).

During the dry season, the lack of precipitation, low atmospheric humidity, and high temperature lead to water stress that weakens the plants, rendering them more vulnerable to pests and diseases. The most destructive disease is stem rot, caused by the fungus *Fusarium oxysporum* f. sp. *vanillae* (Tombe et al., 1997), and is discussed in detail in Chapter 8.

Production begins in the second or third year of planting. Farmers usually retain 8–15 beans/inflorescence, while the number of inflorescences/plant varies. Harvesting is conducted gradually by picking only the ripe beans, as indicated by a slight fading of the green color and yellowing of the tips. This condition occurs at 8–9 months after pollination. The average production is 0.2 kg mature beans/plant, although the production potential is about 1.0 kg/plant (Hadipoentyanti et al., 2006).

Vanilla Curing

Farmers commonly sell the harvested beans at local markets, while exporters proceed with postharvest processes before exporting. Postharvest activities include preparation, withering, aging, and drying (in sunlight and shade) as well as conditioning (Figure 19.2). During preparation, freshly harvested beans are washed with water, followed by selection according to length, thickness, amount of damage, and physical blemishes or defects (Risfaheri and Rusli, 1995). Selected beans are subsequently withered by placing the beans in a wire or bamboo basket and immersed in hot water (between 63°C and 65°C) for 2–2.5 min. The purpose of withering is to terminate vegetative growth and trigger enzymatic activities, which induce the formation of vanillin.

After draining, the beans are placed in a double-walled box for 24 h for the purpose of aging. Coconut fiber or sawmill is inserted between the wall layers as insulation, maintaining the temperature between 38°C and 40°C. The inner wall layer is covered with a relatively thick cloth to absorb moisture from the vanilla beans. The beans eventually become brownish in color with a greasy appearance.

The beans are dried on a black cloth on a drying bamboo tray for 2–2.5 h in the morning with occasional turning over. The beans are then covered with a black cloth and dried in the sun until afternoon when they are placed in a drying room. This process is repeated daily until the water content of the beans reduces to 55–60% and the aroma of vanillin is produced. From this point on the beans are dried away from direct sunlight in order to lower the water content gradually and increase the vanillin aroma. The vanilla beans are arranged in an orderly fashion on trays in a clean and cool ventilated room for 30–45 days, and regularly monitored for incidence of rot or

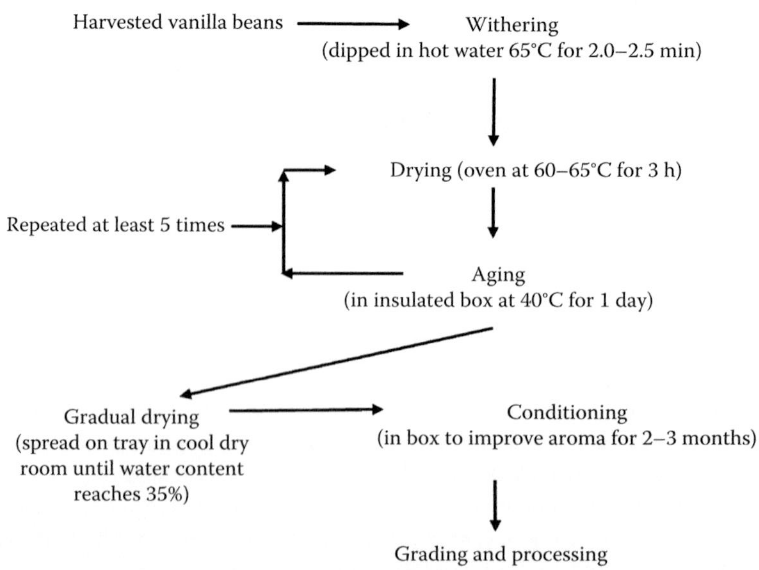

FIGURE 19.2 Schematic representation of the processing of vanilla pods.

mold. When the water content drops to about 30–35%, the beans are conditioned, using an oven maintained at a temperature of 50°C for 3 h. Conditioning is aimed at improving the vanillin aroma. Bundles of 50–100 beans are tied and wrapped in wax or paraffin paper, placed in a box coated with wax paper and stored in a cool dry room for 2–3 months. The beans are again regularly monitored for rot or mold, and if necessary, cleaned with alcohol. Beans with a weak aroma are further dried and returned to the conditioning process. After conditioning, the beans are graded and packed, ready for export.

Commercialization of Vanilla

The vanilla trade in Indonesia lacks structure and organization. In areas where the production is relatively high, exporters commonly have local agents to buy vanilla green beans direct from farmers in order to shorten the marketing chain. But in low production areas, the marketing chain is longer. Vanilla is sold from farmers to village traders, then from village traders to district traders, where it is sold to regency traders before reaching the exporters (Board of National Export Development, 1995).

The price of vanilla is strongly determined by the quality of the beans. Each country determines its own prerequisites. The common and specific sets of prerequisites, determined by the National Standard Board of Indonesia (SNI), are shown in Tables 19.1a and b. The United States prefer vanilla with low water content (20–30%) for its use in the extraction industry. In Europe, vanilla is commonly used in households, therefore, the beans must be whole, with high vanillin content, sharp aroma, and 30–35% water content (Ruhnayat, 2005). The International Standard Organization (ISO) has determined prerequisites for the international market (ISO standard 5565-1:1999).

The price of Indonesian vanilla is generally low due to the low quality of beans, partly a result of early harvesting and suboptimal processing conditions as well as the weak bargaining position of farmers. In the international market, the price of vanilla from Indonesia is in the range of US$15–$48/kg, relatively far below that of Madagascan vanilla, which ranged from US$55–$76/kg.

The annual total production and value of vanilla in Indonesia fluctuate significantly. In 2000, the production was recorded as 350 tons valued at US$8.5 M,

TABLE 19.1a
Common Prerequisites for Vanilla, According to SNI 01-0010-1990

Characteristic	Specification	Testing Method
Aroma	Fragrant, specific vanilla	Organoleptic
Color	Shining, brown to dark brown	Visual
Beans	Compact, greasy, elastic to rather stiff	Organoleptic
Alien thing	Free from alien thing	Visual
Molds	Free from molds	Visual

TABLE 19.1b
Specific Prerequisites for Vanilla, According to SNI 01-0010-1990

Characteristic	Prerequisite				Testing Method
	Quality IA	Quality IB	Quality II	Quality III	
Form	**Whole**	**Whole**	**Whole/cutting**	**Whole/cutting**	**Visual**
Size of whole beans	11	11	8	8	SP-SMP320-1980SP-SMP320-1980
Split beans and cutting	5	None	None	None	SP-SMP7-1975
Water content, max (%)	38	38	30	25	SP-SMP320-1980
Vanillin content, max (%)	2.25	2.25	1.50	1.00	SP-SMP35-1975
Ash content, max (%)	8	8	9	10	

663 tons in 2003 at US$18.4 M (highest in record), and 499 tons in 2006 at US$5.9 M (Ditjenbun, 2006). In 2006, the main importing countries were the United States, accounting for 67% of total export, Germany (20%), and Malaysia (5%). The quality of vanilla exported from Indonesia is relatively low: 27% of grade I, 45% of grade II, and 28% of grade III (Rosman, 2005).

RESULTS OF RESEARCH

There are three important factors that should be addressed in overcoming the constraints in vanilla production: environmental conditions (climate and soil), genetic source (varieties or types of vanilla), and stem rot disease. Research activities are focused on these factors and a summary of the outcomes are as follows.

Climatic and Soil Conditions

Suitable climate is one of the most important conditions for vanilla cultivation. The optimal annual rainfall for vanilla is in the range of 1000–2000 mm/year, and distributed in 8–9 months of wet season, followed by a dry period (rainfall lower than 90 mm/month) of 3–4 months. A distinct dry period, simultaneously with other factors such as looping the stem and removal of the stem apex during the dry period, is required to induce flowering. The formation of flowers is, to a certain extent, affected by stress levels experienced by the vanilla plant (Zaubin, 1994). A rainy season of 150–180 days/year, with temperatures ranging from 20°C to 39°C and relative humidity between 65% and 75%, is preferred.

Vanilla requires only 30–50% of full sunlight (Deinum, 1949). Although vanilla grows and develops well in areas between 0 and 1200 m asl, for practical commercial purposes, vanilla cultivation is recommended at altitudes up to 600 m asl. Vanilla can be grown on a variety of soils, such as andosol, latosol, regosol, as long as the physical properties are good. Soils with a deep solum and rich in organic matter are ideal for vanilla. The soil acidity (pH) should range from 5.5 to 7.0. Although the ideal situation may not always be available, these conditions are used as guidelines for vanilla production and are a useful indication of the amount of cultural input required, given a particular set of existing conditions.

Genetic Source

Recently, the Research Institute for Spices and Medicinal Crops (RISMC) has collected 32 types of vanilla from all over Indonesia. They differ in size and color of leaves and flowers and in the form of leaves, number of stomata, leaf area index, and tolerance to stem rot. Among them only four types seem to have promising prospects (Asnawi and Nuryani, 1995). The four prospective clones have been identified with favorable characteristics, including a relatively high production potential, high vanillin content and relative tolerance to water stress and stem rot (Hadipoentyanti et al., 2007). The characteristics of these clones are shown in Table 19.2. These superior types or varieties of vanilla have not been released until now.

TABLE 19.2
Characteristics of the Four Prospective Vanilla Clones

Characteristics	Clone 1	Clone 2	Clone 3	Clone 4
1. Selected from population	Cilawu–West Java	Gisting–Lampung	Malang–East Java	Ungaran–Central Java
2. Leaf:				
• Length (cm)	20.54 ± 1.48	19.12 ± 1.56	18.15 ± 2.35	21.38 ± 1.69
• Width (cm)	6.42 ± 0.43	7.03 ± 0.51	6.75 ± 1.24	7.00 ± 0.32
• Thickness (cm)	2.32 ± 0.31	2.13 ± 0.16	2.18 ± 0.25	1.98 ± 0.26
• Color	Green	Green	Green	Green
3. Stem	No branches	No branches	No branches	No branches
4. Stem diameter	1.20 ± 0.24	1.17 ± 0.26	1.15 ± 0.38	1.25 ± 0.36
5. Internode length (cm)	12.69 ± 1.13	12.71 ± 1.12	15.38	14.78 ± 1.16
6. Flower bunch	No branches	No branches	No branches	2–4 branches
7. Beans:				
• Length (cm)	19.03 ± 1.87	19.17 ± 1.08	18.51 ± 1.46	19.05 ± 1.38
• Color	Green	Green	Green	Green
8. Number of bunches per plant	5.75 ± 2.75	5.35 ± 2.53	6.50 ± 2.15	7.11 ± 2.32
9. Number of flowers per bunch	20.15 ± 1.76	19.75 ± 2.76	21.35 ± 1.51	22.25 ± 1.46
10. Production of beans (kg/ha)				
• Fresh	9.768	9.768	10.392	10.523
• Dry	2.035	1.918	2.165	2.239
11. Vanillin content	2.80–3.25	2.26–3.16	2.25–3.05	2.35–2.86
12. Others	Relatively tolerant against water stress	Relatively tolerant against stem rot	—	—

Vanilla Diseases

The occurrence of diseases is one of the main constraints in vanilla cultivation. The important diseases are stem rot (*Fusarium oxysporum* f. sp. *vanillae*), shoot rot (*Phytophthora capsici*), Sclerotium rot (*Sclerotium rolfsii*), and anthracnose (*Colletotrichum gloeosporioides*); stem rot being the most devastating disease. Since a significant degree of host resistance to stem rot is not available at present, crop losses are severe under poor management. Although *F. oxysporum* f. sp. *vanillae* is a soilborne pathogen, it is also transmitted via water splash. Two types of dispersal propagules are produced; the macroconidia and the microconidia, which can easily adhere to plant parts, insects, and farming implements (Tombe et al., 1992). The disease is highly infectious and has the ability to spread rapidly and devastate extensive production areas. The disease may attack every vegetative part of the vanilla plant, although it is mostly found on the stem. Under adverse environmental conditions, the fungus produces chlamydospores for long-term survival in the soil and host debris. The pathogen can also survive within the stem without showing any symptoms.

Owing to the general poor performance of the plants and the adverse conditions, which may predispose a variety of diseases, farmers commonly abandon their farms after 5–7 years. For as long as disease-tolerant varieties or other effective control measures are unavailable, the development and production of vanilla in Indonesia remain stymied to a great extent.

Recent research indicates that successful management of stem rot can be achieved by applying the Bio-FOB approach (Tombe, 2008). This approach involves three components, which are detailed in Chapter 8. In brief, the technique consists of:

1. Planting disease-free cuttings tolerant to stem rot such as Bio-FOB vanilla cuttings (produced using nonpathogenic strains of *Fusarium oxysporum*).
2. Using selected bio-control agent and organic matter: *Bacillus pantothenticus* and *Trichoderma lactae* have been shown to be antagonistic to *F. oxysporum* f. sp. *vanillae*. These microbes decompose organic matter as well as have growth-promoting effects on the plant (Tombe, 2008). These two microbes are formulated under the name Bio-TRIBA. Bio-TRIBA can be mixed with compost or manure at a dosage of 2–3 L Bio-TRIBA/ton compost or manure and incubated for 1–2 weeks before application. Bio-TRIBA can also be applied by drenching onto the roots of vanilla at a dosage of 5–10 mL Bio-TRIBA/L water, preferably after the application of organic fertilizers.
3. Applying the botanical fungicide Mitol 20 EC at a concentration of 3–5 mL/L water. The formula contains active ingredients of eugenol extracted from cloves (Tombe et al., 1993; Sukamto et al., 1996) and demonstrated to be antagonistic to several plant pathogens, including *F. oxysporum* f. sp. *vanillae*.

Taking into account the proper climatic and soil conditions, using the best genetic source available, and implementing root and stem rot disease control strategies, the production and quality of vanilla in Indonesia are expected to increase.

FUTURE PERSPECTIVES

Despite the many challenges ahead in developing and improving the production of vanilla in Indonesia, there are nevertheless advantages and sufficient optimism in the country toward ultimately achieving the status of the world's primary producer and exporter of vanilla (Directorate General of Estate Crops, 2008). Indonesia is rich in both human and natural resources, the latter in terms of the availability of suitable land for the expansion of vanilla cultivation. The technology is available for overcoming the most significant production constraint, stem rot, based on an integrated disease management approach, the *Bio*FOB strategy. However, extensive knowledge transfer to production areas and technology adoption by farmers need to be significantly improved (Rosman, 2005). A greater emphasis needs to be placed on the promotion of Indonesian vanilla on the international platform (Directorate General of Estate Crops, 2008). This should be in concert with the establishment of integrative domestic institutions for farmers, entrepreneurs, researchers, and the government to promote farmer knowledge, protection of farmer welfare, and communication among the sectors involved.

REFERENCES

Anonymous. 1993. *Study on the Trade of Vanilla Bean in the Region of Japanese Market.* Department of Trade, Board of National Economic Development, 40pp. (in Indonesian).

Asnawi, R. and Y. Nuryani. 1995. Germplasm of Vanilla. *Proceedings of the Agronomic Status and Post Harvest in Lampung.* Bandar, Lampung, March 15th, 1995. Research Institute for Spices and Medicinal Crops, 59–68 (in Indonesian, English Abstract).

Board of National Export Development. 1995. Export development, prospect and constraint of Vanilla commodity. *Proceedings of the Agronomic Status and Post Harvest in Lampung.* Bandar, Lampung, March 15th, 1995. Research Institute for Spices and Medicinal Crops, 1–3 (in Indonesian).

Deinum, H.K. 1949. Vanille. In: C.J.J. van Hall and C. Van de Koppel, eds. *De Landbouw in de Indische Arschiepel.* II B. Uitgeverij W. Van Hoeve, Nederland, 763–784.

Directorate General of Estate Crops. 1995. Strategy and Development Program of Vanilla in Indonesia. *Proceedings of the Agronomic Status and Post Harvest in Lampung.* Bandar, Lampung, March 15th, 1995. Research Institute for Spices and Medicinal Crops, 15–21 (in Indonesian, English Abstract).

Directorate General of Estate Crops. 2008. *Vanilla Prospects in Indonesia.* Ministry of Agriculture, Jakarta, Indonesia. http://ditjenbun.deptan.go.id/rempahbun/rempah//index.php?option=com_content&task=view&id=101&Itemid=26 (Accessed May 17, 2009. In Indonesian).

Ditjenbun. 2006. *Statistik perkebunan Indonesia 2006.* Panili, Dephutbun, Jakrata, Indonesia. http://database.deptan.go.id/eksim/eksporKomoditi.asp (Accessed July 17, 2009. In Indonesian).

Ernawati, R. 1993. The effect of dosage and time of foliar spray of nutrient on the growth of vanilla. Report of Research, Natar sub-Research Institute for Spices and Medicinal Crops, 8pp. (in Indonesian, English Abstract).

Gusmaini and D.D. Tarigan, 1999. The response of vanilla on time and concentration of nutrition sprays at the nursery. *Science Bulletin GAKURYOKU* V (2):98–104 (in Indonesian, English Abstract).

Hadipoentyanti, E., A. Ruhnayat, and L. Udarno. 2007. *Vanilla Cultivation to support Overcoming the Stem Rot Disease. Prime Technology*. Board of Agriculture Research and Development, Research Center and Development of Estate Crops, 21pp. (in Indonesian, English Abstract).

Hadipoentyanti, E., L. Udarno, D. Seswita, A. Ruhnayat, Sukarman, Emmyzar, and M. Tombe. 2006. The status of technology of vanilla. *Proceedings of the Technology Status of Spices and Other Industrial Crops*. Parungkuda-Sukabumi, September 26th, 2006. Research Center and Development of Estate Crops, Research Institute for Spices and Other Industrial Crops, 58–78 (in Indonesian, English Abstract).

Risfaheri and S. Rusli. 1995. Processing and improving the quality of Vanilla beans. *Proceedings of the Agronomic Status and Post Harvest in Lampung*. Bandar, Lampung, March 15th, 1995. Research Institute for Spices and Medicinal Crops, 93–108 (in Indonesian, English Abstract).

Rosman, R. 2005. Status dan strategi pengembanagan panili di Indonesia. *Perspektif* 4:43–54 (in Indonesian).

Ruhnayat, A., Hadipoentijanti, E., Sukamto, and L. Mauludi. 2005. Vanilla cultivation. Cultivation of vanilla and patchouli. Circular No. 12, 2005. Research Institute for Spices and Medicinal Crops, 1–28 (in Indonesian).

Sukamto, M. Tombe and S. Mugi. 1996. Clove by-product to control stem rot of vanilla. *Seminar on Integrated Control on Main Diseases off Industrial Crops*. Bogor, March 13–14, 1996. AARD–JICA: 77–86 (in Indonesian, English Abstract).

Tombe, M., K. Kobayashi, Ma'mun, Triantoro, M. Oniki and K. Matsumoto. 1993. The role of eugenol in disease suppression of stem rot of vanilla. *Annals of Phytopathological Society of Japan* 59:282 (Abstract).

Tombe, M., K. Matsumoto, A. Nurawan, Sukamto and S.B. Nazarudin. 1992. Strains, morphology, physiology and ecology of causal fungus and disease damages of vanilla. *Proceedings of Final Seminar of the Joint Study Programme RISMC-JICA, ATA-380*. Bogor, Indonesia, 50–57.

Tombe, M., D. Sitepu, and S. Mogi. 1997. Present status of biological control research of vanilla stem rot disease in Indonesia. *Proceedings of the Fourth International Workshop on Plant Growth Promoting Rhizobacteria*. Japan-OECD, Workshop, 13–17.

Zaubin, R. 1994. The effect of climate on flowering of Vanilla. *Communication Media Litbangtri* 15: 12–17 (in Indonesian, English Abstract).

Zaubin, R., R. Asnawi, and R. Kasim. 1994. The effect of NPK compositions on the growth and spreading disease of vanilla. Report of Research, Natar sub-Institute for Spices and Medicinal Crops, 8pp (unpublished).

Zaubin, R. and P. Wahid. 1995. Environmental conditions for Vanilla. *Proceedings of the Agronomic Status and Post Harvest in Lampung*. Bandar, Lampung, March 15th, 1995. Research Institute for Spices and Medicinal Crops, 47–58 (in Indonesian, English Abstract).

20 Vanilla Production in India

Y.R. Sarma, Joseph Thomas, B. Sasikumar, and S. Varadarasan

CONTENTS

INTRODUCTION

Vanilla (*Vanilla planifolia* G. Jackson), a native of southeastern Mexico, is a high-value aromatic orchid spice commercially cultivated in Madagascar, Indonesia, Mexico, Uganda, Comoros, India, and others. It was introduced to India during 1835. As per the documentary evidence (Anonymous, 1992), it was first cultivated at Kallar and Burliar Fruit Research Station, Nilgiris during 1945 and later at Regional Agriculture Research Station, Ambalavayal, Wynad, Kerala. Few enterprising farmers and coffee planters of Wynad took up its cultivation as an intercrop in shade tree plantations under the technical guidance of the Ambalavayal research farm in Wynad and the then Government of Kerala encouraged cultivation of vanilla in the tribal settlement at Cheengeri at Ambalavayal as an alternative income-generating crop during 1960. Similarly, some growers in and around Kallar–Burliar Fruit Research Station, Gudallur and Nilgiris also started cultivating vanilla during the same period. It gradually spread to several parts of Kerala, Karnataka, Tamilnadu, West Bengal, and Assam through innovative farmers. However, almost all these initial attempts did not succeed due to improper care, lack of knowledge, or absence of technical and market support. Nevertheless, these plantations served as a source of planting material for vanilla development programs initiated by various agencies, later on.

Successful production and marketing of vanilla beans was reported from 1 ha of vanilla plantation in Sasthan in South Kanara district in Karnataka. The growers of South and North Kanara districts gradually took up cultivation of this crop and now the state accounts for the largest area under vanilla cultivation/plantation with 58% of the crop in India.

AREA, PRODUCTION, AND EXPORT

Vanilla has become a commercial crop in India, only recently, largely due to the promotional activities taken up by the Spices Board of India since the mid-1990s. Today India is an identified source of quality vanilla beans in the international market. Export of cured beans was 305.1 metric tons at a value of Rs. 2670 lakhs (about $5.8 million) during 2008–2009. Some quantity of vanilla oleoresin was also exported during the corresponding period. The area and production of vanilla in India and its exports are given in Tables 20.1 and 20.2.

The area under production gradually increased. It gained momentum with a price spurt during 2001–2002, probably because of severe crop losses due to a cyclone in Madagascar. Because of the decline in market price of natural vanillin since 2004–2005 coupled with unmanageable disease occurrence, the enthusiasm in growing vanilla has declined dramatically among farmers.

CROP IMPROVEMENT OF VANILLA

Crop improvement of vanilla is in progress in India at the Indian Cardamom Research Institute (ICRI), Spices Board, P.O. Kailasanadu, Idukki, Kerala; Indian Institute of Spices Research (IISR), P.O. Marikunnu, Calicut, Kerala, Kerala Agricultural University, and other institutions. These institutes maintain a good germplasm collection of vanilla as well. A minimum descriptor for vanilla is developed for characterizing the germplasm (Kuruvilla et al., 2000).

Species Diversity

V. planifolia G. Jackson is the cultivated vanilla species in India. In addition to this, the other vanilla native species available in India are *Vanilla pilifera* Holtt., *Vanilla andamanica* Rolfe., *Vanilla aphylla* Blume, *Vanilla walkeriae* Wight, and *Vanilla wightiana* Lindl. The Tahitian vanilla (*Vanilla tahitensis* J.W. Moore) is also conserved in India.

Breeding Behavior

The floral biology of *V. planifolia* is adapted to outcrossing, and hand pollination is resorted to pod set in India, as the natural pollinators are absent in the country (Sasikumar et al., 1992). No self-incompatibility or natural crossing is reported in

TABLE 20.1
State-Wise Assessment of Area and Production of Vanilla in India in 2002–2007

	2002–2003				2003–2004				2004–2005				2005–2006				2006–2007			
State	Total Area	Yielding Area	Prodn (cured)	Yield (cured)	Total Area	Yielding Area	Prodn (cured)	Yield (cured)	Total Area	Yielding area	Prodn (cured)	Yield (cured)	Total Area	Yielding Area	Prodn (cured)	Yield (cured)	Total Area	Yielding Area	Prodn (cured)	Yield (cured)
Kerala	812	239	19	79	1147	342	34	98	1707	575	68	143	1985	883	82	92	2206	1278	122	95
Tamilnadu	268	130	19	146	465	180	18	99	577	186	16	108	732	249	22	89	705	333	23	68
Karnataka	1465	545	54	99	1931	732	82	113	3086	1187	112	101	3098	1751	84	48	2218	1307	88	67
All India	2545	914	92	101	3543	1253	134	107	5370	1948	196	113	5815	2883	188	65	5129	2918	233	80

Source: Spices Board of India.
Area in hectare, Production (Prodn) in Metric Tons and Yield in kg/ha.

TABLE 20.2
Major Country-Wise Export of Vanilla from India 2004–2009

Export	2004–2005		2005–2006		2006–2007		2007–2008(E)		2008–2009(E)	
Country	QTY	VALUE	QTY	VALUE	QTY	VALUE	QTY	VALUE	QTY	VALUE
USA	32.8	2481.0	34.5	555.3	58.7	1108.4	83.3	589.6	111.2	847.0
Netherlands	0.8	10.3	0.0	0.4			0.0	0.7	100.0	689.7
France	5.6	147.4	23.6	433.5	26.4	390.6	27.3	498.7	36.5	405.4
Belgium							1.0	29.2	15.9	344.0
Germany	2.2	52.7	9.0	147.9	22.5	338.3	80.3	577.7	31.8	277.3
Brazil					0.5	6.2	2.1	19.2	6.1	58.0
Others	1.5	184.5	4.5	89.7	18.4	219.2	6.0	59.9	3.7	48.6
Total India	43.0	2875.9	71.6	1226.8	126.4	2062.7	200.0	1775.0	305.1	2670.0
Total World	1259.7	2323.8	1529.6	3117.2	2042.5	4264.3	1300.0	2875.0	2155.1	6074.7

Source: Spices Board of India.
QTY in metric tons, VALUE in lakhs (about 2.000 US$).

V. planifolia from India as observed in Reunion Islands or Mexico (Bory et al., 2008a). However, in *V. wightiana*, natural fruit setting is reported (Rao et al., 1994). Stray fruit set under natural conditions is also seen in *V. aphylla*.

Reproductive biology of *V. planifolia* such as time of pollination, stigma receptivity, and effect of pollen load on the size of the beans were studied (Shadakshari et al., 1996; Bhat and Sudarshan, 1998, 2000). These authors reported that the ideal time for pollination is from 6 a.m. to 1 p.m., and stigma receptivity is up to 24 h. They also observed that complete transfer of pollen results in maximum fruit growth.

Inflorescence initiation in vanilla occurs in late January or early February and the flower opens from mid-February to April–May. Pollination is carried out manually as and when the flower opens and the process is continued for at least 40–60 days. The beans develop quickly in the initial stages and attain their full size within a period of 5–6 weeks under favorable conditions and thereafter slow down. The rate of elongation of beans is maximum during the first 30 days after fertilization (Kuruvilla et al., 1996).

The reproductive mode of *V. planifolia* needs to be studied, thoroughly looking into the rate of outcrossing, self/cross incompatibility, and autogamy.

Cultivar Diversity

Although vanilla was introduced to India about 200 years ago, the present-day genepool in India is derived from this original introduction. Being perpetually propagated vegetatively from these original germplass, a wide genetic base in the primary gene pool of vanilla is very unlikely (Sasikumar, 2004), akin to the situation in some other vanilla-producing countries (Soto Arenas, 1999; Lubinsky, 2003; Bory et al., 2008b). There is no authentic record of any subsequent introductions to the country.

Over the years of domestication and selection by farmers, some new variants (subcultivars) have been recognized in Mexico and Reunion Islands, but no such variants are reported from India, barring a variegated mutant "Marginata" (Minoo et al., 2006a, 2008b) and some accessions with branched inflorescence and varied leaf size. Somatic crossing over (Nair and Ravindran, 1994) reported in *V. planifolia* can give rise to new variation, apart from mutation, sexual recombination, or epigenetic variation.

Genetic diversity analysis of *V. planifolia* accession in India will be worth attempting as it may help to confirm the existence of genetic variability, if any, in the germplasm and help in the breeding program.

Interspecific Hybridization

The variation that exists among the cultivated species of vanilla or even in some related species can be combined to produce new types through hybridization. The secondary gene pool may contain useful genes for self-pollination, root rot and virus resistance, larger fruits, reduced photosensitivity, better aroma profile, and pod indehiscence for incorporating into the cultivated vanilla. Because of incompatibility between some vanilla species, breeders may attempt to raise the progenies of interspecific crosses through *in vitro* seed culture. Interspecific hybrids between

V. planifolia × *V. aphylla* (Minoo et al., 2006b, 2008a) and *V. planifolia* × *V. wightiana* (Rao et al., 1992b) are reported from India.

The xenia/metaxenia aspects too can be looked into inter alia as it is observed that pollen of some vanilla species have a positive effect on the pod size of *V. planifolia* (B. Sasikumar, personal observation).

AGRONOMY

Climate and Soil

Vanilla being a climbing orchid, conditions favorable for its vegetative growth as well as flowering are to be provided for the commercial production of vanilla beans. The plant requires warm and moist conditions of humid tropics for proper growth and sustainable production. It thrives well between 10°N and 20°S latitudes, having well-distributed moderate rainfall between 150 cm and 300 cm. The crop tolerates a wide range of temperature, but ideally, the mean minimum temperature during winter months should not go below 12°C and should not exceed 35°C during summer, and the optimum temperature range being of 25°–32°C with a mean relative humidity of 80%. The plant does not tolerate any prolonged duration of drought or waterlogging, and exposure to hot sun or to strong winds.

According to Potty and Krishnakumar (2003), vanilla can be successfully cultivated in areas nearer to the equator, where warm and humid climate prevails throughout the year and up to an altitude of 1100 m asl.

Vanilla prefers land with gentle slope and light porous soil with adequate drainage. Soil with high humus content is preferred, although the plants can thrive well in sandy loam to even lateritic soils. The humus-rich soils of Western Ghats and the northeastern states of India are highly suited for its cultivation. Vanilla is grown successfully as an intercrop in coconut and arecanut gardens (Sasikumar et al., 1993).

Propagation

Vanilla is amenable to both sexual and asexual methods of propagation. The stem cuttings are capable of striking roots at nodes when they come in contact with soil or any other rooting media. Vegetative propagation through stem cutting is, by and large, the accepted method because it is easy and quick to establish. However, vanilla being a monopodial orchid, collection of sizeable quantities of stem cuttings from the main plantations could lead to the arrest of vegetative growth of mother plants (Ayyappan, 1990) and will be at the expense of subsequent years' crop. Hence, production of planting materials in nurseries is resorted to under commercial farming. In areas where viral disease is rampant, virus-free certified micropropagated materials are recommended for reviving the plantations.

The primary source of cuttings for the nurseries may be collected from disease-free, healthy, and vigorous mother plants from yielding plantations. The length of the vine used for planting varies from place to place, but has profound influence on further growth and time taken to attain maturity. In situations where there is scarcity of mother vines, cuttings of three to four nodes are used. It is recommended to use such cuttings in the nurseries to raise longer vines rather than directly planting in fields because shorter stem cuttings take longer time for establishment and yielding.

Trench Multiplication

A simple and rapid multiplication procedure for planting material production was described (Kuruvilla et al., 2003). The site for the rapid multiplication nursery should be ideally located with respect to accessibility, availability of water source, gentle slope, and deep and fertile soil of loamy nature having optimum natural shade. Wherever the optimum shade is not available, it is necessary to provide adequate shade using agro-shade nets permitting 50% light intensity. Trenches of 60 cm wide and 60 cm deep are opened at convenient length, leaving 40 cm in between. The trenches are filled with topsoil, well-decomposed farmyard manure (FYM), and sand in the ratio 3:1:1. Standards of 2–2.5 m length for trailing of vines are planted at a distance of 1 m along the trenches at least two months before planting vanilla. Fast-growing trees possessing low branching habit, small leaves, and rough bark are preferred. *Plumeria alba*, *Erythrina lithosperma*, and *Glyricidia maculata*, and others are the common live standards used in India. One-meter vanilla vine cuttings with at least 10 nodes are to be used for planting in the rapid multiplication nursery. Three or four basal leaves of the vines are cut retaining one-third of leaf blades and dipped in 1% Bordeaux mixture for 15 min and then kept in shade for about one week for partially losing the moisture.

The basal portion of the vine is laid on the soil surface near the standards and covered with a thin layer of soil in such a way that the cut end of the vine is bent upward above the soil surface to avoid contact with soil. The rest of the vine is then tied on to the standards so that the nodes are pressed to the standard. The plant base should be mulched with partially decomposed wood debris or leaf litter. The vines are allowed to grow on the supports to a height of 1.5–2.0 m and later coiled loosely around the branches of the supports. The plants are to be irrigated at frequent intervals, as per need. Vanilla responds well to fertilizers. Application of 50 g N, 25 g P_2O_5, and 80 g K_2O per vine in six split doses in a year at bimonthly intervals enhances the growth of vines. Alternatively, foliar application of fertilizer could be given at bimonthly intervals. A well-maintained plant would produce about 5–7 m of growth per year, and therefore the size of the nursery can be planned taking into account the annual requirement of the planting materials. On an average 1 ha of nursery would be sufficient to produce about 40,000 m of planting material annually.

Polybag Nursery

An alternative method to generate planting material when there is shortage of standard vine cuttings of 1 m is to raise the shorter cuttings with two or three nodes in polythene bags and growing it to 8–10 nodded rooted cuttings of standard size. Field establishment of polybag plants is invariably better. Nurseries should be located close to the main field for early to transportation. Black polythene bags of 20 × 20 cm size and 100–150 gauge thickness are normally used for planting stem cuttings. In order to facilitate proper drainage, five to six holes are given at the lower half of each bag. The bags are filled with potting mixture prepared with jungle topsoil, decomposed FYM, and sand in the ratio of 3:1:1. Siddagangaiah et al. (1996) observed that vermicompost and decomposed coir pith are better rooting media for vanilla.

Cuttings with three nodes are planted in polybags and tied to bamboo splits/stakes inserted in the bags for support. The length of the cuttings used has been found to have profound influence on subsequent growth. When two node cuttings were used, vine growth was only 32.6 cm in six months, against 51.8 cm with three node cuttings (Krishnakumar, 1995). The exposed soil in the polybag should be given a layer of mulch, preferably decomposed leaves to protect the soil, to retain soil moisture, and to serve as a source of plant nutrients. Partial shade should be provided either by using agro-shade nets or coir nets.

ICRI, Spices Board of India has standardized a more convenient and economical method of raising polybag nurseries. Vanilla cuttings planted in polybags are allowed to trail in coir yarn as illustrated (Figure 20.1). The fully grown nursery vines along with the coir yarn would be tied to the standard during field planting ensuring least disturbances to the poly bagged seedlings. The moisture holding capacity and rough surface of the coir yarn favored better vine growth (ICRI, 2006).

The plants are maintained with regular watering. Foliar application of 1% diammonium phosphate (DAP) can be given to growing vines two months after planting, which can be repeated at monthly intervals. Alternatively, vermiwash can be used for enhancing the growth of vines. Spraying vermiwash once in two weeks increased the length of vine to 66 cm in six months compared to 42 cm obtained in the control (ICRI, 1999). The cuttings so raised in the polybags will be ready for

FIGURE 20.1 Nursery production of vanilla on coir yarns (ICRI method).

field planting in 6–7 months, by which time they should have reached about 50–70 cm of vine length.

Supports for Vines

Vanilla, being a climbing orchid, requires standards for support and shade. A number of tree species have been recommended as suitable standards for vanilla. Apart from live standards, trellis, latticework, wooden and concrete posts, and wire or bars are used as support for vanilla vines. Wooden posts are subject to decay and to the attack of termites and hence it becomes necessary to replace them at frequent intervals. In case of wire or bar supports, the tender portion of the vines may be easily broken. When nonliving supports are used, it is always necessary to provide some form of partial shade for the vanilla plants. Hence, live standards are preferred for vanilla cultivation.

In general, the support tree species with small leaves, which permit filtered light through the foliage, are useful. Species that can easily be propagated through long stem cuttings and those that grow faster and produce branches sufficiently low (from 1.5 to 2.0 m from the ground) for the vines to hang within easy reach of the workers are found to be the most ideal. The trees should be strong enough to support the heavy growth of vines and beans and should withstand strong wind and rainfall. Trees should never entirely defoliate during summer months, but should be amenable for periodic pruning for shade regulation. It is also important for the support trees to have deep penetrating roots so that they do not compete with the shallow rooted vanilla plants for nutrients. Some of the common suitable support trees used in India are *Gliricidia maculate*, *P. alba*, *Morus* sp. *Casuarina*, Liberian coffee, Coral tree, Lebeek tree, *E. lithosperma*, *Jatropha* sp., and so forth.

Planting

Vines of 1 m length either from plantation/trench nurseries/polybag rooted cuttings are generally preferred for planting. Cuttings are planted close to the base of the support tree by laying the 3–4 basal nodes, from where leaves have been removed, on to the soil surface and gently pressing these nodes to the soil or putting sufficient soil to cover the nodes. While planting, care should be taken to ensure that the basal cut end portion of the cutting is kept just above the soil surface, as otherwise chances of decay are more. The top end of the cutting is to be tied to the base of the support tree gently so that it will eventually climb on to them. Partially decomposed organic materials such as coconut husk, mulch, straw leaves, and so on should be placed over the newly planted cutting at the base of the support tree to a thickness of 10–15 cm or more. If shade is not sufficient from the support tree, palm fronds or other leaves can be used to provide shade to cuttings.

The ideal time for planting vanilla is when the weather is neither too rainy nor too dry. Planting in August or September after the southwest monsoon is recommended under Indian conditions. It takes about 4–8 weeks for the cuttings to strike roots and to show initial signs of growth from any one of the leaf axils. Under irrigated conditions, rooted polybag cuttings could be planted any time of the year.

After Care

A vanilla plantation has to be given constant attention after its establishment. It should be frequently visited to train the vines to grow at a convenient height, prune the growing vines and tree supports, and observe diseased and pest-infected plants. Cultivation of other crops near the roots is not advisable as it may disturb the roots and soil leading to damage of vanilla plants. Periodical removal of dead vines or rotting portion of vines is to be undertaken regularly. Success of the plantation would depend on preventing any serious outbreak of diseases and creating favorable conditions for flowering and beans production.

Nutrient Management

Vanilla, being a hemiterrestrial orchid, providing right soil environment at the plant base through appropriate mulching, rather than soil fertility *per se*, is important for its successful cultivation. The best source of nutrients is the deep layer of decomposed mulch maintained over and around the vanilla roots. The nutrient-supplying capacity of mulch depends on the source and composition of the mulch.

Nutritional studies carried out at the ICRI, Spices Board has indicated that vanilla yield can be enhanced by soil application of 20:10:30 g NPK per vine per year and foliar application of urea, single super phosphate, and muriate of potash at the rate of 1.0%, 0.5%, and 1.5%, respectively, during January, May, and September. The crop responds well to foliar application, and therefore during the initial years of vanilla growth, foliar application of inorganic manures is practiced. DAP (2%) and muriate of potash (1%) or NPK complex (2%) may be sprayed on the lower surface of the leaves when adequate humidity (>60%) is available in the plantation. Foliar application of DAP and micronutrients, particularly Zn and B boost the development of pods. Studies on the effect of nutrient uptake through aerial roots of vanilla revealed that Knop's solution (25%) showed improvement in the growth of vanilla (ICRI, 2003).

The crop is highly amenable to organic cultivation. In addition to chemical fertilizers, various organic manures are applied to vanilla. Decomposed organic matter, bone meal, FYM, vermicompost, and fermented deoiled cakes are applied at least twice in a year, that is, during June–July and September–October. Foliar application of vermin wash (1:5 dilutions) favored the growth and development. Application of *Vyrsha ayurvedic* preparations such as panchakavya, fish amino extract, and egg amino extract is also being widely practiced under organic vanilla cultivation.

Mulching

Mulching with green or dry leaves is to be carried out at least twice in a year. They not only provide food in the form of potassium-rich organic matter but also conserve moisture and hold the layer of mulch in position, creating a deep and rich root zone that vanilla favors. In studies on the growth performance of vanilla with various mulch materials, coir pith favored relatively better vine elongation compared to traditionally used wood debris mulch. The advantage of coir pith is attributed to better moisture conservation, maintenance of conducive microclimate, and supply of micro- and macronutrients for vanilla growth (ICRI, 2005).

IRRIGATION

Vanilla requires a moist climate with frequent but not excessive rains. Under excessive rainfall, there is widespread occurrence of rot diseases. Under prolonged drought conditions, the plants may suffer from physiological damage and the vines may not recover. In extremely dry conditions, irrigation should be provided at least once in 4–5 days. Mulching with mango leaves (25 kg per vine per year) and watering once in four days was found to increase vanilla yield (Muraleedharan, 1975). Depending on the area under cultivation, sprinkler or hose irrigation may be practiced. Microsprinkler/fog irrigation is advocated to maintain relatively high humidity levels in the plantations to ensure high percentage of bean sets. Care should be taken to avoid water stagnation around the plant base or excessive water stress at any point of vine or pod development.

SHADE MANAGEMENT

Careful lopping of branches of the support tree is very important to give shade to vanilla plants. Newport (1910) stated that checkered shade rather than dense shade is to be preferred. Vanilla vines require more sunlight than shade during flowering and at the time when the beans are maturing.

It is the farmers experience that the existing shade trees should be pruned to admit sunlight as uniformly as possible, 30–50%. It may be desirable to regulate shade allowing more light during the blossoming period. Allowing more light is helpful to check vegetative growth and favor flowering. Heavy shade should be avoided because the stem will become thin, leaves small, and flowering and fruiting generally reduced. Under too much sunlight, on the contrary, the leaves not only get sunscald and turn yellow but the plants become weak during a drought period and more susceptible to root rot diseases. The amount of light vanilla can tolerate depends mainly on the water supply to the roots as influenced by the atmospheric humidity and irrigation practices.

Under high rainfall and high relative humidity, vanilla can withstand more sunshine than during low humidity and drought periods. Thus, it is important that support trees maintain much of their foliage during dry periods.

TRAILING AND PRUNING

The trailing nature of vanilla vines has an effect on flowering. The vines are coiled around the lower branches of the supporting tree or over the lattice of trellis so that they may hang down. Care is required not to tear or bruise the leaves, branches, or roots. Bending of vines appears to be an important operation for inducing flowering and fruiting beyond the bend, which may be due to accumulation of carbohydrates and possibly other flower-inducing hormones in such regions of the vine.

Arresting the vegetative growth of vanilla vine by pruning helps to induce flowering. Training and pruning are undertaken to induce flowering and bean production. Here, the hanging shoots of 1–1.2 m long are bent down around the branches of the support tree, slightly twisted in the process, with tips pruned at about 45 cm from the soil. Any vegetative shoots appearing after the bend portion of the hanging shoot are removed, but the shoots appearing on the rest of the plant before the bend portion of the hanging shoots are allowed to grow. These shoots will constitute the bearing branches of the following year. As a result, there is a decreased sap flow toward the bearing branches, which favors flower formation. After the harvesting of beans, the yielded portion of the vine is removed from the plant. The following year's bearing branches are to be prepared by bending and pruning. Plants that have been pruned and readied for flowering require heavy manuring of leaf mold, decomposed leaves, lime, ashes, and manure. The vine architecture after 3–4 years is therefore of vine with a number of shoots hanging down over the branches of the live support, without overcrowding or overlapping. Normally, five to six bearing branches per plant can be prepared per annum, depending on the age and growth of the vine.

The yielding behavior of pruned hanging shoots under different growth stages comprising of 4, 6, 8, 10, and 12 nodes per shoot was studied. It revealed that the formation of inflorescence is irregular in the hanging shoots irrespective of its growth stage. The 10-node hanging shoots had maximum number of inflorescences per shoot. However, the first-grade quality of beans (above 15 cm) was 57% in 10-node shoots, 77% in 6-node, and 88% in 4-node hanging shoots, respectively, indicating that production of more number of beans in a bunch or in a productive vine would be at the expense of size of the beans (Hrideek et al., 2003).

When the pods are ripe and harvest is completed, the yielded portion of the vine is removed and only the new shoots of the previous year are retained. Plants that have been pruned and readied for flowering require heavy manuring of leaf molds, rotting leaves, lime, ashes, and cow dung.

DISEASES AND THEIR MANAGEMENT

Fungal and viral diseases are the main production constraints. The crop losses due to these biotic stresses vary. Fungal diseases such as *Fusarium* and *Phytophthora* rots cause severe crop losses, and occasionally total crop losses are reported. The viral diseases, although not lethal, result in varying degrees of production and productivity losses. Vanilla being an exotic plant for India, the planting material might have come from outside unknown sources where the virus-free nature of the material was never ascertained and quarantine regulation was never applied. Lack of adequate molecular diagnostic facilities, especially for viral diseases, is another major constrain that limits the detection of viruses. This is important in view of the vegetative propagation method of cultivation. The genetic variability with respect to disease/pest resistance is comparatively low, which might be the reasons for high disease incidence and consequent crop loss.

Except for the reporting of new diseases, detailed investigations on the management of various diseases are insufficient. Integrated pest and disease management remains an important priority for crop protection of vanilla, with a greater focus on biological control to reduce pesticide inputs for vanilla cultivation.

TABLE 20.3
Diseases of Vanilla in India

Name	Causal Agent	References
	1. Fungal	
Major:		
1. Stem, root rot, and wilt	*F. oxysporum* sp. *vanillae*	Philip (1980), Joseph Thomas et al. (2003)
2. *Phytophthora* leaf, stem blight, and bean rot	*P. meadii*	Bhai and Thomas (2000)
Minor:		
3. Immature bean yellowing and shedding	*C. vanillae*	Thomas et al. (2003)
4. Bean rot	*Sclerotium rolfsii*	Thomas and Bhai (2000)
5. Brown spot	*C. quinquiseptatum*	Bhai et al. (2006)
6. Shoot tip rot	*C. gloesporides*	Thomas et al. (2003)
	2. Viral	
1. Mosaic	CMV—Cucumovirus	Madhubala et al. (2005)
2. Mosaic and stem necrosis	Potexvirus, potyvirus, and closterovirus	Bhat et al. (2004)

Vanilla being a crop with poor genetic variability for disease resistance, efforts should be made to induce more variation to identify reasonable disease resistance or tolerance for the major diseases. The recent report of variable disease reaction in seedling progenies of vanilla and somaclonal variations is of considerable interest and needs to be pursued further (Minoo et al., 2006b).

The main diseases reported on vanilla in India are given in Table 20.3.

Fungal Diseases

Stem and Root Rot

This is one of the most destructive diseases observed in all the vanilla-growing countries and is severe in India too. It was first reported in India in 1980 (Philip, 1980). Later it was shown to be caused by *Fusarium oxysporum* sp. *vanillae* (Thomas et al., 2002).

Symptomatology and Etiology

The disease is observed during warm humid conditions. The fungus colonizes the root and causes yellowing of the vines with degeneration of the root system. It also causes dark brown rot on the stem as well as on the leaves (Figure 20.2). The rotting starts at any part of the plant. All parts of the plant are prone to infection, but more often at the nodal region. Later it spreads upward causing rotting, resulting in drying up of the distal portion beyond the point of infection. Severely affected gardens

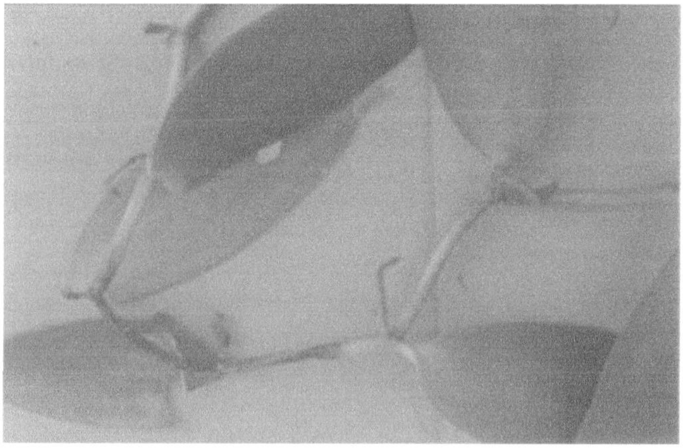

FIGURE 20.2 *Fusarium* stem and leaf infection.

appear dry from distance, typical of wilting. The affected stem exhibits vascular browning, typical of fusarial wilts (Y.R. Sarma, unpubl. data).

Epidemiology

Although detailed epidemiological investigations have not been carried out, ambient temperature around 25–28°C favored the disease incidence and spread. The disease occurs individually and in certain cases overlaps with *Phytophthora* rot. Hence, mixed infections are common. Abundant sporulation is observed in the affected tissues. Disease spreads through rain splashes within the plant and across the vines. Detailed investigation on the disease progression curves in relation to environmental condition is called for. The fungus survives on dead dried-up debris and the dead and the dried-up vines left over from gardens are the perennial source of inoculum.

Disease Management

Since the available cultivars are all highly susceptible and as such, no disease resistance exists. Recent report of variability of seedling progenies of *V. planifolia* to stem rot is important and needs intensive field evaluation (Minoo et al., 2008b).

Cultural Practice

A regular monitoring of the disease is essential for early identification of the disease to take corrective measures. Phytosanitation involving collection and removal of infected vines from the garden is essential to reduce the pathogen inoculum in the soil.

Chemical Control

Phytosanitation followed by spray and soil drenching with 0.2% carbendazim periodically, depending on the intensity of the disease, has been suggested (Thomas et al., 2002).

Phytophthora Leaf, Stem Blight, and Bean Rot

The disease was first recorded in vanilla plantation of Manalaroo Estate in Nelliampathi and Arnakal Estate at Vandiperiyar during 1998 and also in a stray vine at Directorate of Arecanut and Spices Development Office, Calicut (Y.R. Sarma, unpubl. data). It was later observed in Idukki and Kottayam districts of Kerala, specially in Moovattupuzha and Koothattukulam areas (Bhai and Thomas, 2000).

Symptomatology and Etiology

The disease incidence coincides with the southwest monsoon in Kerala. The disease occurs as water-soaked spots. Observations at Manalaroo and Arnakal Estates showed that *Phytophthora* infection occurs on leaves, stems, and beans causing rotting (Figures 20.3 and 20.4). All parts of the vine are prone to infection. On leaves, the water-soaked spot enlarges up to 0.5–1.5 cm with a translucent advancing margin. On stems, it is typically a wet rot unlike that of fusarial rot, and affected portions become soft and rot completely. In vanilla fruit bunches, infection starts at any point of beans either from the tip end or from the stalk region. When the infection starts at the tip, it gradually spreads upward resulting in varying degrees of rotting. If infection occurs at the stalk or peduncle region, either the infected bean drops off or dries resulting in beans hanging to the peduncle region (Y.R. Sarma, unpubl. data).

The causal agent was identified as *Phytophthora meadii*. Under artificial inoculation, infection was recorded 5–8 days after inoculation. The fungus on carrot agar medium appears cottony white. The sporangial size of 39.6×20 μm with an LB ratio of 1.98 was recorded. The cultures were heterothallic. Also, intercalary chlamydospores in the mycelium were observed (Bhai and Thomas, 2000).

Epidemiology

No detailed epidemiological investigations have been conducted. However, the disease coincides with the southwest monsoon with continuous rainfall of 15–20 days

FIGURE 20.3 *Phytophthora* stem and leaf infection.

FIGURE 20.4 *Phytophthora* bean rot.

in a month with intermittent showers during the July–August period of the year. On the infected beans, fungal growth occurs, which harbors abundant sporangia. Disease spreads through rain splashes both in the vine and across the vines.

In Kerala and Karnataka, *P. meadii* has been reported on rubber, small cardamom, and arecanut. It is important to investigate their pathogenic relationship at molecular level and the possibilities of crossing and development of new races/biotypes of *P. meadii*.

Disease Management

Cultural practices: As mentioned earlier in the case of fusarial rot/wilts, phytosanitation measures consisting of systematic collection and pulling out weeds of infected and dried up vines is important.

Chemical control: Spraying of the vines with 1% Bordeaux mixture and soil drenching with copper-oxy chloride (0.25%) was reported effective (Bhai and Thomas, 2000). Prophylactic spraying of the vines with 4–5% of potassium phosphonate (Phytophos 40), a systemic chemical, at 15 days intervals did control the disease but if the wet spells continued, the disease control was poor (Y.R. Sarma, unpubl. data).

Bean Rot

This disease, caused by *Sclerotium rolfsii*, was reported from the Ramamangalam area of Kerala in a survey carried out during 1999 (Thomas and Bhai, 2000).

Symptomatology and Etiology

Rotting of either one or two beans or all the beans in a bunch was observed. Whitish mycelium spreads on the bean in a fan-like fashion. The infected beans showed rotting symptoms with deep sunken areas that are reddish brown. On leaves and stems, the fungus spreads whitish thick mycelial threads and later produces cream-colored to brownish grain-like sclerotial bodies (Y.R. Sarma, unpubl. data).

Brown Spot

This is a minor disease observed in Chempukadavu area of Kozhikode district of Kerala in a mixed cropping system of coconut, arecanut, clove, and vanilla. The disease is caused by *Cylindrocladium quinquiseptatum*. It starts as water-soaked spots on beans, which later enlarge into a sunken spot of 1–10 mm. The lesions coalesce forming reddish-brown sunken lesions resembling anthracnose. Either a single bean or all beans in a bunch get affected. Leaf infection with sunken lesion was also observed (Bhai et al., 2006).

Yellowing and Premature Bean Shedding (YPB)

This is another minor disease of vanilla observed in Kozhikode district of Kerala, caused by *Colletotrichum vanillae*. The disease is characterized by premature yellowing of the beans with dark-brown sunken lesions that gradually expand resulting in rotting of the beans (Figure 20.5). This leads to premature bean shedding (Bhai et al., 2006). In a survey during April–May, 2003, premature bean shedding was recorded. It was reported that high incidence was due to high temperature and low relative humidity (Bhai and Dhanesh, 2008).

Shoot Tip Rot

The disease is generally observed in vanilla plantations during the postmonsoon period of September–December. It starts as a brown patch on the petiole and lower portion of the unfolding youngest leaf. The funnel-like leaf holds rainwater and might predispose to infection. Later infection spreads downward to the tender portion of the shoot causing shoot tip rot. *F. oxysporum* isolated from the affected shoots was found pathogenic. Association of *Colletotrichum gloesporide* was also observed. Spraying the foliage with 0.2% carbendazim was found effective (Thomas et al., 2002, 2003).

Others

Rhizoctonia solani and *Mucor racemosus* have also been recorded as pathogens of vanilla (Bhai and Dhanesh, 2008), but their relevance need to be investigated further.

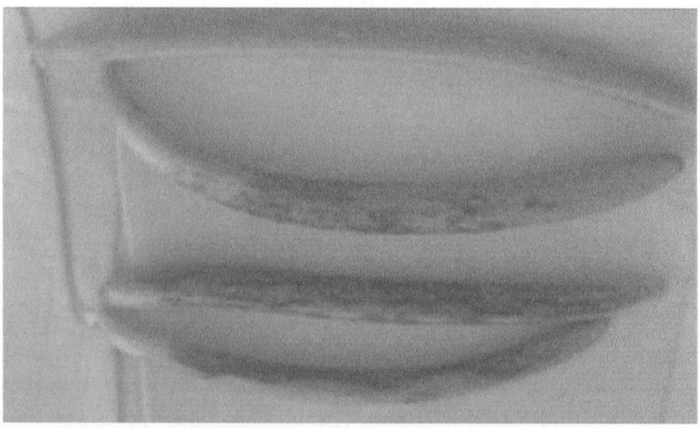

FIGURE 20.5 *Colletotrichum* bean rot.

Biological Control for the Management of Fungal Diseases of Vanilla as a Part of Integrated Disease Management (IDM)

The effectiveness of biocontrol agents for the management of soil-borne plant pathogens has been well established in recent times. Its importance was clearly brought in the disease management of crop spices (Sarma, 2006a, 2006b). Reduction of pathogen inoculum in vanilla plantations by applying *Trichoderma viride, Trichoderma harzianum*, and *Pseudomonas fluorescens* was reported earlier (Thomas et al., 2002). Both *Fusarium* and *Phytophthora* overlap in the field and spatial segregation of these pathogens in a plantation under a set of ecological conditions is not possible. It is imperative and logical to identify biocontrol agents that are suppressive to both these pathogens of vanilla. Studies carried out with different isolates of *P. fluorescens* and *Bacillus* sp. both in *in vitro* and *in vivo* conditions showed the bioefficacy of certain bacterial isolates to suppress *P. meadii* and *F. oxysporum in vitro*. They also provided growth promotion in vanilla. The suppression of fusarial wilt in large-scale field establishment with local strains of *T. harzianum* in Mauritius has been established (Y.R. Sarma, unpubl. data).

Viral Diseases

Importance of vanilla viruses at global level received considerable attention (Wisler et al., 1987; Pearson and Pone, 1988; Pearson, 1990; Benezet et al., 2000; Grisoni et al., 2004). Identification of virus problems of vanilla in India is recent. It was only during 2003 that incidence of viral disease was recognized in the states of Kerala and Karnataka (Bhai et al., 2003; Sudharshan et al., 2003). Systematic investigation on vanilla virus diseases started recently at IISR, Calicut, on identification, characterization, and molecular diagnostics (Bhat et al., 2004; Bhadramurthy, 2008).

Symptomatology and Etiology

Electron microscopic studies revealed the presence of three types of flexuous particles resembling Potexvirus, Potyvirus, and Closterovirus and an isometric particle (Bhat et al., 2004). Investigation on virus incidence in vanilla plots carried out recently revealed about 3–10% incidence in Karnataka and 0.13–5% in Kerala (Bhadramurthy, 2008). These studies revealed that cucumber mosaic virus (CMV, *Cucumovirus*), cymbidium mosaic virus (CymMV, *Potexvirus*), bean common mosaic virus (BCMV, *Potyvirus*), and bean yellow mosaic virus (BYMV, *Potyvirus*) are infecting vanilla in India and mixed infection have been observed. The first authentic investigation on vanilla CMV strains in India was undertaken by Madhubala et al. (2005). The affected vines appear with distorted foliage with small and leathery leaves (Figure 20.6). In artificial inoculation, the members of Chenopodiaceae, Cucurbitaceae, and Fabaciaceae were found infective with CMV strains of vanilla. Isometric virus particles with 28 μm diameter were seen based on the detailed electronic microscopic investigation. Further studies confirmed that vanilla CMV strains belong to subgroup I of CMV. Mild chlorotic mottling and streaks on leaves are the characteristic symptoms of CymMV (Figure 20.7). The investigation revealed that Indian CymMV isolates are closely related to vanilla CymMV isolates from French

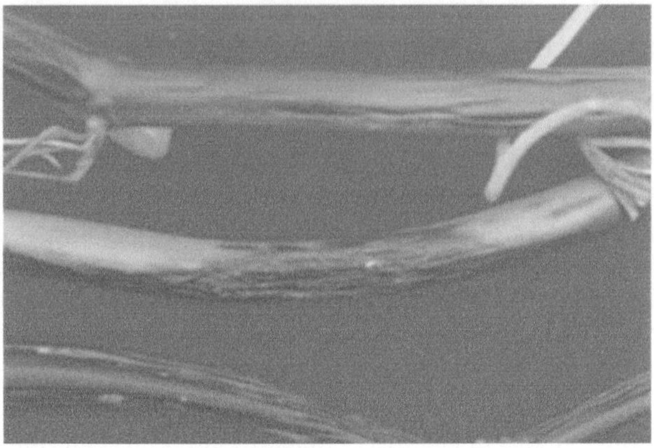

FIGURE 20.6 Stem necrosis caused by cucumber mosaic virus.

Polynesia (Bhat et al., 2004). Studies on the yield losses and the epidemiology are lacking and need to be intensified to develop effective disease management strategies.

Virus Control

Vanilla vines may be symptomless carriers of viruses. The present status of multiplication and new planting would continue as long as yield reduction is not observed. From the practical view point one can go for replanting when the virus infection exhibits declining phase in the yield. It is desirable to build up information on the yield reduction in relation to virus infection to ascertain and identify the exact declining phase with respect to yield. This would guide the management strategy to know when to resort to replanting.

Since these are all sap transmissible viruses, farm operations such as pruning, looping, or pollinating have to be regulated to avoid undue virus transmission leading to disease spread.

FIGURE 20.7 Leaf mosaic caused by Cymbidium mosaic virus.

Elimination of viruses from the planting material through meristem tip culture, production, and distribution of healthy planting material would be a priority for sustainable vanilla production in future.

PESTS AND THEIR MANAGEMENT

In India, pest problems are generally minor on vanilla; however, several insects are recorded to damage vines, shoot tip, flower buds, or roots, by Hemipteran bugs, Lepidopteran caterpillars, and Coleopteran weevils.

Hemipteran Pests

Vanilla Bug

Halyomorpha picus (Pentatomidae) (Figure 20.8) causes serious damage by sucking the sap from the shoot tip and inflorescence. Subsequently, such affected areas become necrotic and rot. Nymphs and adults suck the sap from the peduncle and flower buds. Pin-prick-like punctures at the site of feeding and subsequent necrosis and rotting are the typical symptoms of bug feeding. The affected vegetative buds drop within 3–5 days and the affected inflorescence become rotten. Incidence of the pest is higher during the inflorescence initiation period, that is, January–February.

The bug is reported to occur in Karnataka, particularly where vanilla is intercropped with arecanut plantations. It is also reported from Koothattukulam in Kottayam district of Kerala.

Female bugs lay spherical eggs in clusters on the lower surface of leaf. The eggs are white when laid and turn cream within 3–4 days. Nymphs hatch in 5–6 days. First-instar nymphs are gregarious and remain on top of the egg shells and do not feed. The second- to fifth-instar nymphs are black and actively feed on the shoot tip.

FIGURE 20.8 Vanilla bug *H. picus* (a) shoot tip necrosis, (b) nymph, and (c) adult.

The nymphal duration lasts for about 60 days. The pest causes about 40% damage by way of feeding on the inflorescence (Prakash and Sudharshan, 2002).

Management of *H. picus* includes monitoring of the bug during November–February and removal of egg mass and first-instar nymphs that are seen on the lower surface of leaves. Spray of Monocrotophos at 0.1% a.i. controls the nymphs, if the infestation is high.

Southern Green Stink Bug

Nezara viridula (Pentatomidae) occurs throughout the tropics; the bug lays eggs on leaves and stalks; the nymphs suck the sap of flower buds and stalks. The pest is reported to occur on vanilla in Karnataka and Kerala but its incidence is very low (Varadarasan et al., 2002a).

Riptortus pedestris (Alydidae)

The bug was reported to occur on vanilla in Kerala and Karnataka as a minor pest. The nymph and adult suck the sap from beans and leaves (Varadarasan et al., 2002b).

Scale Insects (Coccidae) (Unidentified)

This insect found in Idukki district, Kerala, sucks the sap from the leaves, vines, and inflorescence. Ants are found to be associated with this sucking insect and the pest incidence occurs during January–February (Varadarasan et al., 2002b). Control measures with Monocrotophos at 0.1% a.i. is warranted only if the pest is serious.

Coleopoteran Pests

Three species of *Coleopteran* pests are recorded in vanilla. Among them a weevil, *Sipalus* sp. occurring in Idukki district of Kerala, causes serious damage to young shoots, vines, and leaves.

Vanilla Vine Weevil, *Sipalus* sp. (Dryothoridae)

The weevil lays eggs singly along the length of the vine and the emerging grubs feed on the inner core of the stem by boring a tunnel, the entire length of vine shreds, rots, and falls down (Varadarasan et al., 2002a).

The adult weevil is 8–10 mm long, 2–3 mm wide, and is light to dark black with two wavy white cross bands on the elytra. They feign death upon approach or touch (Thanatosis). The adult female is larger than the male. Adults mainly feed on shoots by inserting the snout; the injured area becomes necrotic within a day, leading to rotting of the shoot tip/vine. The weevil also feeds on leaves by scrapping upper or lower epidermis with mesophyll tissue, leaving a thin transparent epidermis on the lower or upper surface of leaves (Figure 20.9).

After copulation, females lay eggs singly, 2–4 mm below the epidermis on the vine or shoot tip. The egg is capsule like 5–6 mm long, 0.8–1.0 mm wide, white and subsequently turns to yellow before hatching. The site of egg deposition develops necrosis. Eggs are laid only on tender vines and not on the leaves. The emerging grubs feed on the necrotic tissue by making tunnels in the vine.

FIGURE 20.9 Vanilla vine weevil, *Sipalus* sp. (Left: egg deposited with necrosis; right: grub tunneling the vine).

The first-instar grub is yellow and the final-instar grub is white with brown head capsule. The grub (larva) period lasts for 35–40 days. The grubs tunnel the vine by feeding and the entire length of the affected vine becomes necrotic. The mature grub pupates inside the tunnel with fibrous material; pupation lasts for 19–21 days. The adult weevil emerges from the vine by making a small slit in the dried vine.

Extensive damage by the grubs and the adult weevil on the vine is observed mostly in open areas where shade is less. The adult weevils are seen in the field during November–January, and can easily be located. The weevils are not very active, and, hence, they may be hand picked and destroyed to reduce the damage on the crop (Varadarasan et al., 2002a).

Saula ferriginea Gerst (Endomychidae)

This beetle, found in Wynad and Nilgiris in India, cuts through the leaf from the lower surface, eating the entire leaf tissue, except for the thick translucent cuticle of the upper epidermis. The damaged leaves rot as a result of fungi infection. The pest may be managed by keeping the garden weed-free and in severe cases by application of Malathion 0.1% (Rai and Nayar, 1976).

White Grub

Holotrichia (Scarabaeidae): Roots of vanilla are damaged by white grubs in Idukki and Thiruvananthapuram districts of Kerala. Drenching Chlorpyriphos 0.05% a.i at the plant base during May or application of entomopathogenic nematodes controls the pest.

Lepidopteran Pests

Webber Caterpillar *Archips* sp. (Noctuidae)

The caterpillar of the moth occurring in Idukki, Kottayam, Ernakulam, and Thrissur, districts of Kerala, feeds on the vegetative shoots; these pale green caterpillars are seen in between the shoot bud and the first leaf, forming a web and feeding the shoot that leads to rotting of the terminal bud (Figure 20.10). Incidence of the caterpillar

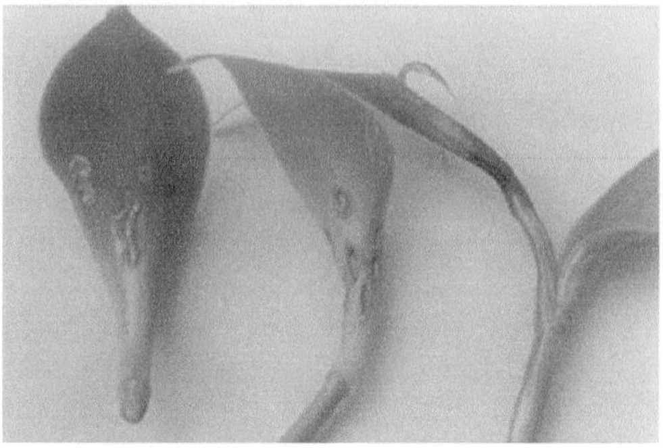

FIGURE 20.10 Shoot tip damage by webber caterpillar, *Archips* sp.

was recorded in January and February. The adult moth is reddish brown with a broad bright yellow band across forewings, gray hind wings, and body. The pest may be controlled by spraying Monocrotophos at 0.1% a.i. if the incidence is high (ICRI, 1996).

Semilooper, *Nemoria* sp. (Noctuidae)

The yellowish ash-colored caterpillars with brown heads are very agile and damage the inflorescence and flower buds by feeding (Figure 20.11). The late instar caterpillar is light brown and pupates in between the damaged inflorescence. The pest was recorded in Chemannar in Idukki district of Kerala during February–March. Monocrotophos 0.075% or Lambda cyhalothrin 0.004% control the pest efficiently.

Noninsect Pests

Among these, mites, snails, and avian pests cause considerable damage.

Mite, *Tyrophagus* sp. (Acarina)

This mite is found to infest beans in storage in India (Sasikumar et al., 1992).

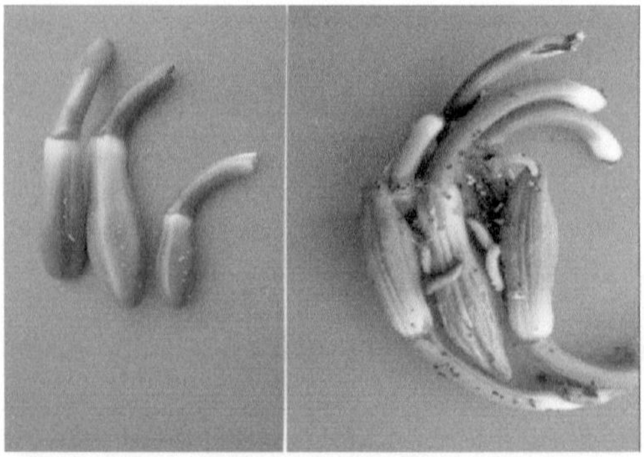

FIGURE 20.11 Left: Egg deposit on flower bud, right: damage on a flower bud by semilooper, *Nemoria* sp.

Molluscan Pests

Snail: Achatina sp: They were found to feed on the chlorophyll tissues from the base of the vine and were recorded in Nagercoil in Tamilnadu, where coconut fronds were used for mulching on the plant base. The snails hide in the mulch during daytime, and come out in the night to feed on vanilla. The snails are also found to feed on leaves (Figure 20.12). The management of this pest includes avoiding mulching with coconut fronds or any other planting materials that do not decompose fast. The snails may be collected from the mulch and destroyed (Varadarasan et al., 2003b).

Avian Pests

Chicken cause much damage by scratching among the mulch and thus exposing and damaging the roots.

Vanilla Scab

Longitudinal corky formations on the beans of vanilla have been observed, which are commonly referred as "vanilla scabs." It was suspected to be caused by insects or fungi or due to abiotic factors such as dew drops or higher light intensity. Experiments conducted at the ICRI, Kerala, indicated that scab formation in vanilla is not due to insect or fungus or due to any other abiotic factors. It was found that the mechanical injury at the early stage of the beans caused scab. Close observation on beans, leaves, and internodes showed that wherever the early stage of plant parts come in contact with hard objects, such as bamboo or dry twig or even mature leaf lamina, scab formation was observed on the point of contact. Subsequently, experiments with nail injury on the ovary during pollination show scabs on the beans. Any minute injury on the early stages of the ovary is stretched longitudinally with the elongation of beans, and thus scabs are formed (Varadarasan et al., 2005).

HARVESTING AND PROCESSING

Fresh vanilla beans are dark green in color and do not impart any aroma because vanillin and other chemical substance are not available in free form at the time of

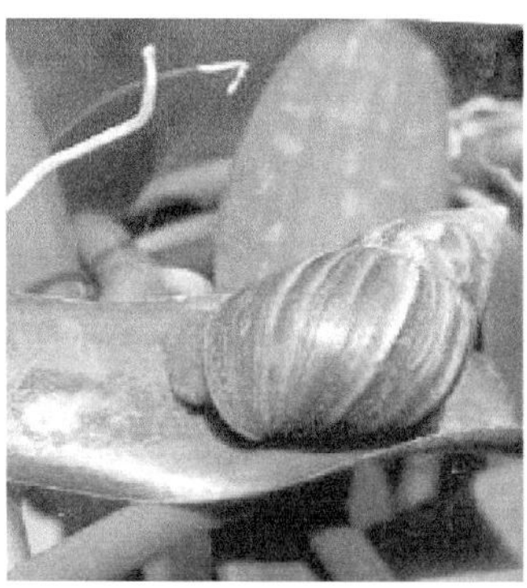

FIGURE 20.12 Damage on leaves by snail, *Achatina* sp.

harvest. The beans are ready for harvest when they attain a mild yellow color at the distal end, and it may take 8–10 months to reach the harvesting stage depending on the conditions. Full ripening of the beans leads to splitting and ultimately affects their quality. The beans are to be harvested at the right time, as the immature ones produce inferior commodity. The nonsynchronized flowering and fruiting behavior of the crop leads to selective harvest extending to 1–2 months. Some of the important factors that determine the vanillin content and beans quality are climatic conditions, stage of harvest, and extent of sweating of the pods during curing. During the process of curing, free vanillin is developed in the beans as a result of a series of enzymatic reactions that provide fragrance.

Many processing methods are followed in vanilla-growing countries. However, the Bourbon method of vanilla processing practiced in Madagascar is followed in India, and modified and standardized to suit local conditions. This method is simple and consists of four stages (killing, sweating, slow drying, and conditioning, see Chapter 11) (Krishnakumar et al., 2007, 2008).

Packing

In the final stage of curing, beans are packed in wax paper or any appropriate material and stored in closed boxes for a period of three months or longer to permit the full development of the desired aroma and flavor. Evaluation of different packing methods and methods for keeping up the quality of beans during storage indicated that the beans stored in polyethylene bottle, glass tubes, polypropylene covers (0.011 and 0.035 mm), acrylic boxes, and waxed paper 0.07 mm plus tin were on par with respect to vanillin content and retention of moisture (ICRI, 2006).

According to Purseglove et al. (1981), the primary quality requirement for cured vanilla beans is the aroma/flavor character. The other traits signifying the quality are general appearance, flexibility, size of beans, and vanillin content. Superior quality beans are long, fleshy, supple in nature, very dark brown to black in color, somewhat oily in appearance, strongly aromatic, and free from scars and blemishes.

The quality standards used in India are as per ISO specifications and there are no separate standards specified by Bureau of Indian Standards (BIS). The local traders/processors have set their own norms based on size of the beans: supergrade beans have a bean size of more than 20 cm, clean without any blemishes; A-grade beans (16–20 cm); B-grade beans (12–16 cm); C-grade (8–12 cm); and low-grade beans comprised of shorter beans, beans with splits, cuts, and wrinkles, and beans with scars/blemishes. In general, from a healthy plantation, 20–40% of beans produced would be of A grade including supergrade; 30–50% B grade, and the remaining C grade or rejects. In order to attain larger proportion of A-grade beans, farmers restrict the pollination to 10 flowers per bunch and 10–12 bunches per vine. The approximate composition of whole vanilla beans is moisture: 25–30%; protein: 2.56–4.87%; fatty oil: 4.69–6.74%; volatile oil: 0.0–0.64%; nitrogen-free extract: 30.35–32.90%; carbohydrates: 7.1–9.1%; fiber: 15.27–19.6%; ash: 4.5–4.7%; vanillin: 1.48–2.90%; resins: 1.5–2.6%; calcium: 19.7 mg%; potassium: 16.2 mg%; sodium: 6.7 mg%; phosphorus: 9.5 mg%; and iron: 0.3 mg%.

Analysis of Indian vanilla beans carried out inside and outside the country have shown that the vanillin content was invariably above 2.5% with an aroma and flavor comparable to Madagascar vanilla.

YIELD AND ECONOMICS OF CULTIVATION

The yield performance of vanilla varies depending on the age and method of cultivation. Although vanilla starts flowering from the third year, the yield is harvested in the fourth year. The yield increases till seventh or eighth year and thereafter declines. Under the moderate management, the yield range of a middle-aged plantation will be around 400–500 kg of cured beans per hectare.

Using biometric characters such as vine length (cm), number of leaves, length of yielding area in the vine (cm), length of nonyielding area in the vine (cm), internodal length (cm), vine girth (cm), leaf area (cm^2), number of inflorescence, number of beans/inflorescence, bean length (cm), and number of beans/vines, a truncated model yield forecast was developed with a precision of about 93% (Priya et al., 2002).

Vanilla in India is grown largely as an intercrop in plantations and homestead and, hence, the cost of production remains competitive. The cost of production under normal farming situation varies from place to place within the country. One hectare with 1600 vanilla plants needs Rs. 53,000 during the first year of establishment and an average Rs. 34,000 for the second and third year of planting. Thus, the total cost of establishment in 1 ha area is around Rs. 120,000. The cost of maintenance would be around Rs. 37,000/annum. The expected yield of cured beans is around 400–500 kg/ha.

RESEARCH AND DEVELOPMENT NEEDS

Since the crop is a recent commercial venture with intensive cultivation under varied cropping systems in arecanut/coconut/coffee plantations, new pest and disease outbreaks are posing challenges. Production and postharvest problems need attention to suit the local needs. A major research effort is made by ICRI, Myladumpara, in Idukki district of Kerala, Spices Board. In view of large-scale cultivation both in Kerala and Karnataka, this institute and its regional station at Saklespur, Karnataka are concentrating on the development of good agricultural practices (GAPs), pest and disease management, postharvest, and storage. Some of the recent processing technologies (Krishnakumar et al., 2007) and integrated pest and disease management (IPM/IDM) initiatives (Thomas et al., 2002; Varadarasan et al., 2003) are logical strategies that are helping the farming community.

Healthy planting material production is a major problem and micropropagation technologies have been developed by the University of Calicut, IISR, Calicut, Kerala Agricultural University, and ICRI (Philip and Nainar, 1986; Rao et al., 1992a; Mary Mathew et al., 1999; Chitra et al., 2007).

Distribution of tissue-cultured plants and their field evaluation were taken up by Spices Board in collaboration with the Department of Biotechnology, Govt. of India, New Delhi.

The efforts to propagate and popularize vanilla have resulted in large numbers of farmers taking to vanilla cultivation, particularly in Karnataka, Kerala, and

Tamilnadu. Research and developmental activities of the Spices Board have given the required impetus and boost in promoting vanilla cultivation. It is estimated that about 3500 ha under vanilla will yield by 2009–2010.

Since virus problems are rampant in vanilla and are vertically transmitted through planting material, the diagnostic techniques for virus detection are essential. The studies carried out on virus diagnostics at IISR are commendable and can be utilized by the vanilla nurseries (Bhadramurthy, 2008).

In order to promote vanilla, the Spices Board undertook several developmental programs, which include vanilla new planting program, vanilla certified nursery scheme, scheme for setting up of processing units, vanilla production award scheme, and scheme for assisting producers for promoting exports of organic spices (including vanilla). These research and developments should continue to sustain vanilla cultivation and production in India.

CONCLUSIONS

In general, vanilla is a crop of small and marginal farmers, grown largely as intercrops and on homesteads, except in a few plantations in the corporate sector. The spurt of price for vanilla during 2001–2002, probably because of crop loss in Madagascar due to cyclone damage during 2000, attracted the small farmers to invest heavily in vanilla cultivation. This led to considerable increase in area expansion, but with extreme substandard planting material. This had its serious repercussion on severe disease outbreaks and consequent crop loss, which was further complicated by violent price fluctuations of vanilla at the global level, offering a poor price structure. Remunerative price ultimately would determine the farmers' long-term interest for large-scale vanilla cultivation. Considering the present subdued global demand, it is likely that the present status quo of vanilla cultivation would be maintained in India. However, the Indian vanilla farmers have the resilience to scientifically cultivate, process, and supply beans of required quality if the right price is offered at the farm gate.

REFERENCES

Anonymous. 1992. *Report of the National Committee on Vanilla*. Spices Board, Kochi, India.

Ayyappan, P. 1990. Vanilla a money making venture for ambitious farmers. *Kisan World* 7:24–26.

Benezet, H., E. Picard, B. Côme, M. Grisoni, F. Leclerq Le Quillec, O. Gambin, and E. Jeuffrault. 2000. Les virus du vanillier à la Réunion. *Phytoma, la défense des végétaux* 526:40–42.

Bhadramurthy, V. 2008. Identification, molecular characterization and development of diagnostics for the virus allocated with Vanilla (*Vanilla planifolia* Andrews). PhD thesis, Mangalore University.

Bhai, R.S. and J. Thomas. 2000. Phytophthora rot—A new disease of vanilla (*Vanilla planifolia* Andrews) in India. *J. Spices Aromatic Crops* 9 (1):73–75.

Bhai, R.S., J. Thomas, S.N. Potty, L. Geethat, and J.J. Solomon. 2003. Mosaic disease of vanilla (*Vanilla planifolia* Andrews)—The first report from India. *J. Spices Aromatics Crops* 12 (1):80–82.

Bhai, R.S., A. Ishwara Bhat, and M. Anandaraj. 2006. Yellowing and premature bean dropping in vanilla (*Vanilla planifolia* Andrews). *J. Plant. Crops* 34 (2):115–117.

Bhai, R.S. and J. Dhanesh. 2008. Occcurence of fungal diseases in vanilla (*Vanilla planifolia* Andrews) in Kerala. *Ind. J. Spices Aromatic Crops* 17:140–145.

Bhat, A.I., M.N. Venugopal, R.P. Pant, and R. Suseela Bhai. 2004. Occurrence and distribution of viral diseases on vanilla (*Vanilla planifolia* Andrews) in India. *J. Spices Aromatic Crops* 13 (2):143–148.

Bhat, S.K. and M.R. Sudharshan. 1998. Floral biology and fruit growth in vanilla (*V. planifolia* Andrews). In: M. Muraleedharan and R. Rajkumar eds. *Recent Advances in Plantation Crops Research*. Allied Publishers, Coimbatore, Tamilnadu, India, 110–112.

Bhat, S.K. and M.R. Sudharshan. 2000. Effect of pollen load on growth and development of vanilla (*V. planifolia* Andrews) fruits. In: *Proceedings of the Centennial Conference on Spices and Aromatic Crops*. Indian Society for Spices, Calicut, Kerala, India, 13–16.

Bory, S., M. Grisoni, M.-F. Duval, and P. Besse. 2008a. Biodiversity and preservation of vanilla: Present state of knowledge. *Genet. Resour. Crop Evol.* 55:551–571.

Bory, S., P. Lubinsky, A.-M. Risterucci, J.-L. Noyer, M. Grisoni, M.-F. Duval, and P. Besse. 2008b. Patterns of introduction and diversification of *Vanilla planifolia* (Orchidaceae) in Reunion Island (Indian Ocean). *Am. J. Bot.* 95:805–815.

Chitra, R., R. Arulozhiyan, M. Jawaharlal, and E. Vadivel. 2007. Micropropagation of vanilla (*Vanilla planifolia* Andrews.). *J. Plant. Crops* 35 (2):111–113.

Grisoni, M., F. Davidson, C. Hyrondelle, K. Farreyrol, M.L. Caruana, and M. Pearson. 2004. Nature, incidence and symptomatology of viruses infecting *Vanilla tahitensis* in French Polynesia. *Plant Dis.* 88:119–124.

Hrideek, T.K., P.P,. Menon, K.M. Kuruvilla, K.J. Madhusoodanan, and J. Thomas, 2003. Yielding behavior of hanging shoot in vanilla. *13th Swadeshi Science Congress*, MSSRF-CabC, 44(Ab).

ICRI (Indian Cardamom Research Institute), 1996. Agenda notes on VIII SRC, ICRI, Spices Board, Ministry of Commerce, GoI, Myladumpara, Idukki, Kerala, India, 25pp.

ICRI (Indian Cardamom Research Institute), 1999. Agenda notes on XI SRC, ICRI, Spices Board, Ministry of Commerce, GoI, Myladumpara, Idukki, Kerala, India, 15pp.

ICRI (Indian Cardamom Research Institute), Annual Research Report, 2003. Spices Board, Ministry of Commerce, GoI, Myladumpara, Idukki, Kerala, India, 31pp.

ICRI (Indian Cardamom Research Institute), Annual Research Report, 2005. Spices Board, Ministry of Commerce, GoI, Myladumpara, Idukki, Kerala, India, 46pp.

ICRI (Indian Cardamom Research Institute), Annual Research Report, 2006. Spices Board, Ministry of Commerce, GoI, Myladumpara, Idukki, Kerala, India, 87pp.

Krishnakumar, V. 1995. Production and planting materials in vanilla. *Spice India* 8:526.

Krishnakumar, V., G.P. Bindumol, S.N. Potty, and C. Govindaraju. 2007. Processing of vanilla (*Vanilla planifolia* Andrews) beans—Influence of storing fresh beans, killing temperature and duration of killing on quality parameters. *J. Spices Aromatic Crops* 16 (1):31–37.

Krishnakumar, V., G.P. Bindumol, and S.N. Potty. 2008. Processing of vanilla (*Vanilla planifolia* Andrews) beans—Effect of method, time and duration of sunning on moisture and vanillin content. *J. Plant. Crops* 36 (1):19–23.

Kuruvilla, K.M., V.V. Radhakrishnan, K.J. Madhusoodanan, and S.N. Potty. 1996. Floral biology of vanilla (*Vanilla planifolia*). *Spice India* 9 (8):20–23.

Kuruvilla, K.M., K. Pradipkumar, and K.J. Madhusoodanan. 2000. A descriptor for vanilla. *Ind. Spices* 37 (1):10–11.

Kuruvilla, K.M., Y.S. Rao, K. Mary Mathew, and K. Pradipkumar. 2003. Vanilla—Planting materials. In: *Vanilla—The Prince of Spices*. Spices Board, Government of India. Cochin, India.

Lubinsky, P. 2003. Conservation of wild vanilla. In: *First International Congress on the Future of Vanilla Business, Protection*, NJ, USA, November 11–12, 2003.
Mary Mathew, K., Y.S. Rao, K. Pradipkumar, K.J. Madhusoodanan, and S.N. Potty. 1999. *In vitro* culture systems in Vanilla. In: P.B. Ravi Kishor, ed. *Plant Tissue Culture and Biotechnology. Emerging Trends.* Universities Press Ltd., Hyderabad, India, 171–179.
Madhubala, R., V. Bhadramurthy, A.I. Bhat, P.S. Hareesh, S.T. Retheesh, and R.S. Bhai. 2005. Occurrence of cucumber mosaic virus on vanilla (*Vanilla planifolia* Andrews) in India. *J. Biosci.* 30:339–350.
Menon, P.P., K.M. Kuruvilla, and K.J. Madhusoodanan. 2002. Yield prediction in vanilla (*Vanilla planifolia* Andrews). *Proc. Placrosym.* XV:414–415.
Minoo, D., G.S. Pillai, K. Nirmal Babu, and K.V. Peter. 2008a. Isolation and fusion of protoplasts in *Vanilla* species. *Curr. Sci.* 94 (1):115–120.
Minoo, D., V.N. Jayakumar, S.S. Veena, J. Vimala, A. Basha, K.V. Saji, K. Nirmal Babu, and K.V. Peter. 2008b. Genetic variation and interrelationships in *Vanilla planifolia* and few related species as expressed by RAPD polymorphism. *Genet. Resour. Crop Evol.* 55:459–470.
Minoo, D., K. Nirmal Babu, and K.V. Peter. 2006a. Conservation of vanilla species *in vitro. Sci. Hort.* 110:175–180.
Minoo, D., K. Nirmal Babu, P.N. Ravindran, and K.V. Peter. 2006b. Interspecific hybridization in vanilla and molecular characterization of hybrids and selfed progenies using RAPD and AFLP markers. *Sci. Hort.* 108:412–414.
Muraleedharan, A. 1975. Effect of mulching on the growth and yield of vanilla (*Vanilla planifolia*). *South Indian Hort.* 23 (3/4):149–151.
Nair, R.R. and P.N. Ravindran.1994. Somatic association of chromosomes and other mitotic abnormalities in *Vanilla planifolia* Andrews. *Caryologia* 47: 65–73.
Newport, H. 1910. Vanilla culture for tropical Queesland. *Queesland Agri. J.* 24:184–189.
Pearson, M.N. and S.P. Pone. 1988. Viruses of vanilla in the kingdom of Tonga. *Australia Plant Pathol.* 17:59–60.
Pearson, M.N. 1990. Virus diseases of vanilla in South Pacific. *Sixteenth Regional Technical Meeting on Plant Protection*, Auckland, New Zealand. South Pacific Commission Noumea, Working paper 16.
Philip, S. 1980. Wilt of *Vanilla planifolia* caused by *Fusarium oxysporum* Schl. Ex. Fries f. sp. *vasinfectum. Agric. Res. J. Kerala* 18 (1):139–140.
Philip, V.J. and S.A.Z. Nainar. 1986. *In vitro* propagation of vanilla through protocorms and root bits. *J. Plant Physiol.* 122:211–215.
Potty, S.N. and V. Krishnakumar. 2003. Agrotechniques. In: *Vanilla—the Prince of Spices.* Spices Board, Kochi.
Prakash, K.V. and M.R. Sudharshan. 2002. Vanilla bug, *Halyomorpha* sp. NR. *picus* (Fabr.) (Hemiptera: Pentatomidae)—A new pest of vanilla (*Vanilla planifolia* Andrews) *Proceedings of XIV PLACROSYM*, Coconut Development Board, Kochi, 493–496.
Purseglove, J.W., E.G. Brown, C.L. Green, and S.R.J. Robbins. 1981. Vanilla. In: *Spices* (Vol. II), Longman Scientific and Technical Publisher, UK, 644–735.
Rai, R. and B.K. Nayar. 1976. *Saula ferruginea* Gerst., a pest of vanilla in South India. *J. Plant. Crops* 4 (1):22–23.
Rao, Y.S., K. Mary Mathew, K.J. Madhusoodanan, and R. Naidu. 1992a. Multiple shoot regeneration in vanilla. (*Vanilla planifolia* Andrews.). *J. Plant. Crops*. 21 (Suppl.):351–354.
Rao, Y.S., K.J. Madhusoodanan, and R. Naidu. 1992b. Detection of genetic variability in vanilla through PAGE studies. *J. Plant. Crops*. 21 (Suppl.):363–365.
Rao, Y.S., K.M. Kuruvilla, and K.J. Madhusoodanan. 1994. Collection of *Vanilla wightiana* Lindl—An endangered wild species from Eastern Ghats of Andhra Pradesh. *Ind. J. Plant Genet Res.* 7 (2):257–259.

Sarma, Y.R. 2006a. Approaches towards biological control of stem and root rot of vanilla in Mauritius—A report, 1–30.

Sarma, Y.R. 2006b. Recent trends in the use of antagonistic organisms for disease management in spice crops. In: B. Ramanujam and R.J. Rabindra, eds *Current Status of Biological Control of Plant Diseases Using Antagonistic Organisms in India*. PDBC, Bangalore, 49pp.

Sasikumar, B., J. Rema, and P.N. Ravindran. 1992. Vanilla. *Ind. Cocoa, Arecanut Spices J*. XVI (1):6–10.

Sasikumar, B., J. Rema, and P.M. Ravindran. 1993. Vanilla—The much fancied export crop. *Indian Perfumes* 37 (4):29–36.

Sasikumar, B. 2004. Vanilla musings. *Spice India* 17 (5):29–30.

Shadakshari, Y.G., H.M. Chandrappa, and K. Umesh, 1996. Flowering and fruit set on Vanilla (*Vanilla planifolia*). *J. Spices Aromatic Crops* 5 (2):148.

Siddagangaiah, B.A. Vadiraj, M.R. Sudarshan, and V. Krishnakumar. 1996. Standardisation of rooting media for propagation of vanilla (*Vanilla planifolia* Andrews). *J. Spices Aromatic Crops* 5 (2):131–133.

Sudharshan, M.R., N.S. Radhika, and K.V. Prakash. 2003. Vanilla viral diseases in Karnataka. *Spice India* 16 (12):28–29.

Soto Arenas, M.A. 1999. Filogeografia y recursos geneticos de las vainillas de Mexico. http://www.conabio.gob.mx/institucion /proyectos/resultados/infj101.pdf.

Thomas, J. and R.S. Bhai. 2000. *Sclerotium* rot—A new disease of vanilla (*Vanilla planifolia* Andrews). *J. Spices Aromatic Crops* 9 (2):175–176.

Thomas, J., A.K. Vijayan, and R.S. Bhai. 2002. Vanilla diseases in India and their management. *Indian J. Arecanut, Spices Medicinal Plants* 4 (4):143–149.

Thomas, J., R.S. Bhai, and A.K. Vijayan. 2003. Diseases and their management. In: *Vanilla—The Prince of Spices,* Spices Board, Cochin, India, 27–41.

Varadarasan, S., M.A. Ansar Ali, and S.S. Chandrasekhar. 2002a. Vanilla vine weevil, a new insect pest on vanilla (*Vanilla planifolia*). *Proceedings of the Placrosym* XV 620–622.

Varadarasan, S., S.S. Chandrasekhar, M.A. Ansar Ali, and K.V. Prakash. 2002b. Status of vanilla pests. In: M. Tamil Selvan, ed. *National Consultative Meeting for Accelerated Production and Export of Spices*. Directorate of Arecanut and Spices development, Calicut, Kerala, 83–87.

Varadarasan, S., B. Gopakumar, S.S. Chandrasekar, M.A. Ansar Ali, and K.V. Prakash. 2003. Pest and their management. In: *Vanilla—The Prince of Spices*. Spices Board, Cochin, 25.

Varadarasan, S., B. Gopakumar, S.S. Chandrasekar, and M.A. Ansar Ali. 2005. Studies on scab formation on vanilla (*Vanilla planifolia* Andrews) beans. *Pest Manage. Hort. Ecosyst.* 11 (1):45–50.

Wisler, G.C., F.W. Zettler, and L. Mu. 1987. Virus infections of vanilla and other orchids in French Polynesia. *Plant Dis.* 71:1125–1129.

21 Vanilla Production in East Africa

Uganda, Tanzania, Kenya, and Eastern Democratic Republic of Congo

Clemens Fehr

CONTENTS

INTRODUCTION

This chapter deals with vanilla production in East Africa.

In East Africa, vanilla production is concentrated around the Lake Victoria basin and around the foothills of the Rwenzori Mountains in the Albertine Rift Valley. Vanilla-producing countries include Uganda, Kenya, Tanzania, and the Eastern Democratic Republic (DR) of Congo, with Uganda being the main producer.

It is notoriously difficult to get accurate information and figures on vanilla production, especially from the smaller producer countries in this region. In addition, a large quantity of vanilla is traded between the countries and is later exported as Ugandan vanilla. This chapter accordingly focuses on Uganda and presents information on an aggregate basis. However, where country-specific information is available, it is mentioned.

HISTORY OF VANILLA GROWING IN EAST AFRICA

Vanilla (*Vanilla planifolia*) was introduced into the then British Protectorate of Uganda from Sri Lanka (then Ceylon) in the 1920s. However, it was not until the 1950s when the first significant plantations were established in Mukono District on the shores of Lake Victoria (Figure 21.1).

In the 1970s, during the Idi Amin regime, the economy of Uganda collapsed, and along with it the vanilla industry. Although a small production of vanilla persisted, the sector did not see a revival until the mid-1990s when the United States Agency for International Development (USAID) initiated a vanilla project. Lasting from 1995 to 2004, the project promoted an increase in vanilla production from a few tons annually in the early 1990s to approximately 200 tons in 2006. It was also

FIGURE 21.1 Situation map of the main vanilla-growing areas in Central Africa.

instrumental to expand the growing area from the lake region to other areas of the country, most importantly to the Rwenzori areas, in particular in Bundibugyo District.

The introduction and promotion of vanilla in other countries of the region is an even more a recent phenomenon. In Tanzania, vanilla is grown along the Western shores of Lake Victoria, in the Kagera region. It started in the mid-1990s and originated from vanilla plants grown in Uganda.

With regard to the DR Congo, although there are reports that vanilla was grown in the west of the country during colonial times, none of these activities can be traced today. Instead, vanilla growing is now concentrated in North Kivu Province, to the west of the Rwenzori Mountains and is a direct offspring of the promotion of vanilla in the western regions of Uganda during the 1990s.

Vanilla growing in Kenya is an even more recent phenomena and the response of the recent high-price phase. Apparently, little progress has been made.

AGRONOMY AND PROCESSING OF VANILLA

The East African highlands are characterized by a mild, equatorial climate with a bimodal rainfall distribution. It permits for two distinct growing and harvesting seasons. This also applies for vanilla and sets the region apart from other vanilla-growing areas worldwide.

Around the shores of the lake, the main vanilla-harvesting season is from June to August, whereas the minor season lasts from December to February. Around the Rwenzori Mountains, it is vice versa.

Although there are flowering and harvesting seasons, due to the distinct microclimate, the main seasons for flowering and harvesting spread over 3–4 months, and even beyond these periods it is possible to find new flowers developing. As a result, although fluctuating, plants usually carry flowers, immature beans, and mature beans at the same time.

AGRONOMY

The production in the entire region is exclusively from *V. planifolia*. In general, vanilla is grown by small farmers and many produce only a few kilos of green beans per year. Some exceptions to this exist, including DML sprl. in the DR Congo, which has established plantations approaching 20 ha.

Planting material usually derives from own fields or from neighbors. During the planting boom that resulted from the high prices obtained for vanilla in 2003 and 2004, the planting material was sold and constituted substantial income for some farmers.

Apart from the mentioned USAID project, the sector receives little input from government institutions, and research remains scanty. As such, the general agronomy is basic and plantations usually produce below their potential. The major shortcomings are poor management of soils and weeds.

Processing

In contrast to other vanilla producing areas in the world, in East Africa most vanilla is purchased by specialized processors as green beans and cured in central curing facilities. In most cases, processing follows the classic Bourbon method, with hot water treatment, and a sunning and sweating phase followed by slow drying indoors. To retain heat, processors usually rely on insulated boxes. To reheat the beans they expose them to the sun. Artificial drying is generally not practiced.

The classic Bourbon approach suffers from two principal shortcomings: it can become quite cool at night and heat retention in the curing containers is inadequate and the vanilla becomes cold. The second problem relates to hygiene—the blankets used to insulate the vanilla constitute a latent risk with regard to contamination and fungal attacks.

Two producers of the region are known to use different curing techniques—in one case, a rapid curing method with previously chopped beans and artificial drying are applied, reducing the curing phase to a few days. The chopped beans are later sold to extractors, and a close cooperation with some exists. The method results in higher vanillin contents than classic Bourbon curing. However, it is claimed that the flavor is also more simplistic—flat and vanillin focussed. Another company in the DR Congo follows a refined Bourbon curing method. It is slower than the classic Bourbon method and results in significantly higher vanillin contents and a very subtle, sweet, balsamic, and complex flavor with elements of dried orange peel, cinnamon, and chocolate.

CHARACTERISTICS OF VANILLA

Centralized and relatively controlled curing is reflected in high vanillin contents. The vanilla from the region often exceeds 3% vanillin content (on dry weight basis). The flavor profile is described as subtle, sweet, and balsamic. A leathery element can be found in the flavor of some vanilla. It can most likely be attributed to vanilla that was cured at very low temperatures.

The typical vanilla is dried at a humidity of 25–33%.

Vanilla from East Africa is sought after by extract makers. The origin is gradually also making its way into the gourmet sector and it competes favorably with vanilla from Madagascar and the Comoros. Vanilla from East Africa is sought after by extract makers. The origin is gradually also making its way into the gourmet sector and it competes favorably with vanilla from Madagascar and the Comoros.

PRODUCTS AND MAIN USES (MARKET SEGMENTS)

The biggest part of the vanilla production from East Africa enters the extraction market and is sold in bulk. Often the vanilla is poorly sorted. At least one company produces cold-milled vanilla powder and another solvent-based extracts.

There is a fair quantity of gourmet vanilla exported, some in retail-packed form.

TABLE 21.1
Main Vanilla Producers and Their Products

Company	Type of Vanilla Produced	Contact
Buiga Farm Industries Ltd.	Processor of conventional vanilla beans; bulk	tamale@buiga.com
Coetzee Natural Products (U) Ltd.	Processor of conventional and certified organic vanilla beans; bulk	cnp@africacentral.net
DML sprl	Grower and processor of vanilla beans; fair trade and certified organic vanilla	lejardinbio@gmail.com
Empire Estates	Trader of conventional vanilla beans; bulk	Abdul214@yahoo.com
ESCO Ltd.	Processor of conventional and certified organic vanilla beans; bulk	esco@africaonline.co.ug
Gourmet Gardens Ltd.	Producer and processor of vanilla beans; powder and extract, certified fair trade and organic; bulk and retail packs	info@gourmet-gardens.net
Ndali Farm	Processor of vanilla beans and extract, conventional and certified organic and fair trade; bulk and retail packs	vanilla@ndali.net
Sekalala	Processor of chopped and rapidly cured vanilla beans, conventional; bulk	aga@simba.fm
Uganda Crop Industries Ltd.	Grower and processor of vanilla beans, conventional and certified organic; bulk and retail packs	Magellan@tenegra.net

In the meantime, a good part of this region's vanilla is organically certified and at least three producers are organically and fair-trade certified, two in Uganda and one in the DR Congo.

In summary, a substantial proportion of the vanilla derived from this region aims at a specific niche in the markets, a characteristic that sets this region apart from other vanilla-producing regions.

The major vanilla producers of this region are listed in Table 21.1.

MARKETS AND TRADE ROUTES

Two types of trade routes are to be distinguished—on the one hand, the local trading routes and on the other, the international trading routes.

Locally, good parts of the vanilla from the Kagera region in Tanzania and from Eastern Congo are purchased green and cured in Uganda and later marketed as Ugandan vanilla. The proportion of Tanzanian vanilla entering the global market as such is in the range of 1–3 tons. Congo currently exports 6–8 tons of homegrown vanilla. It is estimated that at least the same amount enters Uganda and is sold as Ugandan vanilla.

With regard to international markets, most gourmet vanilla is sold to Europe and small volumes of extraction and gourmet beans are sent to Japan. The main market for extraction beans appears to be the United States. The main markets in Europe are the United Kingdom and Germany. The volumes sold in France, which is in overall a big buyer of vanilla, appear small and can be attributed to the French bias towards products from their former overseas territories.

During the last few years, clients from Madagascar regularly bought extract grade vanilla for blending and upgrading vanillin-poor Madagascar vanilla. The trend continues.

Gourmet vanilla is nearly always transported by air. Extract vanilla is transported by air when prices are high. Otherwise, it is transported via the port of Mombasa in Kenya by sea.

PRESENT SITUATION

The Ugandan vanilla market is completely liberalized, without any government interventions or regulations. A similar situation prevails in the neighboring countries. The consequence is that the production levels fluctuate strongly and in line with supply and demand.

With regard to actual volumes, as was mentioned earlier, it is notoriously difficult to get accurate information and figures on vanilla production in this region. The figures presented below are derived from different but usually unconfirmed sources and from direct communication with key stakeholders. Individual figures should therefore be treated with caution.

However, it is undeniable that the revival of the vanilla sector by an USAID-funded project let to an increase in vanilla production from basically nil in 1995 to about 150 tons in 2004.

The end of the project coincided with an unseen period of high prices, which resulted in many more farmers entering vanilla production and increasing the yield to around 200 tons in 2006.

Since then, because prices collapsed, a decline in production can be observed and by 2009 the production declined to about 120 tons (Table 21.2).

Probably, the production will further decline in the coming years. Prices are low and other crops such as cocoa that are grown concurrently with vanilla currently experience a boom. It is not uncommon to see farmers actively uprooting vanilla to give way for cocoa!

OUTLOOK

The conditions for growing vanilla in East Africa are very favorable and the resulting vanilla is of good quality. Over the last 15 years this region has established itself in the global market as a supplier of quality vanilla, which is sought after by extract makers. The gourmet sector is developing too and the vanilla offered in such forms competes favorably with vanilla from the Indian Ocean islands.

The current production potential lies in the range of 100–150 tons/year. However, due to continuous drop in prices, production declines and will continue to do so if no major change occurs in the vanilla market. Should market condition and prices change, the most likely scenario is a delayed reentry into the market, which will

TABLE 21.2
Summary of the Overall Vanilla Production Trend in Eastern Africa

	1994	1995	1996	1997	1998	1999	2000	2001	2002	2003	2004	2005	2006	2007	2008	2009
Volumes (tons)		0.2	16	17	54		70			120		185	195	200	150	120
Value ('000s USD)		8								25000				8400		

contribute to a renewed boom and bust cycle as was observed earlier in this decade.

The outlook for organically and fair-trade-certified vanilla is more promising. Certified farmers are more reluctant to abandon their plantations, primarily because prices are better and the demand more stable. As such, should the market situation change, they should be the major beneficiaries.

In comparison with other vanilla-producing regions worldwide, East Africa is, however, strongly handicapped by its landlocked position. It results in comparatively high transport costs and makes local producers and exporters struggle to remain competitive. The impact is particularly large during low-price phases. As a result, it is very unlikely that vanilla production in this region will increase significantly over the production levels that was seen some years ago.

For East African exporters to stay competitive, many have opted for different types of certification (organic, fair, or both) to add value to their vanilla and some have begun to transfer their vanilla into extract and powder. This trend will most likely continue, at the expense of companies in Europe and the United States that are specialized in transforming vanilla.

Last but not least, although East Africa can produce very good vanilla, it is difficult to compete against the well-established Bourbon brand. Despite the fact that many consumers do not really know what Bourbon stands for they perceive it as superior to vanilla that is not labeled as Bourbon. This is a widespread phenomenon in Europe. French consumers and traders are particularly snobbish and ignorant in this. In the United States, the issue is handled less rigidly and it is not uncommon that vanilla from, for example, Indonesia is advertised as Bourbon beans from Indonesia if they were processed by the Bourbon technique.

22 Vanilla Production in Mexico

Juan Hernández Hernández and Pesach Lubinsky

CONTENTS

INTRODUCTION

The principal vanilla species of commerce ("Mexican vanilla" or "Bourbon vanilla" [*Vanilla planifolia* G. Jackson]) is a rare orchid native to Mexico and Central America (Cameron and Soto Arenas, 2003; Lubinsky et al., 2008; Portères, 1954). The naturally fragrant and cured fruits of *V. planifolia* were used in pre-Hispanic Mesoamerica by various cultures, most prominently by the Totonac of northern Veracruz (the Papantla region), the Aztec of Central Mexico, and the Maya of the tropical lowlands of southern Mexico, Guatemala, and Belize (Bruman, 1948; Lubinsky, 2007; Rain, 2004). These cultures employed vanilla as a medicinal (stomachic) and flavoring agent, usually in the form of an ingredient to spice cacao beverages.

From roughly the mid-1700s to the mid-1800s, Papantla had a global monopoly on vanilla production (Bruman, 1948; Kourí, 2004), earning it the reputation as "The city that perfumes the world" (Rain, 2004). Papantla retains its status as the principal producer of vanilla in Mexico today, although the country itself is a minor supplier globally.

Growers in the Papantla region refer to cultivated *V. planifolia* as "vainilla mansa" (i.e., "domesticated vanilla") to distinguish it from other varieties and species that can still be found growing wild in the forests and which occasionally may be grown by particular farmers (= "vainilla silvestre" or "vainilla del monte"). This genetic wealth emphasizes Mexico's importance as a center of domestication and diversity for vanilla.

Vanilla production in Mexico is situated in the coastal and foothill region of the Gulf of Mexico, from sea level to an altitude of 700 m. The prevailing climate is tropical hot and humid with average temperatures around 24°C, relative humidity of 80%, and rainfall from 1200 to 3000 mm annually. The dry season lasts from March until June. The winter months (November–February) are characterized by cool, humid winds of low intensity called "nortes," which sweep down unimpeded from the Arctic Circle. These low temperatures are critical for stimulating flowering vanilla in the spring.

The major producers of vanilla in Mexico are Veracruz (70%), and Oaxaca and Puebla, the latter two of which are responsible for most of the remaining 30%. Minor vanilla-producing states include San Luis Potosí, Hidalgo, Chiapas, and Quintana Roo.

At the national level, an estimated 4000 families produce vanilla, mostly among indigenous communities. In 2005, approximately 1106.75 ha were used for vanilla production, yielding 189 tons of "green vanilla" and 30 tons of cured beans (SAGARPA, 2005; SIACON, 2005). This supply is managed between six private companies and four farmer cooperatives that produce vanilla, oversee its curing, and arrange its sale for national and international markets (Table 22.1).

VANILLA PRODUCTION SYSTEMS IN MEXICO

There are four systems of vanilla cultivation in Mexico (Hernández Hernández, 2005), which are classified in the following manner.

Traditional or "Acahual"

Vanilla is cultivated in managed secondary forests ("acahual") (Figure 22.1) alone or in combination with coffee (*Coffea*) and/or palms (*Chamaedorea*). The size of these plots is typically less than 1 ha, and are employed by more than 80% of growers. The density of plants in these plots is relatively low as well as the allotment of time the grower uses to manage the growth (e.g., looping and rooting) and care (e.g., fertilizing) of the plants. Consequently, yields are low, averaging 200 kg of green vanilla/ha.

Diversified Citrus Groves

The extensive amount of *Citrus* cultivation in coastal Veracruz has allowed farmers to use these plantings as support trees (Figure 22.2). Most groves consist of Valencia

TABLE 22.1
Purchasing Centers/Curing Facilities in Mexico

Company/Organization	Address
Santa Beatriz	Arturo Tremari No. 104, Papantla, Ver. Tel. 784-8420690
Casa Larios	Gildardo Muñoz, No. 401, Papantla, Ver. Tel. 784-8420160
La Alternativa	Ejido Primero de Mayo, Papantla, Ver. Tel. 728226048
Consejo Nacional de Productores de Vainilla	Benito Juárez No. 202, Papantla, Ver. Tel. 784-8423905
Productores Asociados de Vainilla del Totonacapan	Ejido Cuyuxquihui, Papantla, Ver. Tel. 784-8490548
Gaya Vai-Mex	Av. Hidalgo No. 56, Gutiérrez Zamora, Ver. Tel. 766-8450497
Vainilla del Totonacapan	Benito Juárez No. 14, Gutiérrez. Zamora, Ver. Tel. 766-8450226
The Mexican Vanilla Plantation	Ejido Chacuaco, Tuxpan, Ver. Tel. 783-8346838
Global Fungi SPR de RL de CV.	Xicoténcatl No. 214, Teziutlán Pue. Tel. 231-3121732
Consejo Regional de Productores de Vainilla	Cerro Quemado, San Pedro Ixcatlán, Oax. Tel. 287-8779074

FIGURE 22.1 Vanilla cultivation system, in "*acahual.*"

FIGURE 22.2 Vanilla cultivation system in *Citrus* groves.

oranges, which have excellent characteristics as vanilla support trees, for example, strong branches that grow laterally and that can permit a large quantity of vanilla vines. The *Citrus* trees also provide semifiltered sunlight throughout the year, which encourages vigorous growth and flowering usually in the second year after planting. Ventilation is considered good, resulting in less problems with pathogens and more vegetative growth than is typical of "acahual" systems. Compared to more intensive systems of vanilla cultivation, vanilla grown in *Citrus* groves entails a reduced cost of production since there is no investment required for support trees.

Each proprietor owns between 1 and 5 ha, which obtain yields that fluctuate from 0.5 to 3.0 tons of green vanilla/ha.

Intensive (Monoculture)

This system is characterized by the use of nitrogen-fixing legumes (*Erythrina* sp. and *Gliricidia* sp.) as support trees and planting densities of between 1000 and 5000 support trees/ha. This density allows for some of the highest yields of any system, although generally such a success is obtained in only one year (the fourth or fifth year following planting). After a bumper season, yields in this system often decline drastically because of overcrowding leading to problems with poor light and ventilation leading to pathogen outbreaks. Each grower manages an average of between 0.5 and 2 ha from which 1–2 tons of green vanilla/ha are obtained in rain-fed plantings, or between 2 and 4 tons of green vanilla/ha in irrigated plantings.

Shade Houses

This is the newest and most intensive system of vanilla production in Mexico. The main feature of this system consists of substituting or complementing natural light with plastic shade nets of 50% luminosity. These nets are placed over the support trees, usually at a height of 3–5 m, as well as along the four sides of the plantings. These "shade houses" are generally small (1000 m^2 or 25 × 40 m) because of the high costs entailed in their construction. Supports for vanilla vines are either artificial or living trees, but in either case densities are high (1500–2000 individual vanilla cuttings are planted per shade house, equating roughly to 15–25,000 cuttings/ha).

The first yields from this system were reported in 2007. They ranged from 50–514 kg green vanilla/shade house (1000 m^2). The main factor producing this variation was most likely fruit drop of immature beans caused by adverse weather conditions (i.e., high temperatures and drought) in combination with management practices. The maximum yield of 514 kg suggests that 5140 kg green vanilla/ha can potentially be obtained, a value similar to shade house systems in other producing countries. However, it is still unclear if this can be accomplished in Mexico, especially considering the large financial investment that is needed.

MAIN IMPACTS ON VANILLA PRODUCTION IN MEXICO

Drought and High Temperatures

An estimated 80% of vanilla plantings in Mexico are affected by drought, principally from March to June, during flowering and the early stages of fruit development. Because droughts manifest at such a critical stage, both the overall production and quality of Mexican vanilla are characteristics, which are especially sensitive to drought conditions.

Disease/Pathogens

Stem and root rot, caused by the fungus *Fusarium oxysporum* f. sp. *vanillae*, is the major disease of Mexican vanilla. It results in both the shortening of the reproductive life and mortality of individual plants. The fungus *Colletotrichum* sp. is responsible for causing premature fruit drop, sometimes at levels as high as 50% of a particular planting.

Pests

The insect *Tenthecoris confusus* Hsiao & Sailer (Hemiptera: Miridae) ("chinche roja") is the principal pest on Mexican vanilla (Figure 22.3). It sucks xylem from leaves, stems, and fruits, and can cause the total defoliation of plants. An unidentified species of caterpillar ("trozadores") also consumes flower buds, but the damage is generally minor.

FLOWERING, POLLINATION, AND HARVEST

The first flowering in vanilla occurs in 2–3 years following planting of new cuttings. The flowering season is during March to May. Similar to other countries, pollination in Mexico is achieved manually. Some flowers are still naturally pollinated, and

FIGURE 22.3 (See color insert following page 136.) Adult of *T. confusus* Hsiao & Sailer (Hemiptera: Miridae) and the damages it causes in vanilla.

although the natural pollinator has not been documented, it is most likely to be the orchid bee *Euglossa viridissima* (Lubinsky et al., 2006; Schluter et al., 2007). The method of harvesting in Mexico consists of cutting of entire racemes with the rachis still intact, using shears or knives. The official opening of the harvest in Veracruz is on December 10, and continues through January and February of the following year.

VANILLA CURING IN MEXICO

Curing begins immediately following the harvest. Most curing is done by individual curers rather than by growers. Curing lasts for 3–5 months, generally from January to May.

Vanilla curing in Mexico is distinct from all other vanilla-producing countries, a fact that has helped secure a demand for Mexican vanilla in the world market. This curing process consists of the following steps:

1. Fruit/raceme separation ("Despezonado")
   This step entails the manual separation of each fruit from the rachis ("pezón"). This step begins with the classification process of the fruits by size and type (e.g., whole, split, small, etc.).
2. Oven killing
   This step terminates photosynthesis and cellular activity in green fruits and prevents mature fruits from dehiscing during ripening. Fruits are packaged in small wooden boxes or straw mats ("petates") and then placed in ovens for a period of 1–2 days at 60°C. Afterward, vanilla fruits are removed from the oven and placed in larger boxes for sweating. The fruits are covered

with cotton, sheets, and mats for another 1–2-day period during this first instance of "sweating." In recent years, some curers in Mexico have switched from using ovens for the "killing" stage to using hot water, as is the process in Madagascar.

3. Sun curing and successive sweats
   The vanilla beans are taken out of the sweat boxes and placed in a sunny patio on mats for 3–4 h. Immediately afterward, the beans are returned to the sweat boxes and are covered with more clothes and straw mats to continue the sweating. This process is repeated until the vanilla beans have a moisture content of around 30%, which is usually accomplished after 11–24 cycles of sun curing/sweating, respectively, for smaller and larger fruits.
4. Conditioning
   Vanilla beans that have acquired the desirable moisture content are placed in boxes or wooden racks called "camillas" for observation for 30–45 days.
5. Grading
   Vanilla beans that do not manifest any problems during conditioning are sorted by size and quality according to the Mexican grading system.

COMMERCIALIZATION OF MEXICAN VANILLA

Most growers in Mexico sell their uncured, green vanilla beans to middlemen or directly to curers/exporters. Despite this customary practice, many growers in Mexico are actively trying to learn to cure their own beans, and to make contracts directly with international buyers, or with Mexican curers/exporters looking to fulfill orders.

The centers of commercialization of Mexican vanilla are the cities of Papantla and Gutiérrez Zamora, both in the state of Veracruz. These cities receive green vanilla from growers and middle men from Veracruz, Puebla, Hidalgo, and San Luis Potosi. Curers/exporters are also found in Puebla and in Oaxaca.

INTERNATIONAL TRADE/EXPORT

Mexican vanilla is imported to other countries via multinational companies. The principal brokers of Mexican vanilla are Aust Hatchman, McCormick Co., Eurovanille, Vanipro, and International Flavor and Fragrance. These companies are headquartered in the United States, Germany, France, and Canada.

There also exists a small internal market for vanilla in Mexico. This market mostly entails purchase by extract manufacturers who supply both domestic and foreign consumers.

EXPORT VOLUMES

Almost all of the vanilla produced in Mexico is destined for foreign markets. However, in the last three years, most Mexican vanilla has been stored in warehouses while exporters wait until prices rise above $50 USD/kilo.

In general, Mexico exports 20–30 metric tons of cured vanilla annually. This equates to about 1% of total vanilla production worldwide. The major consumer countries of Mexican vanilla are the United States, Germany, France, Japan, and Canada. A small percent (5%) of the supply of vanilla in Mexico is used nationally to make extracts and handicrafts.

The main demand of buyers of Mexican vanilla is for high-quality beans ("extra" or "gourmet"). Lower-quality beans ("picadura") are also purchased to make extracts.

PRICES FOR UNCURED, "GREEN" VANILLA

The lowest prices for Mexican "green" vanilla were recorded between the years 1984 and 1997, when curers paid less than $2.50 USD/kilo. The highest price paid for green vanilla in Mexico was in 2003, when green vanilla was sold for an average of $45 USD/kilo. Despite the high prices of 2003, green vanilla dropped dramatically in price over the following years. In 2005, green vanilla in Mexico was sold for $12 USD/kilo, and has declined further to $4 USD/kilo during the 2006 and 2007 harvests (Figure 22.4). One exception in 2007 was purchases of green vanilla beans longer than 20 cm for $12 USD/kilo by a private company new to the vanilla trade. The national vanilla growers organization (Consejo Nacional de Productores de Vainilla) also paid growers $6 USD/kilo for green vanilla in 2007.

PRICES FOR CURED VANILLA

Prices for cured Mexican vanilla are set by the international market, and in particular reflect prices offered in Madagascar. The highest prices paid for Mexican vanilla occurred between the years 2001 and 2004, peaking in 2004 at an average of $420

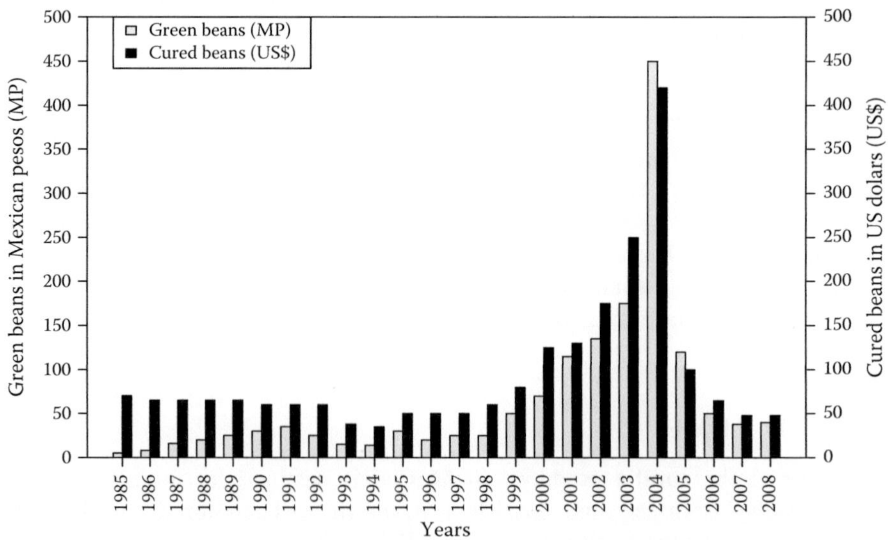

FIGURE 22.4 History of average prices of green and cured vanilla in Papantla, Veracruz, Mexico.

USD/kilo, and a maximum of \$600 USD/kilo. In 2005, the price had declined over 60% to \$120 USD/kilo. Since 2005, small quantities of high-grade vanilla ("extra" or "gourmet") were sold for a maximum of \$60 USD/kilo (Figure 22.4).

GRADING SYSTEM FOR MEXICAN VANILLA

Cured vanilla in Mexico is first classified into three categories: whole ("enteras"), open/split ("rajada"), and chopped/inferior ("picadura"). The "picadura" beans correspond to parts of beans that may have come from immature fruits, or that were damaged or poorly cured. The prices for cured vanilla in Mexico are customarily organized into these three categories.

For whole and split beans, a grading system of five subcategories has been established:

"**EXTRA**": Beans are thick, flexible, lustrous, dark brown/chocolate color ("achocolatado"), sweet/delicate aroma, and with a vanillin content >2.5%.
"**SUPERIOR**": Similar to "extra," but are less thick and lustrous, with a vanillin content of 2.25–2.29%.
"**BUENA**": Beans that are more or less lustrous and flexible, odorous, dark brown with red coloring in longitudinal grooves, and with a vanillin content of 2–2.24%.
"**MEDIANA**": Beans with little flexibility/sheen, weak aroma, mixed dark/light brown, with a vanillin content of 1.75–1.99%.
"**ORDINARIA**": Beans are hard/brittle, not shiny, weak aroma, color mostly light brown with darkened edges, and with a vanillin content of 1.5–1.74%.

In practice, Mexican vanilla may be sold according to other grading systems set by the buyer, usually "gourmet," "splits," and so on.

PACKAGING OF MEXICAN VANILLA

Mexican vanilla was traditionally packaged and distributed in bulk in wax paper and cardboard boxes. Occasionally, high-quality beans ("extra/gourmet") were sold in rolls called "mazos" and distributed in tins. In recent years, due to buyer demand, cured vanilla is now distributed in vacuum packaging in 1 kilo plastic bags.

FLAVOR PROFILE OF MEXICAN VANILLA

The aroma of Mexican vanilla is described as intense, sweet, lightly spicy, and similar to tobacco. The content of the major aromatic compound of vanilla, vanillin, is generally around 2% in Mexican vanilla, but has been found to be as high as 7% (Black, 2005).

Although vanillin content is of principal importance to manufacturers looking to ensure uniform flavor for a mass market, the authentic aroma of vanilla is due to the presence of hundreds of other compounds found in small quantities that develop in cured beans. In Mexican vanilla, 65 volatile compounds have been identified, with acids and phenols figuring prominently (Pérez Silva et al. 2006). Eleven compounds have been found to be unique to Mexican vanilla: hexanoic acid, vanillyl methyl ketone, methyl eicosanoate, 4-butoxy-3-methyl-2-butanone, methoxymethyl acetate, 4-hexen-1-olacetate, 3-ethyl-3-methylpentane, 2,4-dimethyl-1-heptanol, 4-methylene-2-oxethanone, 2-methyl-3-ethylpentane, and 2-ethyl-1,3-dioxolane (Hartman, 2003).

The niche market that persists for Mexican vanilla may in large part be due to the unique aroma/compounds that develop during the curing process that is utilized solely in Mexico.

PERSPECTIVE ON MEXICAN VANILLA

The interest and motivation to produce vanilla in Mexico depends strongly on price. When the international market is favorable, growers in Mexico respond accordingly and organize their labor, and improve the care/maintenance of their plantings, as in the years 2000–2004. When prices fall, growers quickly lose interest and abandon their efforts as is the case currently. This price variable and the international demand for vanilla will continue to be the prime determinants for what visage vanilla production in Mexico will assume.

REFERENCES

Black, J.M. 2005. Vanilla bean volatile analysis, origin and species. Paper presented at the *International Symposium on the Vanilla Business*. Veracruz, Mexico, November 15–16, 2005.

Bruman, H. 1948. The culture history of Mexican Vanilla. *The Hispanic American Historical Review* 283:360–376.

Cameron, K. and M.A. Soto Arenas. 2003. Vanilloideae. In: A.M. Pridgeon, Phillip J. Cribb, Mark W. Chase, and F. Rasmusen, eds. *Genera Orchidacearum. Vol. 3: Orchidoideae (Part 2), Vanilloideae*. Oxford University Press, Oxford, UK, 281–334.

Hartman, T.G. 2003. Composition of vanilla beans from different geographical regions. Paper presented at the *First International Congress*, Princeton, NY, November 11–12, 2003.

Hernández Hernández, J. 2005. Vanilla cultivation in Mexico. Paper presented at the *International Symposium on the Vanilla Business*. Veracruz, Mexico, November 15–16, 2005.

Kourí, E. 2004. *A Pueblo Divided: Business, Property, and Community in Papantla, México*. Stanford University Press, Stanford, CA.

Lubinsky, P. 2007. *Historical and Evolutionary Origins of Cultivated Vanilla*. PhD. thesis, Botany and Plant Sciences, University of California Riverside, CA.

Lubinsky, P., S. Bory, J. Hernández Hernández, S.-K. Chul, and A. Gómez Pompa. 2008. Origins and dispersal of cultivated vanilla (*Vanilla planifolia* Jacks. (Orchidaceae). *Economic Botany* 62 (2):127–138.

Lubinsky, P., M. Van Dam, and A. Van Dam. 2006. Pollination of Vanilla and evolution in Orchidaceae. *Lindleyana* 75:926–929.

Pérez Silva, A., E. Odoux, P. Brat, F. Ribeyre, G. Rodriguez-Jimenez, V. Robles-Olvera, M.A. Garcia-Alvarado, and Z. Gunata. 2006. GC–MS and GC–olfactometry analysis of aroma compounds in a representative organic aroma extract from cured vanilla (*Vanilla planifolia* G. Jackson) beans. *Food Chemistry* 99:728–735.

Portères, R. 1954. Le genere Vanilla et ses espèces. In: G. Bouriquet, ed. *Le vanillier et la Vanille dans le Monde*. Èditions Paul Lechevalier, Paris, 94–290.

Rain, P. 2004. *Vanilla: The Cultural History of the World's Most Popular Flavor and Fragrance*. Jeremy P. Tarcher/Penguin, New York.

Schluter, P.M., M.A. Soto, and S.A. Harris. 2007. Genetic variation in *Vanilla planifolia* (Orchidaceae). *Economic Botany* 61(4):328–336.

SAGARPA (Secretaría de Agricultura, Ganadería y Desarrollo Agropecuario). 2005. Base de datos. Delegación Estatal Veracruz, México.

SIACON (Sistema de Información Agropecuaria de Consulta). 2005. Base de datos, México, D.F.

23 Vanilla Production in China

Hengcang Zhou, Yunyue Wang, Hongyu Wang, Xurui, and Dexin Chen

CONTENTS

HISTORY

Vanilla Introduction and Industrial Development

In the early 1960s, *Vanilla planifolia* Andr. was successively introduced from Indonesia to the Fujian Institute of Subtropical Botany and from Sri Lanka to DanZhou in Hainan Province, and then to the Xishuangbanna region of Yunnan Province. From this point on, the study of its planting, cultivation, and curing began in China.

In the late 1980s and early 1990s, with the support of vanilla research institutions and their achievements in vanilla research, some companies were progressively established, dedicated to the industrial planting, and production of vanilla. By late 1999, the total vanilla planting area reached around 233 ha, mainly using intensive cultivation systems under artificial shading (Figure 23.1), of which 100 ha were distributed in various parts of Hainan Province and belonged to different companies (Chen, 2005), and 133 ha were distributed in the Xishuangbanna region of Yunnan Province (Figure 23.2), which included five plantations belonging to the largest vanilla producer in China, the Yunnan Vanilla Industry Ltd.

Owing to a high input in the early stages, a long production period, poor management, lack of capital support, unstable output, and natural disasters and so forth, many vanilla plantations were ruined. Until 2001, only about 30 ha of the vanilla plantations were running normally in Hainan Province (Mao, 2002). In Yunnan, the situation was similar with the bankruptcy of the Yunnan Vanilla Industry Ltd. in 2007; the intensive cultivation plantations and processing factory in Xishuangbanna were shut down, and only a few farmers were still planting vanilla. In the past, as the planting and production area frequently fluctuated, the volume of cured vanilla fluctuated greatly from very little to several tons each year.

On the basis of a summary of practical experience, Hainan and Yunnan researched and developed a "Company + Farmer Household" agricultural industrialization

FIGURE 23.1 Vanilla growing in Hainan, China, showing an intensive cultivation system under artificial shade houses. (Courtesy of Chen Dexin.)

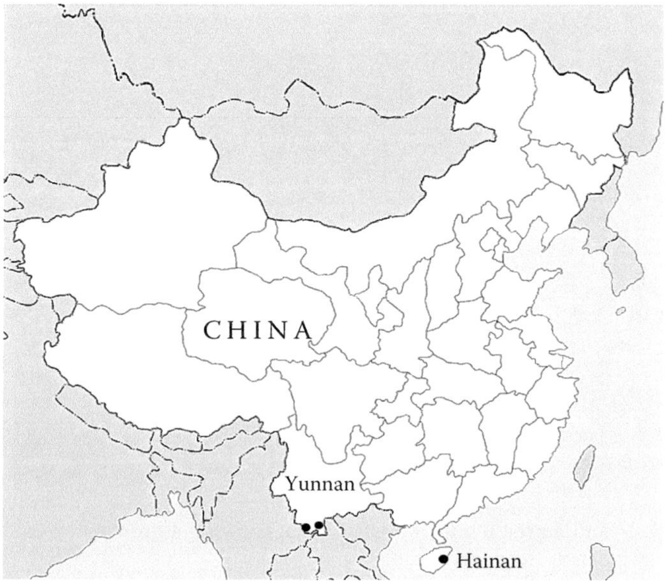

FIGURE 23.2 Map of China showing the regions of vanilla cultivation (black circles) in the Yunnan and Hainan provinces.

model to organize vanilla production. Vanilla producers and local government formulated preferential policies to encourage farmer households to develop market-oriented vanilla cultivation. The producers help farmer households to establish vanilla plantations and provide the relevant technical training and advice, and the farmer households sell their green beans to the producers at fixed prices. From 2002 to 2004, under this new cooperative organization and dispersed cultivation model, more than 70 farmer households in Hainan have planted over 14 ha of vanilla under areca or other trees, or in combination with artificial shade nets (Figures 23.3 and 23.4).

The planting and production area of vanilla in Hainan Island (Figure 23.2) is now around 34 ha, which is divided into eight sections in Anding, Tunchang County, Wanning, and Qionghai City, and belongs to more than 100 farmer households or private companies. Depending on the management and growing systems, its average yield is about 75–300 kg of cured beans per hectare, and the total output is about 5 tons of cured beans per year, which can only satisfy China's domestic consumption demands, with little left for exportation (Chen, 2009; personal communication).

Vanilla is now mainly cultivated and produced in the Hainan Region in China, since the bankruptcy of the biggest vanilla planting and production enterprise, the Yunnan Vanilla Industry Co., Ltd.

A Review of Vanilla Research in China

General Research Situation

With the introduction of vanilla into China and the development of its industrial cultivation, the corresponding science and research academies and production

FIGURE 23.3 Vanilla growing in Hainan, China, showing a growing system under areca trees. (Courtesy of Zhou Hengcang.)

corporations have done a good deal of work in many fields, including planting, manufacturing, pest control, biological characteristic research, and production research and development. By 2008, over 330 academic papers had been published in different periodicals; 7 masters' degree papers and 2 monographs had been published; and over 10 patent applications and 4 national agricultural standards had been issued in China. Furthermore, the vanilla production corporations establish their own rules, regulations and standards on planting, manufacturing, and production.

FIGURE 23.4 Vanilla growing in Hainan, showing farmer households growing vanilla. (Courtesy of Zhou Hengcang.)

Introduction to Main Research Organizations and Research Works

The scientists of the Subtropical Plant Research Academy in Fujian Province were the first to focus on research on vanilla in mainland China. Their early works played a role in the foundation of vanilla research and industrial development in China. Meanwhile, their works initiated and promoted the industrial development of vanilla in China. As regards planting, they researched and established a model of intensive planting in artificial shade houses, which has been widely adopted in commercial vanilla planting in China. The research team published 36 papers, and edited and published the first monograph on vanilla, *Vanilla Planting*, in China.

The Chinese Academy of Tropical Agricultural Sciences (the South China Academy of Tropical Crops before 1994) was also the first to focus on vanilla research. Its team of scientists is the main force in the field of vanilla research in the region of Hainan Province. Its research work directly promotes and supports the industrial development of vanilla in Hainan Province. At the same time, it provides references for scientific research and the industrial development of vanilla in the region of Yunnan Province. It published over 30 research papers about vanilla, and organized and edited national agricultural standards for vanilla in China.

Several organizations are engaged in vanilla research in Yunnan Province. These are mainly the Flavors & Fragrances Research Institute and the College of Plant Protection at Yunnan Agricultural University, the Tropical Crop Research Academy in Yunnan Province, the Kunming Botany Research Institute of Chinese Academy of Science, and so on. A research team in Yunnan Agricultural University has done a good deal of work, especially in pest and disease control for vanilla, and published a monograph on *Pest and Disease in Vanilla*.

Some corporations that plant and manufacture vanilla, especially the vanilla manufacturing corporations in Hainan Province, have also done some research work.

PRESENT SITUATION

Commercial Varieties and Genetic Resources

Vanilla planifolia has been introduced and commercially planted in China. A few other introduced and native wild vanillas are used only for scientific research. The main commercial varieties and genetic resources are as follows:

V. planifolia Andr.: introduced from Indonesia (1960) and Sri Lanka (1962);
Vanilla somai Hayata: Taiwanese vanilla distributed in Taiwan in China;
Vanilla shenzhenica Z.J. Liu and S.C. Chen: found in Shenzhen, South China. It is akin to *Vanilla somai* Hayata (Liu et al., 2007);
Vanilla siamensis Rolfe ex Downie: known as "Big Vanilla," distributed in Guangxi Province and the Xishuangbanna and Hekou areas in Yunnan Province. A relative correlation exists between *V. shenzhenica* Z.J. Liu & S.C. Chen and *V. somai* Hayata. A natural population distributed in the Yachang orchid nature protection region in Leye County, Baise city in northwestern Guangxi Province is the biggest of the wild Big Vanilla group

known so far at the highest latitude and the highest altitude (east longitude 106°19′13″, north latitude 24°48′56″) (Chen et al., 2007).

The Main Growing Regions and Their Climatic Characteristics

Hainan Province and Xishuangbanna in Yunnan Province are major regions in China, and are appropriate for vanilla planting and production. Xishuangbanna, which includes three counties, Jinghong, Mengla, and Menghai, is located in southern Yunnan, E99°55′–101°50′, N21°10′–22°40′, altitude 475–2429 m. This place lies on the northern edge of the eastern and southern Asian tropical region, with tropical and subtropical climate regions, multiple climate types, and complicated terrain. The tropical climate area in Xishuangbanna covers about 400 ha that are located below 800 m altitude, which represents only 20.4% of its area. Total annual precipitation is between 977 and 1655 mm. The annual mean temperature in Xishuangbanna has been 18.3–22.4°C for many years, and the average temperature of the coldest month (January) is 12.0–16.1°C.

Hainan Island lies between north latitude 18°10′ and 20°10′, and east longitude 108°37′ and 110°03′. It is part of the oceanic tropical monsoon climate, which is suitable for vanilla cultivation. Under the influence of the tropical monsoon climate and the terrain, the climate in Hainan is warm in the south and cold in the north, dry in the east and moist in the west. The annual mean atmospheric temperature is 23–25°C in most parts of the province. The average atmospheric temperature in January is 17–21°C. Annual mean precipitation varies between 958 and 2148 mm in parts of Hainan Island, and it is one of the highest of the regions of the same latitude. But it is not evenly distributed, being dry in the east and moist in the west, highest in the middle, higher in the east, northeast and north, lower in the southwest, and lowest in the west. Typhoons frequently hit Hainan Island during the June–October period, and the number of strong wind days per year is around 4.7–6.8.

On the whole, vanilla is more suitable for cultivation in Hainan Island than in Yunnan Province.

Different Cultivation Systems

Brick Stakes under Artificial Shade Houses

In this method, the climbing stakes for the vanilla are made with bricks of 240 mm × 115 mm × 53 mm. Bricks are placed in groups of three, forming a triangle, and each layer of bricks is rotated by a certain angle relative to the previous. In this way, a hollow brick stake is built: the buried part of the stake should measure 0.5 m, the height above ground is 1.5 m, and the space between poles is 2 m. Then substrates should be filled into the empty space of the hollow stakes, and vanilla seedlings planted around the poles. This method was adopted in a 1 ha plantation in Xishuangbanna, and 8.25 kg of fruit were harvested from 100 m^2 of vanilla plants. The method proved effective for vanilla cultivation, but the vines finally became entangled, which influenced the growth, pruning, pollination, and harvesting of the vanilla. The method was therefore replaced by the one described in the next section.

Poles and Strings under Artificial Shade Houses

Nylon netting with a 50–70% shading degree is used for the shade house, which is 2.5 m high. The stakes can be made of cement (15 cm × 15 cm × 2 m), wood, stone, or bamboo. The height above the ground for stakes should be 1.5 m. The field is divided into ridges: each ridge is 1.3 m wide, and the space between ridges is 0.35 m. Stakes are fixed along the middle of the ridges, with a space of 2 m between each stake on the same ridge. The poles on the same ridge are connected with steel bars, iron strings, or other types of strings. Vanilla seedlings are then planted longitudinally along both sides of each pole. All the shade houses in the vanilla plantations in Hainan and Xishuangbanna have adopted this method, which means full use can be made of the land, and the vanilla is easy to manage and convenient for handling. The disadvantage is that the cost is too high.

Courtyard Planting Method

This method can be combined with the family courtyard economic system and makes full use of the scattered land in Xishuangbanna. Furthermore, different families can adopt different methods of planting. Some families build simple shade houses with stakes, some add tall grasses or oil palm leaves to the roof of the shade house; some plant the vanilla directly under trees, whether mango (*Mangifera indica*), litchi (*Litchi chinensis*), *Baccaurea ramiflora*, or pomelo (*Citrus maxima*); some even plant the vanilla in the *Hevea brasiliensis* forest, and others plant them in the secondary forest. These courtyard planting methods require lower input and are easy to build with a sufficient supply of organic matter and good ventilation. Only shading and pruning need special attention. The area using this kind of planting method reached 16 ha in Xishuangbanna. Now only a few families still plant in this way.

Vanilla Cultivation Practices

Vanilla producers in the Yunnan and Hainan regions apply almost the same technologies to vanilla production (Zhou, 2000). The most commonly adopted vanilla cultivation system for commercial purposes is an intensive one under artificial shade house conditions. In order to adapt to this cultivation system and the specific climate in different regions, a whole set of mature cultivation techniques has been developed. Through exchanges and experience sharing, Hainan and Yunnan have developed similar cultivation technologies. However, some differences in skill and the actual implementation of cultivation techniques still exist among different companies. According to the actual production of vanilla under artificial shade house conditions in the Yunnan and Hainan regions, a brief introduction to vanilla cultivation and curing techniques is given below.

Propagation

Generally, vanilla is propagated by means of stem cuttings or tissue culture. Cuttings for propagation should be taken from plants that are healthy, robust, and have not yet flowered. The cuttings are usually 40–60 cm in length with 4–5 nodes, or 20–30 cm in length with at least two nodes, with the length depending on the amount of vines available. If there are not enough cuttings available, vanilla can be propagated rapidly and abundantly by means of tissue culture using shoot tips and nodal segments as explants. After the necessary hardening and acclimatization, seedlings can be transplanted into soil. In order to avoid seedlings becoming entangled and to ensure they are convenient for management, seedlings should be tied to bamboo sticks of around 50 cm in length.

Plantation Establishment and Planting

The construction of a vanilla plantation over a large area with an intensive cropping system includes not only the construction of the plantation itself, but also its supporting infrastructures, such as irrigation systems and living facilities for workers, and so on, which should be considered comprehensively and carefully, and planned and distributed according to actual situations. It is vital to the optimal growth of vanilla in the Xishuangbanna region to develop irrigation systems and to water vanilla plants in the dry season, especially from March to May.

The construction of a vanilla plantation mainly includes excavating fields into planting belts (ridges) and dividing these into units or zones along with the construction of artificial shade nets, vanilla stakes, and pipelines for the installation of the irrigation system, and so on. For ease of operation, the processes should be carried out in turn. A vanilla plantation site requires a slope of a certain gradient, which should not exceed 15°. The planting land should be excavated in planting belts at the same distance apart, so as to prevent soil erosion and to improve soil fertility and water supply, and the width of each planting belt should be 1–1.5 m. Materials such as angle steel, cement or wood columns, which are connected by iron wires or steel bars, are usually used as supports for the artificial shade net with 60–80% shade. Along the planting belt, at intervals of 1.8–2 m, cement or wood poles should be erected and fixed at a convenient height, and connected by iron wires or other materials into a row, which provides a trellis for vanilla plants to climb up. The recommended height of the trellis is around 1.5 m above the ground, and that of the shade net is at least 2.5 m.

After the plantation has been established, rooted seedlings from tissue culture plantlets or cuttings grown in nursery beds can be used for planting. Long cuttings can be planted in an optimal season, when plenty of vanilla planting materials are available. Two cuttings are usually planted beside each trellis column. According to the cuttings available, the density of cuttings can be up to 4000–9000 plants per hectare. Comparatively, at the same distance between rows, if more cuttings are planted along each planting belt, this will help to achieve and sustain greater vanilla plant vegetative biomass in the initial stages. The optimal planting time is immediately before the rainy season, at the beginning of June in the Xishuangbanna region, although the vanilla seedlings could be planted in the other months. The seedlings should be planted 5–10 cm below the surface of the soil, while their rootlets should be situated and distributed at or near the center of the planting belts, stretching along horizontally.

Care and Management

From the initial planting, vanilla plants—especially flowering adult plants—need meticulous care and painstaking management. Only in this way can the desired high yield and sustainable production be obtained. The regular management of a vanilla plantation consists of weeding, fertilizing, mulching, irrigating, vanilla pruning and trailing, and plant protection. A farmer can usually undertake a management workload of 0.4–0.7 ha of intensive vanilla plantation. More workers are of course needed for artificial pollination according to the number of flowers during the flowering season.

Artificial Pollination

Owing to the unusual structure of the flower, in other words the separation of the stamen from the stigma by the rostellum, pollination must be done artificially with a bamboo splinter. There are two methods of artificial pollination. The first one is to push up the rostellum under the stamens and then press the pollen sac into contact with the stigma. The second one is to remove the cap-like rostellum directly and then press the pollen mass to adhere to the stigma. The "pushing method" is commonly used in the Hainan region, and the "removing method" is used in the Xishuangbanna region of Yunnan. Artificial pollination should avoid injuring the stigma or dropping the pollen mass, for both the "pushing method" and the "removing method."

Supplying Nutrients and Fertilization Management

Being a surface rooting plant, vanilla is greatly dependent on favorable mulching to obtain available nutrients. The mulch and decomposed organic materials are the main source of nutrients for vanilla plants, and are helpful in retaining sufficient moisture and forming a loose soil aggregate structure for the roots to spread. Reasonable manuring is vital to obtaining an early, high, and sustainable yield. According to the mulching materials available and the plants' growth, manuring is usually combined with mulching. The fundamental principles of supplying nutrients and fertilization are the application of organic manures combined with chemical fertilizers, foliar spraying of chemical fertilizers, and correctly supplementing plants with trace elements, and so on. Mulches used and available in the main regions of China include the humus litter layer from the rainforest, coconut shells, wood chips, sugarcane bagasse, and rice straw. Some enterprises have researched and developed specialized vanilla fertilizers. In combination with the trimming of planting belts, mulching should be carried out at the end of the rainy season and completed before the cold season begins.

Pruning and Training of Vanilla Plants

The pruning and training of vanilla vines is one of the main agronomic measures required to regulate vegetative and reproductive growth, to reduce the occurrence of vanilla disease and to provide propagation cuttings and so on. Depending on different pruning purposes, the main parts to be pruned may include the infected portion of a plant, or weak and flowered vines.

The disease-infected leaves, stems, and fruits should be removed as early as possible so as to check the spread of the disease. Aging vines need pruning periodically to stimulate the induction of offshoots and the promotion of vegetative growth. The tips of slim vines and juvenile plants reaching a height of over 50 cm should be nipped off. With this method, more than one offshoot will come out and the branches will become stronger. However, more water sprouts and branches should be removed or inhibited by pruning, especially during the flowering and fruit development period. Branches that have flowered and borne fruit and have no living axillary buds can be cut off, since they will not produce any flowers. A specific pruning of hanging branches on adult plants is generally carried out 3–4 months before the flowering season—around November to December in China—to encourage the production of inflorescences in the axils of the leaves on the hanging branches. When the desired number of fruits per inflorescence has been reached after pollination, the remaining buds and the tip of inflorescences should be removed.

As essential as pruning measures are, vanilla vine training is also helpful in facilitating the even distribution of the vanilla vine on the climbing trellis in order to maintain a sustainable biomass and yield. Generally, the vines of the initial planted seedlings attached to the column of the supporting trellis or the erected bamboo sticks beside it always grow upward. When the vine or branch climbs up and reaches the top of the trellis, it should be brought back to the ground, and the portion of the vine with 4–5 nodes should be covered with surface soil or mulch, which will induce the vine under ground to take root and thereby support the whole plant in order for it to grow well. Thus, the vines are looped up and down, and evenly distributed on the supporting trellis, but do not accumulate on the top half of the trellis. Gradually, vanilla plants develop into a row along the planting belt, reaching as high as the trellis. Vanilla vines can be trained along the whole trellis within the same row.

Weeding

Weeds in vanilla plantations do little harm to vanilla plants. The aim of weeding is simply to prevent competition for nutrients with vanilla plants. There are many different species of weeds in newly established vanilla plantations that grow in abundance, especially during the rainy season. Weeds should be cleared before blossoming and fruit bearing, and can be used as mulch and organic fertilizer after fermentation. With manual weeding, the quantity and species of weeds will gradually diminish to a lower level until the third year.

Other Maintenance

In Xishuangbanna and some regions of Hainan Province, the annual alternation of dry and rainy seasons leads to soil compaction, poor root growth, and the incidence of diseases. Depending on the soil water content and moisture, in combination with mulching, adequate irrigation during the dry season is essential for vanilla plants to grow well. Lack of irrigation would lead to the death of aerial roots, and even the death of the plants suffering from water deficits. It is recommended to irrigate plants by frequent and low-volume sprinkling during extended dry periods, and to avoid waterlogging, especially in the rainy season.

Typhoons in the Hainan region from June to October, and especially from August to October, and monsoons in the Xishuangbanna region from February to June, occasionally damage vanilla plantations. It is therefore necessary to strengthen and maintain plantations to prevent damage by typhoons or monsoons.

Furthermore, it is necessary to replant where the plants have died or been destroyed by livestock or poultry, and so on. Measures should be taken to keep them away from plantations located near to dwellings.

Vanilla Diseases, Pests, and Diseases Management

Diseases

In different vanilla-growing regions in China, vanilla diseases, especially *Fusarium* root rot caused by *F. oxysporum* f. sp. vanillae (Tucker) Gordon has been a limiting factor to vanilla production. The integrated control measures for vanilla plantations in practice put prevention first, combined with treatment.

The measures recommended to prevent and counteract vanilla diseases are as follows:

- Avoid choosing low-lying and clay soil for growing vanilla plants, but choose gently sloping land that is well drained with a thick surface layer of humus.
- Use healthy seedlings for planting.
- Create a favorable environment for vanilla plants, especially better soil conditions by organic mulching in the rhizosphere.
- Do not trample on the planting belts when carrying out pollination or other operations.
- Regularly remove the disease-affected portions and apply necessary medicament treatments in emergency cases.
- Compartmentalize the plantation into different sections to prevent the disease spreading once it has occurred.
- Apply comprehensive measures, such as pruning and training plants, regulating shade, water drainage, and controlling flower numbers for pollination and so on, in order to rejuvenate vanilla plants and enhance their resistance to stresses and pathogens.

By applying the above measures, it is possible to reduce disease occurrence to an acceptable economic level and to guarantee long-term production.

Pests

Comparatively, insect pests do less damage to vanilla plants and are easier to prevent. Among these pests, coccidia must be promptly removed once they appear in the plantation, as their large-scale spread would destroy the vanilla plantation; it is hard to kill them since they are covered and protected by a waxy layer. Furthermore, terrestrial molluscs, snails and slugs, may cause certain damage to the tender parts of vanilla plants. Their population size can be efficiently controlled by attractant trapping and manual mass killing.

Harvesting

After planting, vanilla usually starts flowering in the third year. Vanilla plants growing in China bloom once a year, reaching the peak flowering stage in April, and are harvested during November to January of the following year. Depending on the degree of maturity of the beans, the right time for harvesting is when the beans turn a pale yellow color at their distal end. Harvesting earlier or later would have negative effects on the quality of cured beans.

Curing, Use, and Marketing

Curing

The harvested green beans have no aroma or flavor; this will develop after curing. Careful processing of beans will take 4–5 months depending on the climatic conditions. Various companies in China have developed different methods and capacities for processing vanilla beans. Although, different methods are used in different companies, they are all fairly similar.

The following takes the method employed in the former Yunnan Vanilla Company as an example to present an overview of the processing method in China, which consists of cleaning, killing, fermenting, slow drying and conditioning, grading and packaging.

- *Sorting*: According to length, appearance, and degree of maturity, the green beans are sorted into four types.
- *Cleaning*: Applying special mechanized high-pressure cleaning equipment with a vibrating screen, batches of different types of beans are cleaned and loaded onto stainless-steel mesh plates.
- *Killing*: The plates full of beans are immersed into an automatic temperature control water pool for several minutes. The temperature and duration of the "killing" stage are dependent on the size of the different types of beans.
- *Fermenting and rapid drying*: After killing, the beans are transferred to a drying chamber compartment, in which the specific temperature and humidity conditions are controlled. During this operation, the beans lose some water, turn deeper brown, and become supple with a perceptible aroma. This is the so-called "sweating" of the traditional method, rather than a true microbial fermentation.
- *Slow drying and conditioning*: This procedure is carried out in a clean room at ambient temperature. It is a process to continue the aroma development and browning, and the beans slowly lose more water. Ultraviolet sterilization is used to prevent microbial damage to beans at this stage.
- *Grading and packaging*: On the basis of the sorting of green beans, cured beans are sorted into grades according to their length, appearance, vanillin,

and water content. The vanillin or water content is mainly affected by curing, therefore, the length of cured beans constitutes the real difference between grades. Cured beans of different grades are packed and labeled, and then stored in a special warehouse for sale or delivery.

The production capacity of this semiautomatic line in the former Yunnan Vanilla Company was at least 300 tons of green beans per year. It was built in 1997 and unfortunately dismantled when the Yunnan Vanilla Company went bankrupt in 2007. In Hainan, several companies own vanilla production lines or equipment with different capacities, and they usually cure their own green beans and those they buy from local farmers.

Use and Marketing

Vanilla growers in China include companies and farmer households. Usually, green beans are cured by a few companies that own their equipment or production line for vanilla processing. These companies sell cured beans directly to customers or market their own processed products from vanilla. Some companies are therefore not only vanilla growers, but also the main users of cured beans.

Vanilla beans and their derivatives are mainly used in the cigarette and food industries, such as in coffee, wine, sauces, baked foods, and confectionery, especially tea products. Furthermore, vanilla beans are also used for cooking, although there are few people in China who use vanilla for cooking as they are unfamiliar with the exact usage and unaccustomed to using it for cooking in Chinese foods. With the development of information technology and the rise in the living standards of the people, more and more common people are becoming acquainted with vanilla and beginning to learn to use it for cooking or other purposes. At scenic spots in the vanilla-growing region, a lot of vanilla beans are directly marketed as local tourism products and sold to tourists.

Owing to the increasing domestic demand and limited output, vanilla beans produced in China are seldom exported. In the past, only a few batches of cured beans produced in Yunnan were exported to Japan and other countries by the former Yunnan Vanilla Industry Ltd. In recent years, Chinese production has only just covered domestic demand for vanilla bean.

PROSPECTS

The rapid development of vanilla industries in the past years has met with difficulties and problems in many respects. However, the industrial development of vanilla in China has great potential and many advantages. Over the past 49 years of research and development and in over 10 years of vanilla industrialization, China has fostered and established the market for vanilla and its related products, and has developed a whole set of mature cultivation and curing techniques. In recent years, the vanilla industry has begun to show a new trend and has maintained steady and sound growth. With the support of vanilla companies and local government, an increasing number of farmers are beginning to plant vanilla in Hainan Province. We have full confidence in the prospects for vanilla development in China. The vanilla industry will have broad horizons and bright prospects for development.

NATIONAL AGRICULTURAL STANDARDS FOR VANILLA IN CHINA

For promoting the development of the vanilla industry, the following agricultural standards for vanilla were issued by the Chinese Ministry of Agriculture in 1999. They were drafted by the Chinese Academy of Tropical Agricultural Sciences.

NY-T 362-1999 Vanilla seedlings
NY-T 483-2002 Vanilla
NY-T 713-2003 Determination of vanillin in vanilla beans
NY-T 968-2006 Technical rules for vanilla cultivation

REFERENCES

Chen, D. 2005. About vanilla bean, topic 111 Survey papers on present situation of Chinese vanilla bean industry. *Flavor and Fragrance Cosmetics* 5:32–36.

Chen, D., Liu, S., Deng, Z., Lan, Y., and Ling, C. 2007. Exploration of the largest wild population of *Vanilla siamensis* Rolfe in the world, and its implication for the development of vanilla industry in China. *Flavor and Fragrance Cosmetics* 5:49–52.

Liu, Z., Chen, S., and Ru, Z. 2007. *Vanilla shenzhenica* Z. J. Liu & S. C. Chen, the first new species of orchidaceae found in Shenzhen, South China. *Acta Phytotaxonomica Sinica* 45:301–303.

Mao, G. 2002. Suggests of promoting vanilla production in Hainan island. *Journal of Yunnan Tropical Crops Science & Technology* 25:22–26.

Zhou, H. 2000. *Technique and Operating Regulations of Vanilla Producing*. Enterprise Standard, Yunnan Vanilla Industry Ltd.

24 Vanilla Production in French Polynesia

Sandra Lepers-Andrzejewski and Michel Dron

CONTENTS

CULTIVATION AND PRODUCERS

Vanilla has been cultivated in French Polynesia since the nineteenth century. Tahitian vanilla, the genuine black gold of French Polynesia, is cultivated for its famous beans, which develop an intense fragrance of anise notes (see Chapter 13). The vanilla plantations are mainly located in the archipelago of the Society Island (Figure 24.1). The surface dedicated to vanilla production covers 306 ha (SDR, 1995). The island of Tahaa is also called the "Vanilla Island." The island has a land area of 88 km^2 and harbored 573 vanilla producers in 2002. Its nearby islands, Raiatea and Huahine, comprised 206 and 300 vanilla producers (census of the SRD, 2002). Vanilla is also widely cultivated throughout French Polynesia. From Tahiti and its sister island Moorea (10.3 ha of vanilla production) to the Marquesas Islands (2.4 ha), the Tahitian people meticulously cultivate their invaluable Tahitian vanilla. Vanilla is cultivated by two techniques: the first one known as the "traditional way" and the second called the "under shade house technique."

Traditional Way of Cultivation

The traditional way of cultivation has been used for several generations, where the vanilla vine is planted at the base of a tree. The tree provides stake and shade for the climbing orchid. The vanilla plantation is generally located on a hillside, within a cleared forest, chosen for its high content in organic matter (Wong, 1999). The plantation

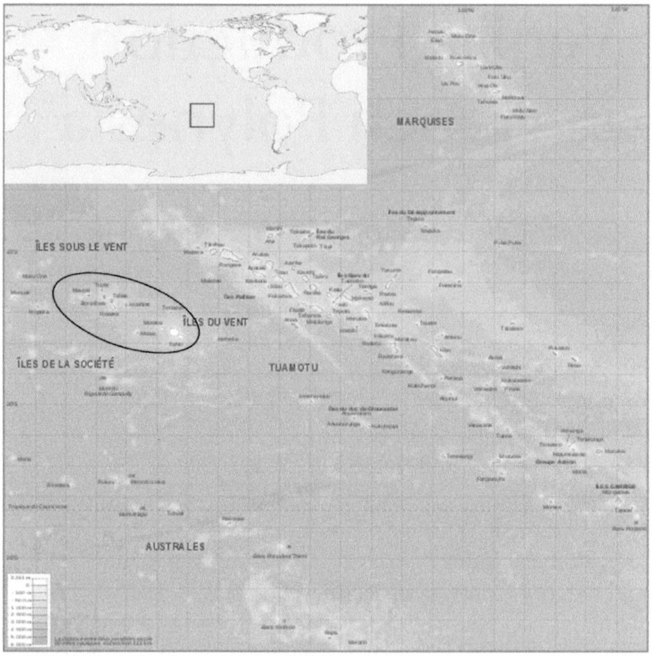

FIGURE 24.1 Map of French Polynesia. The Society Islands (Iles de la Société), where vanilla production is the most important, are circled in black.

is sloped, below wooden trees to benefit the humus generated by the trees at the top. If the ground is flat, the soil needs to be drained, digging channels to clear up the excess of water. The choice of the stake is important that it must protect the vanilla from the heat of the sun or wind and not compete for water supply and compost. The two favorite stakes are the "piti" (*Gliricidia maculata*) and the "pignon d'Inde" (*Jatropha curcas*) (Figures 24.2a and c). *Gliricidia* is a leguminous plant, which does not compete with vanilla for nitrogen supply. It grows quickly and requires frequent pruning. *Jatropha* requires less pruning to limit its growth, as leaves fall naturally during the dry season when flower induction occurs in vanilla. Thus, it does not need pruning. Other natural stakes are also used—the "purau" (*Hibiscus tiliaceus*), the "aute" (*Hibiscus rosa-sinensis*), or the "nono" (*Morinda citrifolia*) (Figures 24.2b, d, and e).

The traditional type of cultivation is the one most currently used. In this system, vanilla is in fact associated with other species such as food crops or fruits. In 1995, the average size of traditional vanilla plantations was about 3000 m^2 and the average age of the producers was over 50 years old. This traditional type of cultivation displays many constraints for the farmer. The plots are generally very sloppy and far from the house. It is regularly necessary to cut the branches of the stake tree to maintain constant shade, neither too intense, nor too weak, in order to preserve the development and bloom of vanilla plants. Finally, the vanilla plants can compete with the stake tree for water and its nutritive requirements. This will hamper the development of vanilla plantations.

FIGURE 24.2 Different plants used as stake in traditional vanilla plantation (a) "piti" (*Gliricidia maculata*), (b) "purau" (*Hibiscus tiliaceus*), (c) "pignon d'Inde" (*Jatropha curcas*), (d) "aute" (*Hibiscus rosa-sinensis*), and (e) "nono" (*Morinda citrifolia*).

Owing to this, some producers using the "traditional way" choose techniques involving the use of a shade house. Of course, they maintain their natural plantation and expect the first harvest through their new "shade house" plantation. It is therefore likely that, in forthcoming years, the quantity of ripe vanilla from traditional cultures will decrease due to the desertion of "traditional plantations."

The Under Shade House Culture

In recent years, vanilla plantations have been set up in shade houses in order to increase its production. Shade is ensured by insect-proof canvas thus avoiding virus transmission through insect vectors (Figure 24.3). The stake is generally made of concrete, and therefore is not in competition with the vanilla plant for the compost supply, which can be easily controlled. The production per hectare is strongly increased and daily work is less intensive. According to traditional plantations, the geographical location of shade houses must be taken into account. Ventilation plays an important role for shade houses. It is preferable to build a shade house on the coast

FIGURE 24.3 (a) Vanilla plant under shade house, two years after plantation; (b) vanilla plants flowering, under shade house, second year of production.

on the sheltered side of the island. This facilitates ventilation and diminishes risks of fungal disease. The soil must always be drained and slightly sloped.

This new method of culture is more productive and facilitates the task of farmers; however, the cultural technique, traditional, or under a shade house, has little influence on the aromatic composition of the vanilla beans. Healthy vanilla vines will give high-quality beans whatever the cultural technique used.

Production and Agricultural Value

Production

French Polynesia was once among the big vanilla producers. Between the years 1900 and 1960 more than 100 tons of ripe vanilla beans were produced per annum (Figure 24.4). The production dropped during the 1960s, following the installation of the Center of Nuclear Experimentation of the Pacific, which attracted many Polynesian

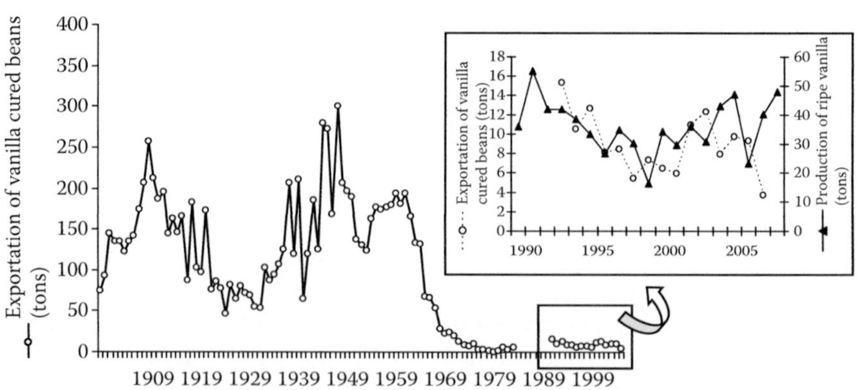

FIGURE 24.4 Evolution of the exportation of vanilla since 1905 and detail of the production and of the exportation of ripe and cured vanilla during the last 18 years.

people (Petard, 1986). Since the 1990s, the production of vanilla is making a come back, and is averaging 35 tons of ripe vanilla per annum. However, the production remains subjected to the weather. The cyclone of 1998 generated a drop in production by 16 tons in 1999. High temperatures in 2005 also had a negative impact on the production of 2006. In order to raise vanilla production in French Polynesia, governmental programs were set up to enhance its cultivation.

Etablissement Vanille de Tahiti

The Service du Développement Rural (SDR), a public service in charge of the agricultural productions and livestock in French Polynesia, originally in charge of the "vanilla program" developed a cultivation system under shade houses and techniques of composting. In order to popularize these innovations, the Polynesian government established the Etablissement Vanille de Tahiti (EVT, Figure 24.5) in 2003. This public establishment with commercial interest was mandated to restimulate the production of vanilla in French Polynesia, in the archipelago of the Leeward Islands, where a "traditional way" had been maintained. In addition, the culture of vanilla "under shade house" has been developed in the other French Polynesian islands.

This program is primarily devoted to human resource management. The aim is to increase the income of farmers and perpetuate their production.

The EVT assists growers through the sale and installation of shade houses and the supply of virus-free certified cuttings of vanilla. It ensures training and provides technical support on cultivation and sanitary control to the farmers.

The Sanitary Control

Cuttings sold to the farmers with the installation of shade houses are certified to be virus-free. The sanitary certification of these vines consists in the repeated controls of the cutting sources in the nurseries of the EVT, using ELISA detection of three types of viruses (Potyviruses, Cucumber mosaic virus, and Cymbidium mosaic virus) (Richard et al., 2009). Cuttings are collected only on vines in which no virus was detected over the past 9 months. At planting, a map of the new plantation is established with indication of the origin of each cutting (nursery, cutting source name, and location). At 3–4 months after planting of the cuttings, an ELISA test for the detection of the three viruses is carried out in the newly planted shade house in order to check if contamination did not occur during planting.

The Research and Development Department

As a priority, the EVT laboratory has undertaken research to characterize the traditional vanilla vines for their yield, their aromatic and lipidic composition, and their

FIGURE 24.5 Logo for Etablissement Vanille de Tahiti.

genome (see Chapter 13); in order to select the best cultivars and then create new varieties to be proposed to the farmers. The objectives were also to set up an "Appellation of Origin" "Vanille de Tahiti." This would enable consumers and industrialists, to have a better visibility of the quality of Tahitian vanilla.

Since the installation of new plantations under shade houses (20 ha including 16 in the Leeward Islands and 4 ha in the Windward Islands) by the EVT and despite the relative decline of traditional culture, the production should increase over the next few years to reach an estimated production of 60 tons of ripe vanilla.

Agricultural Value

The agricultural value of vanilla in French Polynesia, more precisely the value of marketed vanilla mature beans, has strongly increased since 2000, starting from 65 million Pacific Francs (547,000 euros) in 2000 to 116 million in 2006 (972,100 euros). This agricultural value of ripe vanilla represents 2% of the total Polynesian agricultural value.

After a peak in 2004, the average price of a kilogram of ripe vanilla beans drew back to the level originally observed of 4000–5000 FcfP (33–42 euros/43–54 USD). Export prices of a kilogram of cured vanilla hovered around 22,000 FcfP (184 euros/236 USD) (Figure 24.6). This price is much higher than that of other cultivated vanillas, including Bourbon vanilla. This is due to the aromatic quality of the Tahitian vanilla, which is very rich in flavor and thus has become a top-of-the-range product, in particular, in pastry products and catering.

REGULATIONS AND QUALITY CONTROLS OF TAHITIAN VANILLA

Harvesting the beans at the stage of maturity is necessary to obtain excellent quality of cured beans and long shelf-life preservation. In order to guarantee the quality of vanilla to the professionals and buyers in the vanilla sector, regulations have been in place for several years. The Polynesian regulation precisely defines the harvesting and sale of ripe vanilla (deliberation no. 77-119 of 10/11/1977 on regulation of the

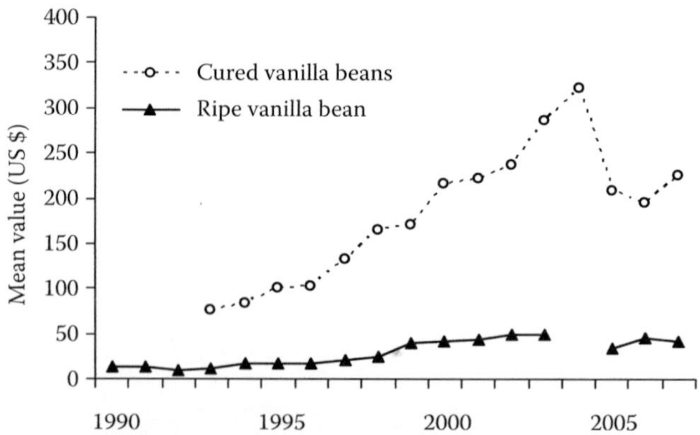

FIGURE 24.6 Evolution of the mean price of a kilogram of ripe vanilla and of exported cured vanilla beans.

harvesting, curing, packaging, and exporting of vanilla produced in French Polynesia; decree no. 1198 CM of 03/11/1992 on the production and the marketing of the vanilla produced in French Polynesia). The beginning and end of the harvest season are determined to allow the harvest only when the vanilla beans are at the best stage of maturity. Several public sales are organized during the harvest season. These sales of vanilla are conducted under the control of the "committee of ripe vanilla inspection" designated in each vanilla-producing district. These "committees of ripe vanillas inspection" control the level of ripening of the pods, their appropriate size (more than 14 cm), verify the absence of splitting and defaults (Figure 24.7). If some pods are not in agreement with the regulations, they will be removed from the batch and destroyed. The producer receives a certificate, which enables him to sell his vanilla to a curer or to cure it by himself.

The curing of the vanilla is also regulated (decree no. 1424 C of December 28, 1949). This regulation stipulates that the pods must be cured strictly by natural processes. Moreover, the curing of vanilla is prohibited to nonholders of a curers' patent (deliberation no. 91-120AT of October 25, 1991 on the establishment of commissions of control of vanilla and regulating access to the curing and the expertise of vanilla). The curers' patent is issued by the Polynesian Ministry of Agriculture. It depends on the candidate's technical ability, his theoretical knowledge of vanilla culture and curing, and the ongoing regulations. To date, there are only 10 effective vanilla bean curers.

Three categories of cured vanilla are defined, according to their length and their external aspect and quality.

1. The "Extra" category concerns vanilla of higher quality, healthy, complete, unsplit, flexible, and fleshy vanillas, of uniform dark brown color, presenting an oiled and shining aspect, with a fine and perfect odor of vanilla.

FIGURE 24.7 Control of vanilla beans during a sale of vanilla.

These beans must measure at least 16 cm and must present neither defect nor gall. Only a few scars on beans can be accepted.

2. The "First" category is defined by a vanilla of good commercial quality, healthy, complete, unsplit, flexible, and fleshy, of uniform dark brown color, presenting an oiled and shining aspect, with a fine and perfect odor of vanilla. These beans measure less than 16 cm and must present neither defect nor gall. Only a few scars are permitted.
3. There is also a so-called vanilla of second category. These vanilla beans can have different lengths and may present some defects on their surface.

The last quality control of the beans intended for export is carried out by the "territorial vanilla experts" (deliberation no. 91-120AT of October 25, 1991). These experts are named for two years following an examination testing their general knowledge on vanilla, notably curing, and regulations on vanilla. They are sworn in by the judicial authorities and can implicate antifraud measures. The "vanilla experts" are also involved in the jury which issues the curers' patent.

Thus, all the stages, from harvesting to curing and exporting of the Tahitian vanilla beans, are regulated and controlled in order to guarantee a high-quality vanilla to the buyers.

ACKNOWLEDGMENTS

The authors would like to thank E. Reva, K. Leoce-Mouk-San for productive discussions, L. Panie and J. Serven for invaluable comments on the manuscript.

REFERENCES

Petard, P. 1986. *Plantes utiles de Polynésie française et raau Tahiti*. Haere po no tahiti, Papeete, 354pp.

Richard, A., K. Farreyrol, B. Rodier, K. Leoce-Mouk-San, M. Wong, M. Pearson, and M. Grisoni. 2009. Successful control of virus diseases in intensively cultivated vanilla plots of French Polynesia. *Crop Protection* 28:870–877.

Service du Développement Rural (SDR). 1995. *Census of Agriculture*. Pirae, Tahiti, French Polynesia.

Wong, C. 1999. La culture de la vanille Tahiti en Polynésie française. Doc. Service du Développement Rural, 2ème secteur agricole.

Index

G

M

Q

R

S

T

U

V

W

X

Y

Z